AF594961

The Conformational Universe of Proteins and Peptides: Tales of Order and Disorder

The Conformational Universe of Proteins and Peptides: Tales of Order and Disorder

Editor

Marilisa Leone

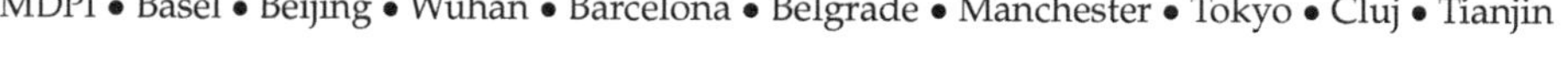
MDPI • Basel • Beijing • Wuhan • Barcelona • Belgrade • Manchester • Tokyo • Cluj • Tianjin

Editor
Marilisa Leone
Institute of Biostructures and
Bioimaging
National Research Council of
Italy (IBB-CNR)
Naples
Italy

Editorial Office
MDPI
St. Alban-Anlage 66
4052 Basel, Switzerland

This is a reprint of articles from the Special Issue published online in the open access journal *Molecules* (ISSN 1420-3049) (available at: www.mdpi.com/journal/molecules/special_issues/Conformational_Analysis_Proteins).

For citation purposes, cite each article independently as indicated on the article page online and as indicated below:

LastName, A.A.; LastName, B.B.; LastName, C.C. Article Title. *Journal Name* **Year**, *Volume Number*, Page Range.

ISBN 978-3-0365-2352-1 (Hbk)
ISBN 978-3-0365-2351-4 (PDF)

© 2021 by the authors. Articles in this book are Open Access and distributed under the Creative Commons Attribution (CC BY) license, which allows users to download, copy and build upon published articles, as long as the author and publisher are properly credited, which ensures maximum dissemination and a wider impact of our publications.
The book as a whole is distributed by MDPI under the terms and conditions of the Creative Commons license CC BY-NC-ND.

Contents

About the Editor

Marilisa Leone

Marilisa Leone earned a Master Degree in Chemistry (110/110 cum laude) in 1999 and a PhD in Chemical Sciences (Subject: protein NMR) in 2003 from the University "Federico II" of Naples, Italy. She was mostly trained in the field of NMR-based drug discovery at the Burnham Institute for Medical Research in San Diego (USA) (2002–2003 and 2005–2009). In 2006, she was awarded the ITALFARMACO PRIZE for young scientists from the Italian Chemical Society.

In 2009, she was recruited as a permanent researcher by the National Research Council of Italy (CNR-IBB). She is, to date, the co-author of more than 75 articles on international peer-reviewed journals (Scopus H-index: 24). Her scientific interests are protein/peptide structure and dynamics, protein/protein and protein/ligand (i.e., peptide and small molecules) interactions, and drug discovery, largely centered on cancer-relevant targets.

MDPI

Editorial

Special Issue—The Conformational Universe of Proteins and Peptides: Tales of Order and Disorder

Marilisa Leone

Institute of Biostructures and Bioimaging (IBB), National Research Council of Italy (CNR), Via Mezzocannone 16, 80134 Naples, Italy; marilisa.leone@cnr.it

Citation: Leone, M. Special Issue—The Conformational Universe of Proteins and Peptides: Tales of Order and Disorder. *Molecules* **2021**, *26*, 3716. https://doi.org/10.3390/molecules26123716

Received: 15 June 2021
Accepted: 16 June 2021
Published: 18 June 2021

Publisher's Note: MDPI stays neutral with regard to jurisdictional claims in published maps and institutional affiliations.

Copyright: © 2021 by the author. Licensee MDPI, Basel, Switzerland. This article is an open access article distributed under the terms and conditions of the Creative Commons Attribution (CC BY) license (https://creativecommons.org/licenses/by/4.0/).

Among biological macromolecules, proteins hold prominent roles in a vast array of physiological and pathological processes. The protein sequence-structure-function paradigm establishes that the amino acid sequence governs the structure that in turn determines the function [1]. Thus, the knowledge of the 3D structure of a protein, along with the possible conformational transitions occurring upon interaction with diverse ligands, are essential to fully comprehend its biological function.

Apart from globular well folded proteins, during the past years intrinsically disordered proteins (IDPs) have attracted a lot of attention. IDPs present a general tendency to aggregate and may form toxic amyloid fibers and oligomers associated with many human pathologies [2–4]. Intensive studies have been lately devoted to structural characterizations of aggregates formed by IDPs, along with the search for novel techniques to easily perform such analyses. Interestingly, IDPs are also known to often undergo a disorder to order switch following binding to their targets that generates specific outcomes in a cellular context.

Peptides, which are characterized by a smaller size than proteins, represent key elements of cells as well. Peptides can even find diagnostic and therapeutic applications, for instance working as tumor markers [5]. In the drug discovery field, the structural features of bioactive peptides are employed to design potential novel drugs acting as selective modulators of specific receptors or enzymes. Nevertheless, synthetic peptides reproducing different protein fragments have frequently served as model systems in folding studies relying on structural investigations in water and/or other environments.

This Special Issue comprehends contributions (i.e., seven original research articles and five reviews) on the above-described topics and, in detail, it includes structural studies on globular folded proteins, IDPs and bioactive peptides. These works were conducted by utilizing different experimental (including solution, solid states and high-pressure NMR and mass spectrometry) and/or computational approaches (mainly molecular dynamics simulations and bioinformatic tools).

A broad range of structural biology topics are covered by the Special Issue as summarized below.

Proteins under biological conditions unfold and refold several times in vivo showing a marginal structural stability. A detailed molecular-level knowledge not only of the native but also of the diverse non-native conformational states, which are accessible to a protein in solution (i.e., its denatured state ensemble (DSE)), is necessary to fully comprehend its function. Several investigations employed short peptides as models to obtain the canonical features of the DSE [6]. Short peptides are advantageous within this context as, being too short to assume a compact fold, they can sample unfolded states under folding conditions. Different peptide structural studies showed the strong tendency for the polyproline II (PPII) backbone conformation, which consequently could be a dominant component of the DSE [6]. Another structural model for the DSE is represented by the protein coil library that was built up from the segments of protein structure in the Protein Data Bank (PDB) that are located outside the α-helix and β-strand domains. Overall, coil libraries exhibit structural trends that are in good agreement with the results from peptide structural studies underlying a high preference for PPII that can be also linked to the amino acid type. In order to assess structural preferences in unfolded states under non-denaturing conditions, IDPs can also be employed as an additional experimental model system. An interesting review by Steven T. Whitten and collaborators critically analyzes spectroscopic and calorimetric works on short peptides, structures in the protein coil library and sequence and temperature-based investigations of IDP hydrodynamic sizes; they demonstrate how the three model systems used for describing unfolded proteins under folding conditions deliver a consistent structural and energetic view of the DSE [6]. Results from analyses of the three model systems (i.e., peptides, the coil library and IDPs) highlight that the structural and energetic features of the DSE at normal temperatures can be predicted by a PPII-dominant ensemble. At cold temperatures, the DSE can undergo a transition in population toward the α-helix backbone conformation as revealed by the analyses of both peptides and IDPs.

The Special Issue further includes interesting reviews describing structural and functional features of different proteins. Leonardo L. Fruttero and collaborators focus on intrinsically disordered polypeptides from plant ureases (i.e., Jaburetox and Soyuretox). Jaburetox represents a recombinant peptide derived from the jack bean (*Canavalia ensiformis*) urease that possesses entomotoxic and antimicrobial functions [7]. NMR studies point out that Jaburetox possess only low amounts of secondary structure and behaves as an IDP. Nevertheless, Jaburetox can undergo a disorder to order transition after binding lipid membranes. Soyuretox is another IDP homologous to Jaburetox and it is derived from the soybean (*Glycine max*) ubiquitous urease. Compared to Jaburetox, Soyuretox contains a higher secondary structure content but preserves similar entomotoxic and fungitoxic properties. Due to the positive toxicity profile, both peptides find biotechnological applications and have in fact been already used to generate transgenic crops giving rise to plants active against insects and nematodes [7].

Nicholas J. Bradshaw and collaborators focus, instead, on the TRIOBP (TRIO and F-actin Binding Protein) isoforms and describe within their review the proteins structural characteristics along with their function in actin stabilization and the relationship to pathological conditions (deafness, mental illness and cancer) [8]. In detail, the TRIOBP gene encodes multiple proteins, TRIOBP-1 represents principally a structured protein that interacts with F-actin and avoids its depolymerization. TRIOBP-1 has been linked to schizophrenia, as it can give rise to protein aggregates in the brain. TRIOBP-4 is, on the other hand, a completely disordered protein and a few of its mutations are related to severe or profound hearing loss. TRIOBP-1 and TRIOBP-4 have both been related to cancer [8].

Another interesting review by Christian Roumestand and collaborators is centered on High-Pressure (HP)-NMR. The authors summarize recent advances of HP-NMR and describe how this technique can be employed to characterize, at a quasi-atomic resolution, the protein folding energy landscape [9]. Globular proteins can be perturbed in several ways and high-hydrostatic pressure represents an alternative destabilizing method. At difference from heat or chemical denaturant, which generate a uniform protein destabilization, pressure produces local effects on protein regions or domains provided with internal voids. HP-NMR spectroscopy allows one to follow the structural transitions occurring upon unfolding and to study the kinetic properties of the process [9].

Beat H. Meier, Anja Böckmann and collaborators describe within their review the lessons learned from studies on two ATPases (the bacterial DnaB helicase from *Helicobacter pylori* and the multidrug ATP binding cassette (ABC) transporter BmrA from *Bacillus subtilis*) [10]. The authors report on NMR approaches that can be employed to examine proteins binding to ATP-mimics. In order to reveal conformational and dynamic changes occurring upon interaction with ATP-mimics, carbon-13, phosphorus-31 and vanadium-51 solid-state NMR spectra of the proteins or the bound molecules are shown.

The reported information can surely be relevant to researchers conducting studies on other NTPases (Nucleoside TriPhosphatases) [10].

An interesting research article by Michael O. Glocker and collaborators highlights how mass spectrometry techniques have enlarged their horizons by becoming additional tools to study intermolecular interactions [11]. In detail, the authors report on an original approach relying on electrospray mass spectrometry (called ITEM-TWO (Intact Transition Epitope Mapping-Thermodynamic Weak-force Order)) that requires only a small sample amount and permits the simultaneous identification of epitopes (such as peptide segments that are recognized by a certain antibody) and gas phase binding strengths for the antibody–epitope peptide interactions [11].

In this pandemic era, most research efforts are centered on SARS-CoV-2 related studies. Thus, a contribution to the field is also included in this Special Issue. Vladimir N. Uversky and collaborators, by utilizing different bioinformatic tools, analyze amino acid sequences of ACE2 (Angiotensin-converting enzyme 2) receptors from eighteen non-human species and compare them with the human ACE2 sequence pointing out the degree of variability [12]. Results indicate that many non-human species have, in the binding site of ACE2 receptor, similar amino acid types to humans letting speculate that the RBD (Receptor Binding Domain) of the SARS-CoV-2 Spike (S) protein could be involved in similar interactions with ACE2 receptors from different species. Consequently, this detailed investigation of the ACE2 protein let speculate that, indeed, interspecies SARS-CoV-2 transmission could be quite possible and allows to formulate a possible transmission flow. Nevertheless, the authors also examine the per-residue intrinsic disorder tendency of the ACE2 proteins from several species pointing out a certain degree of similarity among disorder profiles and highlighting regions where the most striking differences are evident [12].

Authors Alberto Perez and Lijun Lang report on the p53-MDM2 (Mouse Double Minute 2 homolog) interaction, which is highly relevant in cancer research. In fact, p53 triggers programmed cell death when cells misbehave, while MDM2 downregulates p53 anticancer activity; inhibitors of the p53-MDM2 interaction thus possess anticancer potentials [13]. The authors present, in their research article, a computational approach that could result really useful for drug design. Nowadays computational tools such as virtual screening techniques have to face a huge challenge when dealing with the binding of flexible peptides that could fold following interaction to specific receptors [13]. The authors in their research article investigate the binding of five peptides, including three intrinsically disordered ones, to MDM2 with a Bayesian inference approach (MELDXMD (Modeling Employing Limited Data Accelerated MD)). The method is able to capture the folding upon binding mechanism, showing the most likely bound conformations and highlighting the differences in the binding mechanisms [13].

Work by Rajni Verma, Jonathan M. Ellis and Katie R. Mitchell-Koch focuses instead on molecular dynamics simulations of the enzyme YqhD. YqhD represents an *E. coli* alcohol/aldehyde oxidoreductase that, beginning from a wide range of materials, is able to generate relevant bio-renewable fuels and fine chemicals [14]. The computational work sheds light on the conformational dynamics of the enzyme upon interaction with oxidized and reduced NADP/H cofactor [14]. The study highlights how YqhD complexed with NADP may fluctuate between open and closed conformations, while interaction with NADPH induces a slower opening/closing dynamic of the cofactor-binding site. This dynamical view let speculate that the frequent opening of the binding cleft is necessary to favor release of NADP, while a more closed conformation is necessary to enhance NADPH interaction along with aldehyde reductase activity [14]. This work clearly points out how molecular dynamics simulations may provide access to structural details that could help better understand how enzymes work.

In their research article, Richard P. Cheng and collaborators analyze, largely by using NMR analyses, cross-strand lateral ion-pairing interactions and their importance for antiparallel β-sheet stability [15]. The authors provide interesting insights for the design of functional peptides provided with lateral ion-pairing interactions across antiparallel-strands. In detail, they perturbed the cross-strand lateral ion-pairing interactions in a β-hairpin peptide by swapping the position of ammonium and carboxylate containing residues with different side-chain lengths [15]. Chemical shift data permit gaining the fraction folded population and folding free energy. Results point out that, similarly to the unswapped peptide, the most stabilizing cross-strand contacts occur between short residues, although an increase in folded populations upon swapping is detected [15].

David A. Snyder and collaborators tells us about protein flexibility and its importance for proper protein functioning by performing a comparative investigation of protein flexibility captured by crystallographic B-factors, molecular dynamics and NMR studies [16]. This work highlights that NMR- structural ensembles present a pattern in the coordinate uncertainties of backbone heavy atoms that also recurs in coordinate variances across MD trajectories but not in crystallographic B-factors [16]. This evidence let speculate that either MD trajectories and NMR structures are able to detect the motional behavior of peptide bond units not contemplated by B-factors or could highlight a deficit linked to the force fields employed in both NMR and MD calculations [16].

Finally, Gennaro Esposito and collaborators in their research article report on the structure characterization of a nanobody by utilizing solution state NMR techniques coupling to molecular modelling investigation [17]. Nanobodies derive from heavy chain-only antibodies, which can be found in camelids, with their smaller molecular size and higher stability represents an alternative to mAbs for therapeutic use. Two nanobodies, Nb23 and Nb24, are able to block similarly self-aggregation of highly amyloidogenic variants of β2-microglobulin. The authors carried out a structural characterization of Nb23. The study points out peculiar structural features of Nb23 with respect to Nb24, which could be linked to diverse target antigen affinity [17].

In conclusion, this ensemble of studies clearly stresses how the knowledge of structural features enables the understanding of the multifaceted roles of protein and highlights the importance of flexible regions and disorder in dictating binding events and directing protein functions under normal physiological and pathological conditions. Coupling of experimental and computational work is always necessary to reach a complete and detailed structural portrait of a protein in its isolate state and when it is bound to a ligand (intended either as a small cofactor or a bigger peptide). Researchers are still eager to develop better computational tools to keep into account proteins/peptide flexibility and to predict interactions of proteins with large flexible systems, such as peptides, and in the next few years much efforts will likely be devoted towards reaching this goal and optimizing existing tools. Nevertheless, the reported studies also highlight the continuous search by the scientific community for novel improved experimental techniques to a protein/peptide structure, dynamics and interactions.

Given the variety of topics embraced by this Special Issue, a great interest from researchers working in the protein/peptide structural biology field is expected.

Funding: This research received no external funding.

Acknowledgments: The guest editor wishes to thank all authors for their contribution to this Special Issue. All reviewers are greatly acknowledged for supporting the editor in selecting the most scientifically sound articles. MDPI staff is also kindly acknowledged for the collaboration and helpful suggestions.

Conflicts of Interest: The author declares no conflict of interest.

References

1. Redfern, O.C.; Dessailly, B.; Orengo, C.A. Exploring the structure and function paradigm. *Curr. Opin. Struct. Biol.* **2008**, *18*, 394–402. [CrossRef]
2. Uversky, V.N. Intrinsically disordered proteins and their (disordered) proteomes in neurodegenerative disorders. *Front. Aging Neurosci.* **2015**, *7*, 18. [CrossRef] [PubMed]
3. Uversky, V.N.; Oldfield, C.J.; Dunker, A.K. Intrinsically disordered proteins in human diseases: Introducing the D2 concept. *Annu. Rev. Biophys.* **2008**, *37*, 215–246. [CrossRef] [PubMed]
4. Coskuner, O.; Uversky, V.N. Intrinsically disordered proteins in various hypotheses on the pathogenesis of Alzheimer's and Parkinson's diseases. *Prog. Mol. Biol. Transl. Sci.* **2019**, *166*, 145–223. [PubMed]

5. Lau, J.L.; Dunn, M.K. Therapeutic peptides: Historical perspectives, current development trends, and future directions. *Bioorg. Med. Chem.* **2018**, *26*, 2700–2707. [CrossRef] [PubMed]
6. Paiz, E.A.; Lewis, K.A.; Whitten, S.T. Structural and Energetic Characterization of the Denatured State from the Perspectives of Peptides, the Coil Library, and Intrinsically Disordered Proteins. *Molecules* **2021**, *26*, 634. [CrossRef] [PubMed]
7. Grahl, M.V.C.; Lopes, F.C.; Martinelli, A.H.S.; Carlini, C.R.; Fruttero, L.L. Structure-Function Insights of Jaburetox and Soyuretox: Novel Intrinsically Disordered Polypeptides Derived from Plant Ureases. *Molecules* **2020**, *25*, 5338. [CrossRef] [PubMed]
8. Zaharija, B.; Samardzija, B.; Bradshaw, N.J. The TRIOBP Isoforms and Their Distinct Roles in Actin Stabilization, Deafness, Mental Illness, and Cancer. *Molecules* **2020**, *25*, 4967. [CrossRef] [PubMed]
9. Dubois, C.; Herrada, I.; Barthe, P.; Roumestand, C. Combining High-Pressure Perturbation with NMR Spectroscopy for a Structural and Dynamical Characterization of Protein Folding Pathways. *Molecules* **2020**, *25*, 5551. [CrossRef] [PubMed]
10. Lacabanne, D.; Wiegand, T.; Wili, N.; Kozlova, M.I.; Cadalbert, R.; Klose, D.; Mulkidjanian, A.Y.; Meier, B.H.; Bockmann, A. ATP Analogues for Structural Investigations: Case Studies of a DnaB Helicase and an ABC Transporter. *Molecules* **2020**, *25*, 5268. [CrossRef] [PubMed]
11. Danquah, B.D.; Opuni, K.F.M.; Roewer, C.; Koy, C.; Glocker, M.O. Mass Spectrometric Analysis of Antibody-Epitope Peptide Complex Dissociation: Theoretical Concept and Practical Procedure of Binding Strength Characterization. *Molecules* **2020**, *25*, 4776. [CrossRef] [PubMed]
12. Hassan, S.S.; Ghosh, S.; Attrish, D.; Choudhury, P.P.; Aljabali, A.A.A.; Uhal, B.D.; Lundstrom, K.; Rezaei, N.; Uversky, V.N.; Seyran, M.; et al. Possible Transmission Flow of SARS-CoV-2 Based on ACE2 Features. *Molecules* **2020**, *25*, 5906. [CrossRef] [PubMed]
13. Lang, L.; Perez, A. Binding Ensembles of p53-MDM2 Peptide Inhibitors by Combining Bayesian Inference and Atomistic Simulations. *Molecules* **2021**, *26*, 198. [CrossRef] [PubMed]
14. Verma, R.; Ellis, J.M.; Mitchell-Koch, K.R. Dynamic Preference for NADP/H Cofactor Binding/Release in *E. coli* YqhD Oxidoreductase. *Molecules* **2021**, *26*, 270. [CrossRef] [PubMed]
15. Huang, C.H.; Wong, T.W.; Yu, C.H.; Chang, J.Y.; Huang, S.J.; Huang, S.L.; Cheng, R.P. Swapping the Positions in a Cross-Strand Lateral Ion-Pairing Interaction between Ammonium- and Carboxylate-Containing Residues in a beta-Hairpin. *Molecules* **2021**, *26*, 1346. [CrossRef] [PubMed]
16. Reinknecht, C.; Riga, A.; Rivera, J.; Snyder, D.A. Patterns in Protein Flexibility: A Comparison of NMR "Ensembles", MD Trajectories, and Crystallographic B-Factors. *Molecules* **2021**, *26*, 1484. [CrossRef] [PubMed]
17. Percipalle, M.; Hunashal, Y.; Steyaert, J.; Fogolari, F.; Esposito, G. Structure of Nanobody Nb23. *Molecules* **2021**, *26*, 3567. [CrossRef]

MDPI

Article

Structure of Nanobody Nb23

Mathias Percipalle [1,2], Yamanappa Hunashal [1], Jan Steyaert [3,4], Federico Fogolari [5,6] and Gennaro Esposito [1,6,*]

1 Science Division, New York University Abu Dhabi, Abu Dhabi 129188, United Arab Emirates; mp5604@nyu.edu (M.P.); yh45@nyu.edu (Y.H.)
2 Department of Chemistry and Magnetic Resonance Center, University of Florence, 50019 Florence, Italy
3 Structural Biology Brussels, Vrije Universiteit Brussel, B-1050 Brussels, Belgium; jan.steyaert@vub.be
4 VIB-VUB Center for Structural Biology, Vlaams Instituut voor Biotechnologie, B-1050 Brussels, Belgium
5 Dipartimento di Scienze Matematiche, Informatiche, e Fisiche, Udine University, 33100 Udine, Italy; federico.fogolari@uniud.it
6 Istituto Nazionale Biostrutture e Biosistemi, 00136 Roma, Italy
* Correspondence: ge22@nyu.edu

Abstract: Background: Nanobodies, or VHHs, are derived from heavy chain-only antibodies (hcAbs) found in camelids. They overcome some of the inherent limitations of monoclonal antibodies (mAbs) and derivatives thereof, due to their smaller molecular size and higher stability, and thus present an alternative to mAbs for therapeutic use. Two nanobodies, Nb23 and Nb24, have been shown to similarly inhibit the self-aggregation of very amyloidogenic variants of β2-microglobulin. Here, the structure of Nb23 was modeled with the Chemical-Shift (CS)-Rosetta server using chemical shift assignments from nuclear magnetic resonance (NMR) spectroscopy experiments, and used as prior knowledge in PONDEROSA restrained modeling based on experimentally assessed internuclear distances. Further validation was comparatively obtained with the results of molecular dynamics trajectories calculated from the resulting best energy-minimized Nb23 conformers. Methods: 2D and 3D NMR spectroscopy experiments were carried out to determine the assignment of the backbone and side chain hydrogen, nitrogen and carbon resonances to extract chemical shifts and interproton separations for restrained modeling. Results: The solution structure of isolated Nb23 nanobody was determined. Conclusions: The structural analysis indicated that isolated Nb23 has a dynamic CDR3 loop distributed over different orientations with respect to Nb24, which could determine differences in target antigen affinity or complex lability.

Keywords: nanobody; protein structure; immunoglobulin domain; NMR

Citation: Percipalle, M.; Hunashal, Y.; Steyaert, J.; Fogolari, F.; Esposito, G. Structure of Nanobody Nb23. *Molecules* **2021**, *26*, 3567. https://doi.org/10.3390/molecules26123567

Academic Editor: Marilisa Leone

Received: 13 April 2021
Accepted: 21 May 2021
Published: 11 June 2021

Publisher's Note: MDPI stays neutral with regard to jurisdictional claims in published maps and institutional affiliations.

Copyright: © 2021 by the authors. Licensee MDPI, Basel, Switzerland. This article is an open access article distributed under the terms and conditions of the Creative Commons Attribution (CC BY) license (https://creativecommons.org/licenses/by/4.0/).

1. Introduction

Single-domain antibodies, or nanobodies, are derived from heavy-chain only antibodies (HcAbs) found in camelids [1]. Essentially, they can be used for the same therapeutic purposes as monoclonal antibodies (mAbs) and single-chain variable fragments (scFvs) but with some advantages brought about by their inherent properties. For one, the small molecular size of nanobodies (~15 kDa) facilitates penetrance to target sites, as nanobodies are half as large as scFvs and five times smaller than human conventional antibodies [2]. This, in combination with more extended loops of the complementarity determining regions 1 and 3 (CDR1 and CDR3), enables binding to a wider range of epitopes with different shapes at sub-nanomolar affinity, potentially increasing the application of nanobodies as drugs. The lack of a light chain in HcAbs also allows nanobodies to exist as a single domain with less susceptibility to aggregation through hydrophobic interactions, as is the case for scFvs [3–5]. Due to their small size and high similarity to the human immunoglobulin variable domain, they provoke little to no immune response [5] which often makes humanization unnecessary.

Amyloidogenic proteins have previously been targeted with nanobodies to inhibit the course of amyloidogenesis [4]. Nanobodies have been shown to inhibit the formation of amyloid β (Aβ) fibrils formed in Alzheimer's disease patients, and also to recognize non-conventional epitopes on Aβ fibrils for diagnostic use [6], although the clinical trials to validate antibody drugs have been unsuccessful so far.

Non-neurodegenerative amyloidoses may prove more amenable for nanobody treatment. A paradigmatic amyloidogenic protein, β2-microglobulin (β2m), which is a component of class I major histocompatibility complex (MHC-1), accumulates as amyloid deposits in the joints of patients undergoing long-term haemodialysis [7]. The deposits contain some 30% of ΔN6β2m, the proteolytic variant of β2m devoid of the N-terminal hexapeptide, that forms fibrils also by mild stirring at neutral pH [8]. This amyloidogenic propensity, much stronger than the parent protein, was also observed with D76Nβ2m, a naturally occurring variant of β2m that causes progressive bowel dysfunction and systemic amyloidosis, i.e., deposits in several vital organs [9].

Several nanobodies were raised against wild-type (WT) β2m and ΔN6β2m by immunization of both a camel and a llama. Nb24, a camel-derived nanobody raised against WT β2m has been shown to inhibit the self-aggregation of the very amyloidogenic ΔN6β2m and D76Nβ2m variants in vitro and, indirectly, also in vivo, and the binding thermodynamics and kinetics along with the epitope mapping of the D76Nβ2m-Nb24 complex were characterized [10,11]. In this case, D76Nβ2m self-aggregation was inhibited despite the fact that Nb24 was raised against the WT β2m. The crystal structure of Nb24 complexes with ΔN6β2m (PDB ID 2X89) and P32Gβ2m (PDB ID 4KDT) are known [11,12] whereas no structure is available for the isolated nanobody. Nb23, which is instead llama-derived and raised against ΔN6β2m, inhibits self-aggregation of its raising antigen, but fails to inhibit D76Nβ2m self-aggregation, despite it being raised against a very amyloidogenic variant of β2m. In order to characterize the interaction of Nb23 with a target other than the original antigen, structural information is crucial. In this study, the solution structure of Nb23 has been determined using nuclear magnetic resonance (NMR) spectroscopy, as a first step of a general project aimed at rationalizing the determinants of nanobody performance with β2m variants. In particular, structure knowledge enables systematic analysis of the conformational, thermodynamic, and kinetic properties of the binding to the β2m variants in order to improve the affinity between nanobody and antigen or attenuate their complex lability through rational design.

2. Results

2.1. Nb23 Sequence Inferences

The Nb23 construct characterized here consists of 136 amino acids, including an initial methionine residue introduced as a start codon and therefore referred to as Met0, and a $(His)_6$ tag at the C-terminus of the protein for expression in *E. coli* and purification, amounting to a molecular weight of 15.1 kDa. There are two cysteines at position 22 and 96 which form the disulfide bond between the two β-sheets of the expected immunoglobulin domain. Nb23 and Nb24 are of equal lengths with 71% identity, and 75% positive identity. This level of homology indicates structural and functional similarity [13]. The fact that the main variation in sequences between Nb23 and Nb24 coincides with the CDRs (located between residues 26–32, 52–57, and 100–116), together with a general consensus on the typical structural similarity of the framework regions of immunoglobulin variable domains, suggests that the frameworks of both nanobodies are similar.

2.2. NMR Spectroscopy Results and Chemical Shift Assignment Completeness

The ^{15}N-^{1}H HSQC spectrum of Nb23 is shown in Figure 1. The resonance spreading already appears quite satisfactory, and TROSY pulse schemes further enabled the resolution of certain overlapping peaks in the regular ^{15}N-^{1}H HSQC. Apart from the two prolines which lack amide protons and excluding Met0 and the $(His)_6$ tag, amide connectivity assignments are missing for Gln1, Arg27, Thr28, Ser63, and Ser105, which include residues of the expectedly mobile CDR1 (Arg27 and Thr28) and CDR3 (Ser105) loops. The occurrence of conformational mobility at intermediate rate on the chemical shift scale leading to signal broadening seems confirmed by the fact that neighboring residues in CDR1 and CDR3 (Gly26 and Gly102) exhibit below-average intensities and by the $^{15}N\{^1H\}$ NOE data, where residues in conformationally rigid regions show a close-to-average ratio of peak intensity with and without hydrogen saturation (Figure 2). It is thus plausible that an unfavorable conformational exchange rate in the CDR regions could affect the detectability of some signal in ^{15}N-^1H HSQC and TROSY spectra. On the other hand, the unassigned peaks other than sidechain resonances that were observed in the ^{15}N-^1H HSQC or TROSY maps—namely three cross-peaks highlighted by blue boxes and letter labels in Figure 1—were addressed, but no conclusion could be achieved through the correlation patterns of the 3D triple resonance experiments acquired for backbone assignment, suggesting again that some slow conformational exchange occurring over the ms-to-μs time scale accelerates relaxation, thereby hindering the propagation of the coherence transfer pathway. The extent of population transfer from $^{15}N\{^1H\}$ NOE data (Figure 2) enables, however, a tentative assignment. The negative heteronuclear NOE of boxed peak (a) is very likely to arise from Gln1. The close-to-average NOE value of boxed peak (c) could be consistent with the mobility expected at Ser63. Finally, the NOE value observed for boxed peak (b) suggests a possible attribution to Thr28, given the similar NOE value measured at Phe29. This dipolar-coupling-based assignment leaves only Arg27 (CDR1) and Ser105 (CDR3) without observable ^{15}N-^1H connectivity signal that, in turn, corresponds to the signature of a conformational exchange process at the start of CDR1 and CDR3.

Typical TROSY-based 3D triple resonance spectra [14,15] (see Section 4) were used to assign the backbone and sidechain atoms. The sidechain assignment was arduous especially for residues with very long sidechains, due to the relaxation attenuation ensuing from many magnetization transfers combined with the relatively low sample concentrations, leading to noisy data with reduced intensity. The low sample concentrations were in turn due to poor protein solubility, at least for the particular sample conditions used here, and concentrations were further reduced by the subsequent protein precipitation occurring during the data acquisition.

The aromatic sidechain hydrogen atoms of Tyr, Phe, and Trp residues were assigned using the 2D experiments correlating the Hδ and Hε to the Cβ (2D CBHD and CBHE [16]) with samples in 100% D_2O. The corresponding aromatic carbons were identified in the ^{13}C-^1H HSQC. Due to extensive overlap of the aromatic carbon atoms in the spectra, only 32% of them could be assigned unambiguously.

The total percentages of chemical shifts assigned are reported in Table 1. Excluding Met0, the $(His)_6$ tag and two Pro residues, the backbone assignments (Cα, C′, HN, N and Hα) were 95% complete, the sidechain residue assignments (including Cβ and Hβ) were 67% complete, and the aromatic residue assignments were 50% complete. Overall, the chemical shift assignment was achieved to an extent of 77%. The majority of the unassigned chemical shifts for both backbone and sidechain belong to residues of the CDR1 and CDR3 regions, which are expectedly less rigid than the remaining structure, thereby leading to inherently poor frequency spreading and/or broad line widths when unfavorable mobility rates are also involved. The completeness limits of the aromatic residue assignment could instead be totally ascribed to extensive resonance degeneracy from high mobility, for which characterization was mostly ambiguous and hence peaks unassignable, especially for carbons.

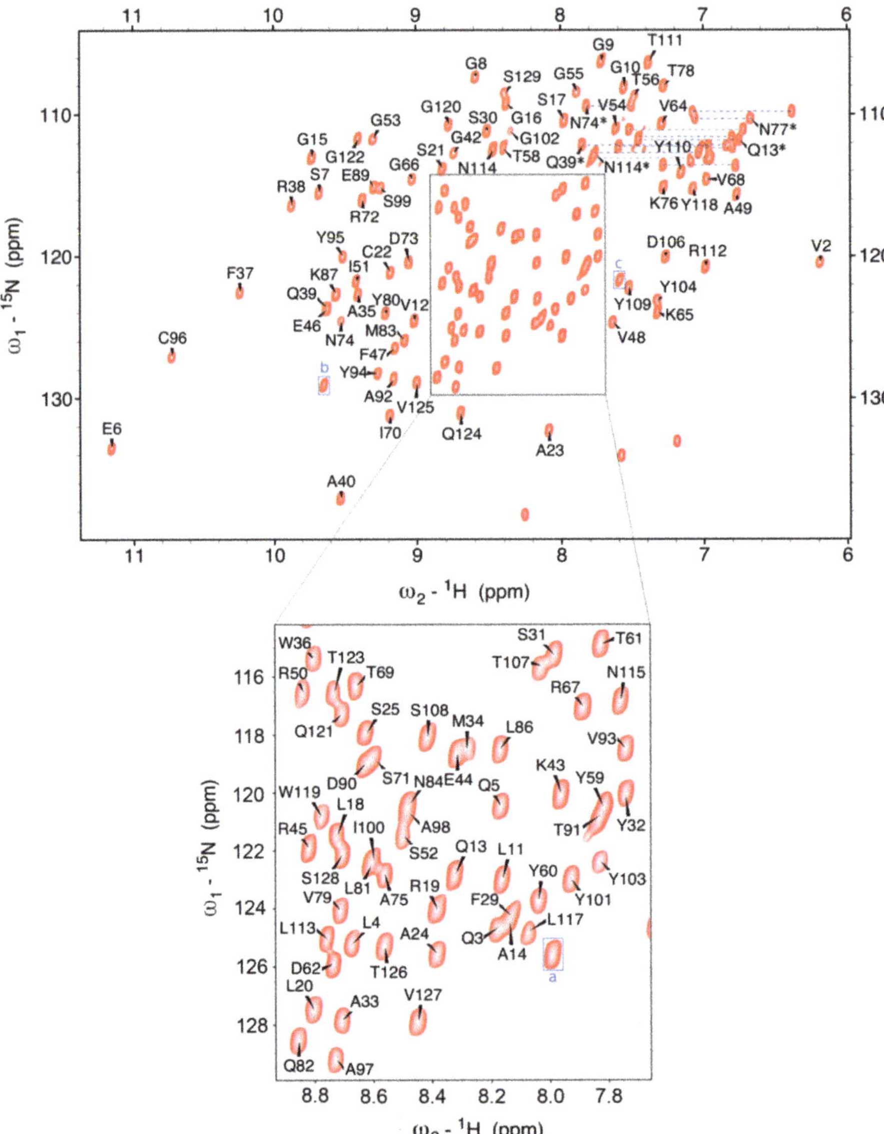

Figure 1. The $^{15}N-^{1}H$ HSQC of Nb23 from a freshly prepared sample (247 μM in 19.5 mM bis-Tris and 21 mM NaCl). The good signal−to−noise of the spectrum allowed the application of a squared sine−bell shifted by $\pi/6$ to achieve complete resolution. Excluding Met0 and the C−terminal $(His)_6$ tag used for expression, five N−H connectivities could not be assigned (Gln1, Arg27, Thr28, Ser63, and Ser105). Only the three blue-boxed connectivities, labeled a, b, and c, out of those that were observed, could not be attributed through scalar correlation. A tentative assignment is proposed based on heteronuclear NOE (see main text). The central area highlighted with a box has been enlarged for better visualization (lower panel) to limit the assignment annotation crowding given the high density of peaks. The Asn and Gln sidechain carboxyamide pairs could be connected from the slow exchange cross−peak of 2D $^{1}H-^{1}H$ NOESY, which also enabled the identification in a few cases from intra−residue NOE. The pairs are connected with blue dashed lines and the assigned ones are marked with an asterisk. The dispersion of peaks indicates a well−structured protein. The remaining peaks without labels belong to sidechain NHs, i.e., Arg, His, and Trp.

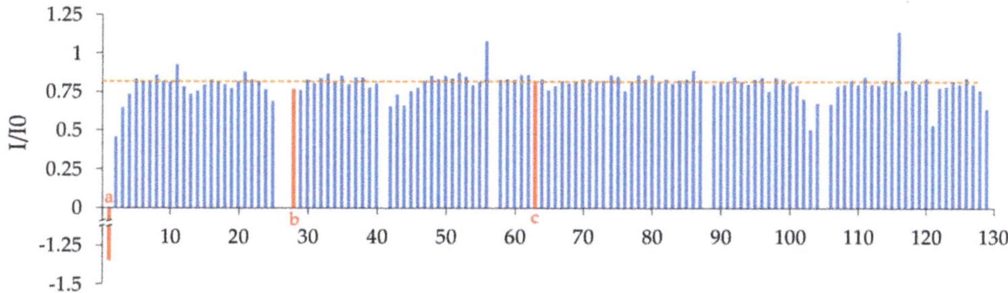

Figure 2. $^{15}N\{^{1}H\}$ NOE values, with I/I0 ratios representing the individual amide signal intensity with and without hydrogen saturation. The horizontal dotted line marks the average ratio value. Ratios below the average line indicate regions of mobility in the protein. The main regions of flexibility correspond to the supposed CDR1 (positions 26−31), a supposed loop between positions 42 and 45, and the supposed initial part of the CDR3 (positions 102−106). Residues with no bar correspond to either prolines (Pr041 and Pro88) or residues which were missing NH assignment. Based on the NOE values obtained for peaks (a), (b), and (c), that did not show scalar correlation in 3D spectra (Figure 1), a tentative assignment is proposed, respectively Gln1, Thr28, and Ser63, as indicated by the positions of the red bars.

Table 1. Chemical shift assignment completeness.

	Total	^{1}H	^{13}C	^{15}N
Backbone	95%	96%	94%	96%
Sidechain	67%	73%	69%	0%
Aromatic	50%	68%	32%	0%
Overall	77%	80%	75%	71%

2.3. Secondary Structure Content Assessment

An assessment of secondary structure content was made by looking at the difference of the deviations from random conformation chemical shifts of the assigned Cα and Cβ resonances ($\Delta\delta^{13}C\alpha - \Delta\delta^{13}C\beta$) [17]. To identify secondary structure elements using the individual carbon resonances, the chemical shifts are compared to the random coil chemical shift of the corresponding residue. A difference larger than ±0.7 ppm from the random coil chemical shift for several consecutive residues indicates the presence of secondary structure elements. Four consecutive downfield shifted Cα resonances beyond the 0.7 ppm threshold with respect to the random coil shift indicate α-helical structure, while three consecutive upfield shifted resonances in a row indicate β-strand presence. The opposite is true for Cβ resonances (downfield shift indicates β-strand, upfield shift indicates α-helix) [18]. The difference between the $\Delta\delta^{13}C\alpha$ and $\Delta\delta^{13}C\beta$ eliminates any possible chemical shift reference error on the individual deviations, with a positive $\Delta\delta^{13}C\alpha - \Delta\delta^{13}C\beta$ difference indicating α-helix and a negative difference indicating β-strand. Here, a cumulative approach to identify secondary structure elements from the $\Delta\delta^{13}C\alpha - \Delta\delta^{13}C\beta$ difference was employed by using an error threshold derived from the individual ± 0.7 ppm deviations of $\Delta\delta^{13}C\alpha$ and $\Delta\delta^{13}C\beta$, i.e., $\sqrt{(0.7^2 + 0.7^2)} \cong 1$ ppm. The results are illustrated in Figure 3, with the expected secondary structure elements highlighted in the figure. Overall, nine β-segments could be identified, a number consistent with the typical β-strand content of a canonical immunoglobulin variable domain, with a percentage of residues involved in β-strands of 49.6%. In comparison, Nb24 has a β-strand content of 50.4% when bound to antigen [11]. One possible α-helical tract was identified in the supposed CDR3 loop between residues 107 and 109.

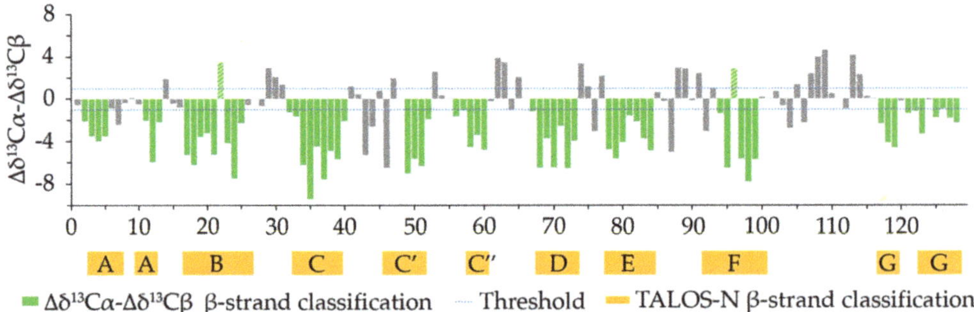

Figure 3. The chemical shift indexing analysis (CSI), computed by taking the difference between the experimentally determined Cα chemical shifts and the Cα random coil chemical shift ($\Delta\delta^{13}C\alpha$) minus the difference between the experimentally determined Cβ chemical shift and the Cβ random coil chemical shift ($\Delta\delta^{13}C\beta$). Three negative $\Delta\delta^{13}C\alpha - \Delta\delta^{13}C\beta$ values in a row indicate the presence of β-strand. A cumulative threshold error based on the individual Δδ deviations of ±0.7 ppm, i.e., $\sqrt{(0.7^2 + 0.7^2)} \cong 1$ ppm, was used as a threshold to include only significantly varying consecutive negative values. Residues predicted to be in β−strands are highlighted in green in the graph. The chemical shift differences of Cys22 and Cys96 (highlighted by green hatched bars) deviate because of upfield shifts induced by aromatic sidechains. As a consequence, especially for the Cβ chemical shifts, typical values of the reduced cysteines were observed despite the presence of the disulfide bridge with the associated β structure content. Control CD spectra of oxidized and reduced Nb23 are reported in Supplementary Materials to illustrate the issue, showing that Cys22 and Cys96 form a disulfide bridge. Yellow blocks indicate the position of residues that were estimated to be in β−strands by TALOS-N.

For an alternative assessment of secondary structure content, TALOS-N [19] was also used to infer φ and ψ torsion angles of Nb23 sequence from its backbone and Cβ chemical shift assignments. Torsion angles are in turn characteristic for certain types of secondary structures. The secondary structure content obtained by TALOS-N assessment is also illustrated in Figure 3. Here β-strand content was also 50.4% (as for Nb24), marking a difference with the chemical shift indexing analysis.

Circular dichroism (CD) data collected for Nb23 and uploaded to the Beta Structure Selection (BeStSel) server, a CD data analysis server especially useful for identification of β structures [20], show that Nb23 is mainly composed of antiparallel β-strands with different twists. No α-helical segments were identified. The overall β-strand content of the structure was 55.2%, which is slightly exceeding the content from the chemical shift indexing and TALOS-N estimations. This is not surprising as BeStSel assessment also includes relaxed β-strands. The results from the BeStSel analysis can be found in the Supplementary Materials.

2.4. Constraints and Nb23 Structure Calculation

Given the lack of assignment for a number of Nb23 sidechain resonances, an alternative strategy was employed to collect necessary constraints for restrained modeling. The CS-Rosetta server was used to provide a model for Nb23 in order to facilitate the search for experimental constraints. CS-Rosetta uses chemical-shift-constrained homology modeling to outline a 3D protein structure, based on the prediction of backbone and side-chain dihedral angles from the amino-acid sequence and the analogy of the experimental chemical shifts with those of a characterized model ensemble derived from PDB and BMRB [21]. The CS-Rosetta run generated 40,000 models of Nb23. The Cα-Root Mean Square Deviation (Cα-RMSD) was calculated for all of the models with respect to the lowest energy structure, yielding an averaged Cα-RMSD of 1.53 ± 0.99 Å for the ten best structures, calculated over the fragments 1–102, 117–122. Residues 103–116, coinciding with the tentative location of CDR3 loop, were considered as a flexible region. The CS-Rosetta run was deemed as successful as it achieved a Cα-RMSD below 2 Å for non-flexible regions for the ten lowest energy structures and the run converged towards a single structure.

The average β-structure content of the CS-Rosetta models was 49.2%, comparable to the β-structure content of TALOS-N and CD. The β-strand positions also coincided well with the TALOS-N β-strand positions except between residues 57 and 60, where β-secondary structure was consistently absent in the models.

Given the good agreement between the TALOS-N estimates, CD spectroscopy results, and the CS-Rosetta models regarding the β-secondary structure content, as well as the satisfactory Cα-RMSD for the ten best structures, the CS-Rosetta models were deemed as representative of Nb23 for the residues 1–102 and 117–122, and used as prior knowledge for NOE-constraint identification. The conformation of the CDR3 (residues ~101–116) was however not defined for the CS-Rosetta models and was not used for the same purpose.

A 3D ^{15}N-^{1}H NOESY HSQC spectrum, and aliphatic and aromatic 3D ^{13}C-^{1}H NOESY HSQC spectra, were acquired in order to extract NOE constraints for structure determination. Complementary 2D ^{1}H-^{1}H NOESY spectra were also acquired using unlabeled protein samples. Besides the attribution difficulties deriving from the missing sidechain assignments, the NOE identification was also hampered by resonance overlap and critical signal-to-noise ratio due to progressive decrease of protein concentration. The total number of NOE constraints extracted from the spectra using automated and manual assignments, handled by means of the software PONDEROSA [22,23], with prior knowledge from CS-Rosetta models was limited (619), first because of the lack of extensive assignment for the aliphatic and aromatic sidechains, and second because of selection of only unequivocal correlations. This apparently "minimalist" approach was adopted because the structural restraining was already based on the experimentally constrained models of CS-Rosetta, that included 734 chemical shift values constraining 353 dihedral angles. Nonetheless, very characteristic NOE patterns for β-secondary structure types [24] concerning backbone atoms were identified for most residues expected to be found in β-strands as per the chemical shift indexing analysis. Hydrogen bonded amides were also identified by recording a ^{15}N-^{1}H HSQC spectrum one week after transferring the protein to D_2O. This allowed for identification of slowly exchanging amide protons which are involved in secondary structure formation or are otherwise hydrogen bonded [25]. In that spectrum, the backbone NHs of 18 residues were characterized as slowly exchanging, all of which were expected to occur in secondary structure elements as per the chemical shift indexing analysis. The corresponding H-bonds were thus added as distance restraints (the relative list is reported in Supplementary Materials, Table S1). The 20 best NOE-restrained structures were validated with the tools of the PDB Validation Service [26–28] (see Supplementary Materials) and subjected to energy minimization as described in the Materials and Methods section. The ensemble of the ten lowest energy and most similar structures was retained. The relative validation report can be found in the Supplementary Materials.

A summary of the structural features and violations of the CS-Rosetta ensemble, the 20 NOE-restrained structures, and the ten NOE-restrained energy-minimized ensemble is shown in Table 2.

Table 2. Summary of features and violations for the CS-Rosetta ensemble, the NOE-restrained ensemble, and final NOE-restrained and energy minimized ensemble.

Nb23 CS-Rosetta (10 Structures)	
Clashes	
van der Waals clashes	8 (0.8 clashes/structure)
Average clash	0.48 ± 0.05 Å
Ramachandran plot distribution	
Residues in favored regions	97%
Residues in allowed regions	2%

Table 2. *Cont.*

Nb23 CS-Rosetta (10 Structures)	
Outliers	1%
χ outliers per structure	0.1
Cα-RMSD 1–129 w.r.t. lowest energy structure *	3.442 ± 2.212 Å
Cα-RMSD 1–102, 117–122 w.r.t. lowest energy structure *	1.531 ± 0.994 Å
Nb23 NOE-Restrained (20 Structures)	
Distance Constraints	
Short-range	417
Medium-range	16
Long-range	186
Hydrogen bonds	18
Total	637
Violations	
Distance constraint violations	115 (5.75 violations/structure)
Short-range	4
Medium-range	6
Long-range	76
Hydrogen bonds	29
Average violation	1.13 ± 0.61 Å
Clashes	
van der Waals clashes	189 (9.45 clashes/structure)
Average clash	0.48 ± 0.08 Å
Ramachandran Plot Distribution	
Residues in favored regions	93%
Residues in allowed regions	6%
Outliers	1%
χ outliers per structure	2.45
Cα-RMSD 1–129 w.r.t. least violation structure	1.98 ± 0.68 Å
Cα-RMSD 3–100, 118–128 [Å] w.r.t least violation structure	1.70 ± 0.68 Å
Nb23 NOE-Restrained Energy-Minimized (10 Structures)	
Clashes	
van der Waals clashes	0
Ramachandran Plot Distribution	
Residues in favored regions	96%
Residues in allowed regions	4%
Outliers	0%
χ outliers per structure	0.6
Cα-RMSD 1–129 w.r.t least violation structure *	1.57 ± 0.32 Å
Cα-RMSD 3–100, 118–128 w.r.t least violation structure *	1.23 ± 0.30 Å

* The pariwise Cα-RMSD for the respective ensembles, as well as the pairwise Cα-RMSD between the CS-Rosetta ensemble and the final NOE-restrained and energy minimized ensemble, are reported in Table S2 in the Supplementary Materials.

2.5. Nb23 Structural Features

The ten best Nb23 structures from energy minimization were deposited in the PDB (PDB ID 7EH3) and will be henceforth referred to as NOE-restrained best cluster. The first structure of the NOE-restrained best cluster is shown in Figure 4. The dispersion of the structures within this cluster was assessed by Cα-RMSD. The averaged Cα-RMSD with respect to the best structure was 1.57 ± 0.32 Å. Excluding the CDR3 (residues 101–117), which is expectedly more mobile and is the most variable part of immunoglobulin domains, and residues 1, 2, and 129, the Cα-RMSD was instead 1.23 ± 0.30 Å, highlighting the extent of the CDR3 contribution. An overlay of the backbone of the NOE-restrained best cluster is shown in Figure 5a. The corresponding β-structure content detailed in Table 3 for each element of the cluster can be compared to the experimental data from the $\Delta\delta^{13}C\alpha - \Delta\delta^{13}C\beta$ chemical shift indexing analysis and the TALOS-N assessment of secondary structure content shown in Figure 3. The superposition of the CS-Rosetta ensemble displayed in Figure 5b highlights the much larger dispersion of the CDR3 region with respect to the NOE-restrained best cluster. A visualization of the positions of the β-strands is shown in Figure 5c. The average β-structure content of the NOE-restrained best cluster is 40.9%, which is lower with respect to the CSI and TALOS-N estimations. Structure 3 (43.4% β-structure content) and Structure 8 especially (46.5% β-structure content) exhibit better and very similar overlap with the CSI, TALOS-N and CS-Rosetta models, while the remaining conformers of the ensemble have a more lacking β-structure content to the one inferred from the CSI and TALOS-N. It is possible that proper β-structure did not appear in the fragments highlighted in Figure 4 due to the relatively low number of constraints found for Nb23. Given that both the β-strand content scores from CSI, TALOS-N and CS-Rosetta modeling indicate higher values, in analogy with the evidence from CD, the β-structure content of the NOE-restrained best cluster may be underestimated. However, the absence of inter-strand NOEs, especially at the edges of the sheets, concerning primarily backbone residues, also suggests the occurrence of loose geometry in solution, as observed with isolated immunoglobulin motifs in solution [8,10].

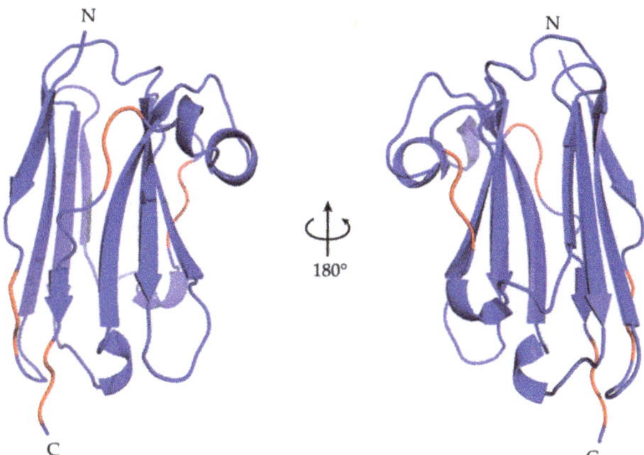

Figure 4. The best Nb23 structure from energy minimization of the NOE-restrained PONDEROSA C/S models. The structure is the lowest energy conformer of the NOE-restrained best cluster deposited in PDB (7EH3). It has the general features of a variable immunoglobulin domain, with the characteristic extended CDR3 of nanobodies which for Nb23 shields the solvent-exposed hydrophobic sidechains of Phe37, Phe47, Ile51, and Trp119. The β-strand content in the NOE-restrained best cluster is under-represented with respect to the analogous content of the CS-Rosetta structure ensemble. The red color highlights the location of the fragments extended but devoid of regular β-structure. Table 3 shows the positions of the β-strands for each structure of the NOE-restrained best cluster.

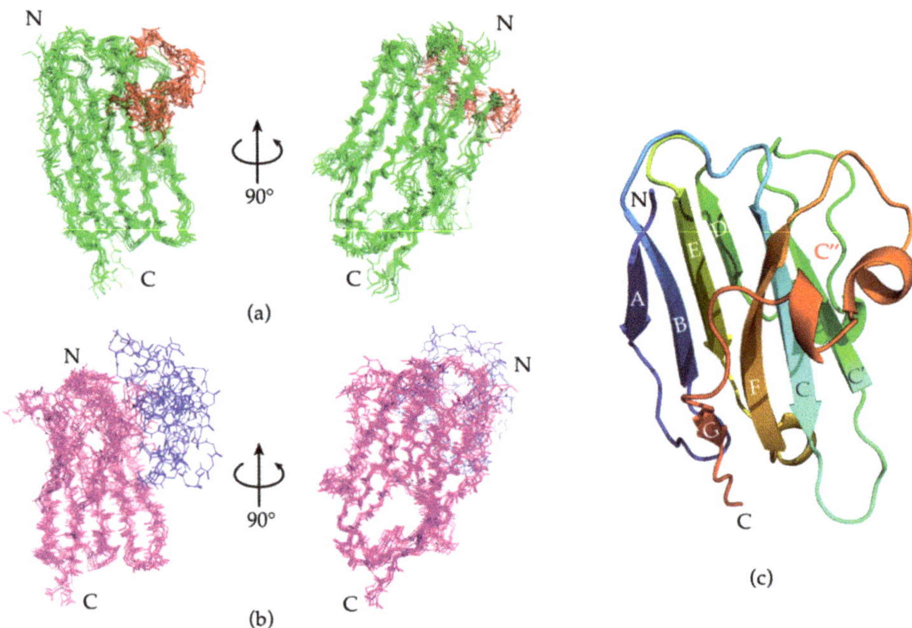

Figure 5. (**a**) An overlay of the Nb23 backbone of the NOE-restrained best cluster. The Cα−RMSD with respect to the best structure was 1.57 ± 0.32 Å, with substantial conformational dispersion localized in the CDR3 (highlighted in red). By excluding the CDR3 and residues 1, 2, and 129 from the alignment, the Cα−RMSD was 1.23 ± 0.30 Å. (**b**) An overlay of the Nb23 backbone of the CS−Rosetta ensemble. The Cα−RMSD with respect to the lowest energy structure was 3.42 ± 2.12. By excluding the CDR3 (highlighted in blue), the Cα-RMSD was 1.53 ± 0.99 Å, calculated over the fragments 1−102, 117−122. The conformational dispersion at the CDR3 is much more pronounced than the spread of the corresponding region in the NOE-restrained best cluster. (**c**) A visualization of the positions of the β−strands, lettered in white or grey. The only whole strand missing (C″) is highlighted in red, and protein terminals in black.

Table 3. β-structure content of the calculated Nb23 structures.

β-Strand	A	A *	B	C	C′	C″	D	E	F	G	G *
Structure 1	3–7	-	17–25	32–39	46–51	-	69–73	77–84	92–100	-	123–125
Structure 2	3–7	-	17–25	32–39	46–51	-	69–73	77–84	92–100	117–119	-
Structure 3	3–7	-	17–25	32–39	46–51	-	69–73	77–84	92–100	117–119	123–125
Structure 4	3–7	-	17–25	32–39	46–51	-	69–73	77–84	93–100	-	-
Structure 5	3–7	-	17–25	32–38	46–51	-	69–73	77–84	93–99	-	-
Structure 6	3–7	-	17–25	32–39	46–52	-	69–73	77–84	93–100	117–119	-
Structure 7	3–7	-	17–25	32–39	46–51	-	69–73	77–84	93–101	117–119	-
Structure 8	3–7	-	17–26	32–39	46–51	59–61	69–73	77–84	92–100	117–119	123–125
Structure 9	3–7	-	17–25	32–39	46–51	-	69–73	77–84	93–100	117–119	-
Structure 10	3–7	-	17–25	32–39	46–51	-	69–73	77–84	93–100	117–119	-

* The A and G strands are composed of two separate β-segments as per the CSI and TALOS-N analyses. A dash (-) indicates the absence of a particular segment in the corresponding NOE-based Nb23 structures.

A different assessment of this scenario may come from an evaluation of the structural data that were obtained by CS-Rosetta or NOE-restrained and energy minimization modeling, based on the recently proposed ANSURR method [29]. According to this validation approach, the accuracy of an NMR structure cannot be inferred from the spread of the final conformation ensemble, which reflects only the precision of the determination. The structural dispersion must be coupled to the correlation between the CSI and the flexibility of the molecule, as scored by software suites that exploit prior knowledge from data banks and/or neural networks. The ANSURR evaluation tested on decoys and real structures shows an interesting diversification between prevalently helical proteins and prevalently β proteins, with the former exhibiting a much higher flexibility-CSI correlation score than RMSD score, and the latter showing the opposite, i.e., a higher RMSD score than flexibility-CSI correlation. The ANSURR evaluation of the CS-ROSETTA ensemble appears to feature somehow the characteristics of the prevalently β-structured proteins, with average correlation and RMSD average scores of 24 ± 15 and 89 ± 11. Conversely, the NOE-restrained energy-minimized models exhibit unsatisfactory average correlation and RMSD scores of 9 ± 6 and 12 ± 6. A graphical presentation of the ANSURR results is reported in Supplementary Materials (Figure S3). The close Cα-RMSD values of the CS-Rosetta ensemble (1.53 $\pm$ 0.99 Å) and the NOE-restrained best cluster (1.57 $\pm$ 0.32 Å) seem to conflict with the RMSD scores of ANSURR that appear satisfactorily high, as expected for β-rich proteins, only with the CS-Rosetta ensemble. Also, the CSI-flexibility correlation score shows an appreciable difference between the CS-Rosetta and the NOE-restrained ensembles. Given the identity of the sequence and the associated chemical shift list, with the consequent flexibility estimates, the difference of CSI-flexibility correlation of the ANSURR assessments must be related to the different β-structure content of the two ensembles, namely the small deviations from regular geometry of the NOE-restrained ensemble shown in Figure 4 that prevent classification as β-structure and therefore conflict with local CSI. Even with a modest CSI-flexibility correlation score and a structural dispersion equivalent to that of the NOE-restrained best cluster, the CS-Rosetta cluster reaches the typically large RMSD score of the β-rich proteins.

No helical segments were identified from the $\Delta\Delta\delta^{13}C\alpha - \Delta\Delta\delta^{13}C\beta$ chemical shift indexing analysis, although TALOS-N predicted four helical segments. Four of the NOE-restrained minimized structures have a right-handed helical fragment between residues 29 and 31. This fragment coincides with the putative CDR1 loop, and the recurrent three-residue helix in the structures could be an indication of a 3_{10}-helical segment, which has a characteristic three-residue turn. The carbonyl oxygen of Thr28 (i) seems to face the HN of Ser31 (i + 3) at an average distance of 2.4 Å. The remaining structures have a helically-shaped loop at the same location; however, no secondary structure element came out for those structures. A similar helical segment is formed in eight of the ten structures of the NOE-restrained best cluster, between residues 62 and 64, with the carbonyl oxygen of Thr61 facing the HN of Val64. There is also a three-residue helix tract, i.e., a helical turn, where the carbonyl oxygen of Lys87 (i) seems to face the HN of Asp90 (i + 3) at an average distance of 2.1 Å, the residues completing a full turn. This is possibly also a 3_{10}-helix. One segment in helical conformation is present in all of the NOE-restrained best cluster structures, in the supposed CDR3 loop, from position 107 to 111 (107–109 for one structure). This segment is in right-handed α-helix conformation, where the carbonyl oxygen of Thr107 (i) faces the HN of Thr111 (i + 4), at an average distance of 2.4 Å. The residues complete a full turn consistent with an α-helical segment. Another segment in helical conformation can be found in five of the structures between positions 113 and 115. This segment shows that the carbonyl oxygen of Arg112 (i) faces the HN of Asn115 (i + 3) at an average distance of 2.1 Å, i.e., a geometry that is consistent with a 3_{10}-helix.

Figure 6 shows the orientation and surface of the CDR loops for the first structure of the NOE-restrained best cluster. The orientation of the CDR3 is of particular interest, given its length and the degree of mobility at the beginning of the loop evidenced by the $^{15}N\{^{1}H\}$ NOE analysis. Hence, several different orientations for the CDR3 were, in principle, possible. This is also reflected in the CS-Rosetta-generated models, where the β-core of the structure is very similar for each model while the CDR3 has a different conformation for each model. The CDR3 of the PONDEROSA-C/S energy-minimized structures included in the cluster has instead a more consistent conformation, with limited variations in the CDR3 relative to the CS-Rosetta models (Figure 5a,b). Fundamental to the orientation of the CDR3 in the NOE-restrained best cluster are the NOEs between Arg50 in β-strand C′ and Tyr104 of the CDR3. This well detectable interaction in the NOE spectra suggests a possible cation–π electrostatic interaction [30] between the Arg50 sidechain and the aromatic ring of Tyr104, which would partially keep the loop in a more defined orientation. Interestingly, position 104 of Nb24—the mentioned nanobody with similar binding properties to the β2m mutants as Nb23—is occupied by a cysteine which forms a disulfide bond with Cys33 of the β-strand C, essentially freezing the loop in a rigid conformation in Nb24. Position 33 is structurally arranged to be adjacent to position 50. Therefore, the cation–π interaction of Nb23 could vicariate the Cys33-Cys104 disulfide bridge of Nb24. One possible orientation of the sidechains of Arg50 and Tyr104 in Nb23 is shown in Figure 7, where the Arg50 sidechain faces the aromatic ring making the cation–π interaction possible [30].

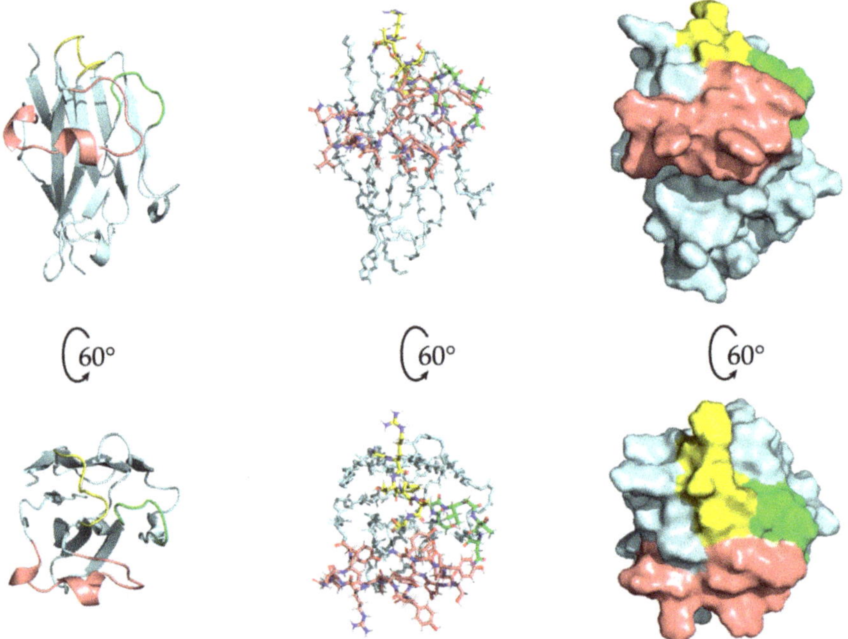

Figure 6. The CDRs of Nb23, with CDR1 in yellow, CDR2 in green, and CDR3 in salmon. The left column shows the cartoon representation of Nb23 without any sidechains. The central column shows the CDRs with sidechains (and only backbone for the β−core). The right column shows the surface of the protein with the CDRs highlighted. The predominance of the CDR3 in the antigen−binding site is evident, highlighting its importance in interacting with the antigen(s). Its orientation affects the size and shape of the antigen-binding site for the unbound nanobody, although the flexibility in residues 102−106 suggests that the CDR3 conformation may change as the nanobody binds its antigen(s).

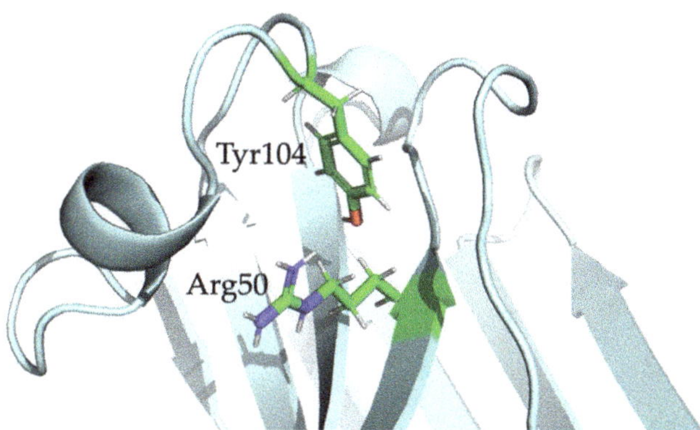

Figure 7. The Tyr104 phenolic ring in the CDR3 and the Arg50 βCH2 in β−strand C″ of Nb23 show proximity as per the assigned NOE constraints. This indicates the possible presence of a guanidinium−π interaction, partially keeping the CDR3 in a defined orientation. The cartoon shows one of the arrangements of the residues in the NOE−restrained best cluster.

2.6. Molecular Dynamics Simulations

The possible conformations for the CDR3 were investigated with molecular dynamics (MD) simulations, starting from representative of the six different clusters including all the best 18 energy-minimized structures from PONDEROSA C/S modeling. All simulations show an initial increase of the RMSD from the first structure of the specific NOE-restrained cluster, followed by rather stable equilibration at the value of about 2.5 Å (Figure 8a). During the simulation, most of the structures fluctuate about an average conformation with lower RMSD with respect to the initial structure, as witnessed by the much lower residue root mean square fluctuations (RMSFs) on the superimposed residues (Figure 8b). Large RMSF values are observed at loops and in the region 100–120 encompassing the CDR3. This is observed in most simulations, although in one of the simulations the region 50–70 is also showing large fluctuations.

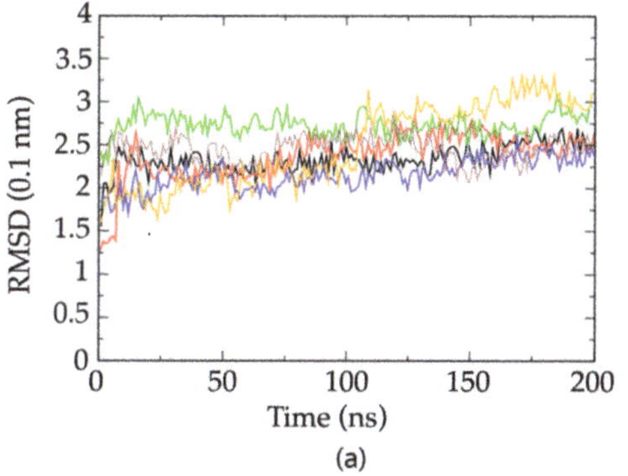

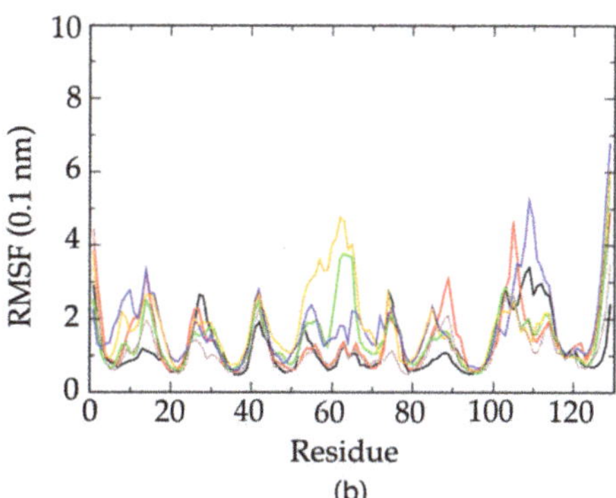

Figure 8. (**a**) RMSD with respect to the lowest energy structure of the NOE−restrained clusters as a function of time of the six MD simulations that were carried out starting from the minimized representative structures from the six clusters of the NOE−restrained PONDEROSA C/S models of Nb23. Black trace = cluster 1 (11 members); red trace = cluster 2 (2 members); green trace = cluster 3 (2 members); blue trace = cluster 4 (1 member); orange trace = cluster 5 (1 member); pale brown trace = cluster 6 (1 member). (**b**) RMSF in the same six MD simulations as in panel (**a**), as a function of the residue number of Nb23. The color code of the traces is the same as in panel (**a**).

MD confirms the proximity of Arg50 and Tyr104 sidechains in all of the simulations originating from the different clusters of PONDEROSA C/S energy-minimized conformers, with a geometry of either cation–π or π–stacking interaction in the snapshots of the simulation concerning the NOE-restrained best cluster.

An interesting observation is that the simulations starting from different minimized conformers of the PONDEROSA C/S clusters sample different regions of the conformational space, as can be seen by comparing the average RMSD at each residue for the ensemble of MD snapshots from each pair of simulations and for the ensemble of the pooled snapshots. An example is provided in Figure S4 with the pooling (dashed curve) of two of the MD snapshot ensembles depicted in Figure 8B. The large increase in RMSD upon pooling the two ensembles is indicative of large differences in the conformations about which the two MD simulations are fluctuating (see Figure S4).

3. Discussion

Nb23 was raised against ΔN6β2m to inhibit its amyloid formation, and could potentially be used for inhibiting fibril formation of other amyloidogenic β2m-variants. By using typical TROSY 3D experiments for backbone and aliphatic sidechain assignments, and 2D aromatic sidechain experiments for aromatic assignments, the chemical shifts of Nb23 were assigned. These chemical shift assignments were used for chemical shift-based homology modeling with CS-Rosetta giving a representative protein model as output. The model was in turn used together with the chemical shifts for NOE-restrained structure calculation supported by prior-knowledge of the structure. Relying on the experimental character of this prior knowledge, the choice was deliberately made to include only the unambiguously assigned NOEs to determine the solution structure of Nb23. Despite using what is considered a low number of NOE constraints (619) for structure determination—usually one would need ten NOEs per residue and Nb23 has ~130 residues—the resulting structures showed the general features of a single variable immunoglobulin domain and the general features of a nanobody. This minimalist approach was employed because of extensive signal overlap (especially for sidechains) making the unambiguous assignment not possible. Unfortunately, the issue of ambiguity could not be addressed because the necessary improvements of signal-to-noise and resolution conflicted with (i) the solubility and stability limits of Nb23 samples, which form precipitate in a matter of hours after dissolving the protein, and (ii) the current difficulties of accessing higher magnetic field facilities. Strictly speaking, the adopted minimalist approach is more rigorous than assigning NOEs, even when they are ambiguous, and then minimizing the constraints violations by progressive refinement with repeated trial-and-error calculations. When the spectral quality is not sufficient to remove assignment incompleteness or/and ambiguity, managing to reach the minimal restraint violation level with arbitrary release or retain of the internuclear distance attribution may only improve the precision of the determination, but definitely not its accuracy, as recently pointed out [29]. Thus, instead of relying on the number of NOE constraints as a quality determinant, the structures restrained with only unambiguous NOEs were evaluated on their similarities to the CS-Rosetta modelled ensemble, that was anyway based on the experimental chemical shifts (CS-Rosetta modelling included more than 700 chemical shift values constraining more than 350 dihedral angles).

The structures resulting from this protocol were subjected to energy minimization to adjust energetically unfavored sidechain conformations and to reduce the number of too-close contacts between adjacent atoms. A cluster of ten similar structures, deemed as representative of the structure of nanobody Nb23, was deposited in the PDB. The overall quality of this deposited ensemble was ranked to be far above average by the PDB validation server with respect to the deposited NMR structures (see Supplementary Materials).

The clustered structures were subjected to MD simulations to assess the conformational space available to the CDR3. The CDR3 showed particularly high values in RMSF, conforming that this functionally crucial region indeed could possibly have a range of conformations.

The deposited Nb23 structures (PDB ID 7EH3) have the main structural features observed in nanobodies: a β-core structure, and an extended CDR3, both for shielding solvent exposed hydrophobic sidechains (in particular Phe37, Phe47, Ile51, and Trp119) and for binding cryptic epitopes [1]. A comparative superposition of the solution structure of free Nb23 and the Nb24 structure to explain their activity differences can be misleading at the present stage. For Nb24, in fact, no structure of the free protein in solution is available as of now, whereas the crystal structures of the complexes with β2m variants were reported [11,12] to exhibit peculiar aspects that may be related to the crystalline state [11] or to the specifically selected β2m variant [12].

Structural characterization is fundamental to uncover subtle conformational differences that lead to changes in thermodynamic and kinetic parameters for the complexation of different nanobodies such as Nb23 and Nb24 with the β2m-mutants. In this respect, the lack of some fragments of secondary structure elements in the β-core of Nb23 is not of concern, because the departure from the canonical geometry amounts to small deviations that are consistent with loose arrangements and absence of inter-strand NOEs, especially at strand edges. This contributes to decreasing the number of employed NOE contacts, barely half of the required minimum threshold of ten contacts per residue. It was reasoned that the β-core of those immunoglobulin domains, so well represented in the PDB and in literature, would be well evidenced by the convergence of the CS-Rosetta models that guided the NOE search and could therefore determine a satisfactory result.

The impact of the 'lacking' β-strand content on the function of the nanobody should not be of great relevance, considering that the paratope of the nanobodies and immunoglobulin domains in general lies in the CDRs. Moreover, some loosening of the β-scaffold in the solution structure of isolated immunoglobulin domains is not surprising [8,10]. Of much more importance is instead the definition of the interactions that shape the CDR3 conformation, partially uncovered in this study. The structure and orientation of the CDR3 in Nb23 was found to both satisfy one of its principal tasks, i.e., shielding of conserved hydrophobic residues in the isolated protein, and be similar to that of the best CS-Rosetta model. In particular, Nb23 shows an interesting series of contacts between the sidechains of Arg50 and Tyr104 which could reflect the occurrence of a cation–π electrostatic interaction between the guanidinium and the phenolic ring. This interaction may vicariate for the disulfide bridge of Cys33 and Cys104 that occurs in camel-derived nanobodies such as Nb24. Besides the canonical disulfide linking the two β-sheets of immunoglobulins, camel-derived V_HH domains exhibit in fact an additional cystine in the CDR3 region, that of course affects the local conformational options. Llama-derived V_HH domains such as Nb23 do not possess this additional covalent constraint, but the occurrence of an energetically non-labile interaction such as a cation–π electrostatic one could help to modulate more precisely the available conformational repertoire. Importantly, the non-trivial character of this interaction should not conflict with the mobility in other regions of the CDR3, as suggested by the pattern of $^{15}N\{^1H\}$ NOE histogram (Figure 2) and the hypothesized conformational exchange that prevents the observation of the Ser105 NH signal.

In conclusion, Nb23′s structure determination is a first characterization step that will enable a more holistic assessment of its performance in inhibiting amyloidogenic β2m variants, once the solution structure of the isolated Nb24 and those of the complexes of both nanobodies with their antigens are also available. One possible outcome for this type of comparison could be the rational design of new hybrid nanobodies that perform better in fibril inhibition than the already existing ones.

4. Materials and Methods

4.1. Nb23 Expression and Labeling

Nb23 was previously obtained by immunization of a llama with a truncated version of β2-microglobulin, ΔN6β2-m (a β2-m variant devoid of the first six residues), as reported by Domanska et al. [11]. Nb23 was obtained uniformly doubly labeled with ^{13}C and ^{15}N by growing the transgenic *E. coli* strain containing the expression vector previously described [11] on ^{13}C and ^{15}N enriched medium. Expression and purification were performed by ASLA Biotech AB (Riga, Latvia), that also provided the unlabeled Nb23. Nb23 consists of 136 amino acids, including an initial Met introduced as a start codon for expression in

E. coli, and a His6 tag at the C-terminus of the protein for purification purposes, amounting to a molecular weight of 15.1 kDa.

4.2. Nb23 Sample Preparation, NMR Data Acquisition, and Peak Assignment

All the NMR spectra were collected at the NMR facility of the Core Technology Platform at New York University Abu Dhabi on a 14 T Bruker Avance III spectrometer operating at 600, 150, and 60 MHz for ^{1}H, ^{13}C, and ^{15}N, respectively, with a triple resonance cryoprobe. The acquisition temperature was always set to 298.2 K. All samples for backbone and sidechain assignment or homonuclear correlations were prepared at labeled or unlabeled protein concentrations ranging from 190 to 291 μM in 95/5 H_2O/D_2O and 10 mM phosphate buffer, pH 6.95, with or without NaCl (6.3–21 mM). Occasionally 19.5 mM bis-Tris aqueous buffer was also used, always at pH 6.95. The samples for aromatic sidechain assignment were prepared in D_2O, at protein concentrations in the range 100–190 μM with 10 mM phosphate buffer, pH 6.98 (uncorrected pH-meter reading), without or with 20 mM NaCl. Importantly, the heteronuclear fingerprint of the ^{15}N-^{1}H HSQC spectra overlapped satisfactorily regardless of the mentioned buffer mixture. Protein concentrations were determined by UV absorption at 280 nm with an IMPLEN nanophotometer based on calculated molar extinction coefficients of 30,495 for Nb23. The sample concentrations were unstable over long time intervals. The initial concentration values invariably decreased by some 50% after 7–10 days as a consequence of protein precipitation. This proved detrimental for the sensitivity of the collected data sets, especially the later acquired ones, that could not be re-acquired due to labeled protein shortage.

A summary of the collected spectra with corresponding acquisition parameters is shown in Table 4. Pure phase detection in t1 and t2 dimensions of 3D data sets were obtained via gradient-based echo-antiecho selection and States-TPPI scheme [31–33]. The States-TPPI scheme was also employed for homonuclear NOESY and TOCSY spectra, whereas 2D heteronuclear spectra pure phase detection in t1 was obtained using echo-antiecho selection. The solvent was typically suppressed with a flip-back pulse [34], whereas in homonuclear spectra WATERGATE elements [35] applied in the excitation sculpting mode [36] were employed.

All 3D matrices were acquired with non-uniform sampling schemes by collecting 10%–20% of the whole datasets and by reconstructing the matrices with the dedicated routine of the Bruker Topspin 4.05 software [37]. The same software was used for processing all of the spectra with standard processing routines.

The NMR data were analyzed using NMRFAM-SPARKY [38], including peak assignment which was performed in a semi-automated manner using NMRFAM-SPARKY incorporated tools. The assignment list is available in BMRB, accession number 50808. Table 1 lists the overall assignment percentages.

Table 4. List of the collected spectra for backbone and side-chain nuclei assignment of Nb23, with the corresponding acquisition parameters. Experiments denoted with tr indicate the use of TROSY pulse schemes.

Spectrum	Time Domain Dimensions	Transients (NS)	Carrier (ppm)	Spectral Width (ppm)	References
2D ^{15}N-^{1}H HSQC	t_2 (^{1}H): 2048 t_1 (^{15}N): 128	8, 16	t_2 (^{1}H): 4.7 t_1 (^{15}N): 118	t_2 (^{1}H): 16 t_1 (^{15}N): 50	[39]
2D tr-^{15}N-^{1}H HSQC	t_2 (^{1}H): 2048 t_1 (^{15}N): 80	16	t_2 (^{1}H): 4.7 t_1 (^{15}N): 118	t_2 (^{1}H): 16 t_1 (^{15}N): 50	[40]
3D tr-CBCANH	t_3 (^{1}H): 1024 t_2 (^{15}N): 50 t_1 (^{13}C): 128	576	t_3 (^{1}H): 4.7 t_2 (^{15}N): 118 t_1 (^{13}C): 43	t_3 (^{1}H): 14 t_2 (^{15}N): 50 t_1 (^{13}C): 80	[14,41]

Table 4. *Cont.*

Spectrum	Time Domain Dimensions	Transients (NS)	Carrier (ppm)	Spectral Width (ppm)	References
3D tr-CBCA(CO)NH	t_3 (^{1}H): 1024 t_2 (^{15}N): 50 t_1 (^{13}C): 128	96	t_3 (^{1}H): 4.7 t_2 (^{15}N): 118 t_1 (^{13}C): 43	t_3 (^{1}H): 14 t_2 (^{15}N): 50 t_1 (^{13}C): 80	[14,42]
3D tr-HNCA	t_3 (^{1}H): 1024 t_2 (^{15}N): 50 t_1 (^{13}C): 96	32	t_3 (^{1}H): 4.7 t_2 (^{15}N): 118 t_1 (^{13}C): 54	t_3 (^{1}H): 18 t_2 (^{15}N): 50 t_1 (^{13}C): 80	[15]
3D tr-CC(CO)NH	t_3 (^{1}H): 1024 t_2 (^{15}N): 50 t_1 (^{13}C): 128	256	t_3 (^{1}H): 4.7 t_2 (^{15}N): 118 t_1 (^{13}C): 43	t_3 (^{1}H): 14 t_2 (^{15}N): 50 t_1 (^{13}C): 80	[14,43]
3D tr-H(CCO)NH	t_3 (^{1}H): 1024 t_2 (^{15}N): 50 t_1 (^{1}H): 128	256	t_3 (^{1}H): 4.7 t_2 (^{15}N): 118 ^{1}H: 4.7	t_3 (^{1}H): 14 t_2 (^{15}N): 50 ^{1}H: 14	[14,43]
3D tr-HBHA(CO)NH	t_3 (^{1}H): 1024 t_2 (^{15}N): 50 t_1 (^{1}H): 128	96	t_3 (^{1}H): 4.7 t_2 (^{15}N): 118 t_1 (^{1}H): 4.7	t_3 (^{1}H): 14 t_2 (^{15}N): 50 t_1 (^{1}H): 8	[14,44]
3D tr-HNCO	t_3 (^{1}H): 1024 t_2 (^{15}N): 50 t_1 (^{13}C): 96	32	t_3 (^{1}H): 4.7 t_2 (^{15}N): 118 t_1 (^{13}C): 173.5	t_3 (^{1}H): 14 t_2 (^{15}N): 50 t_1 (^{13}C): 22	[15]
3D tr-HN(CA)CO	t_3 (^{1}H): 1024 t_2 (^{15}N): 50 t_1 (^{13}C): 96	96	t_3 (^{1}H): 4.7 t_2 (^{15}N): 118 t_1 (^{13}C): 173.5	t_3 (^{1}H): 14 t_2 (^{15}N): 50 t_1 (^{13}C): 22	[45]
2D ^{1}H-^{1}H TOCSY	t_2 (^{1}H): 4096 t_1 (^{1}H): 768	192	t_2 ($_1$H): 4.7 t_1 ($_1$H): 4.7	t_2 ($_1$H): 14.4 t_1 ($_1$H): 14.4	[36,46,47]
3D ^{15}N-^{1}H NOESY HSQC	t_3 (^{1}H): 1024 t_2 (^{15}N): 50 t_1 (^{1}H): 400	96	t_3 (^{1}H): 4.7 t_2 (^{15}N): 122 t_1 (^{1}H): 4.7	t_3 (^{1}H): 14 t_2 (^{15}N): 42 t_1 (^{1}H): 14	[31,39,48]
2D CBHD (D2O)	t_2 (^{1}H): 2048 t_1 (^{13}C): 98	1024	t_2 (^{1}H): 4.7 t_1 (^{13}C): 36	t_2 (^{1}H): 16 t_1 (^{13}C): 28	[16]
2D CBHE (D2O)	t_2 (^{1}H): 2048 t_1 (^{13}C): 98	1024	t_2 (^{1}H): 4.7 t_1 (^{13}C): 36	t_2 (^{1}H): 16 t_1 (^{13}C): 28	[16]
2D ^{1}H-^{1}H NOESY (D2O)	t_2 (^{1}H): 4096 t_1 (^{1}H): 400	192	t_2 (^{1}H): 4.7 t_1 (^{1}H): 4.7	t_2 (^{1}H): 14.4 t_1 (^{1}H): 14.4	[36,46,49]
3D ^{13}C-^{1}H NOESY HSQC aliphatic (D2O)	t_3 (^{1}H): 1024 t_2 (^{13}C): 80 t_1 (^{1}H): 160	96	t_3 (^{1}H): 4.7 t_2 (^{13}C): 43 t_1 (^{1}H): 4.7	t_3 (^{1}H): 14 t_2 (^{13}C): 80 t_1 (^{1}H): 14	[31,39,48]
3D ^{13}C-^{1}H NOESY HSQC aromatic (D2O)	t_3 (^{1}H): 1024 t_2 (^{13}C): 80 t_1 (^{1}H): 160	96	t_3 (^{1}H): 4.7 t_2 (^{13}C): 105 t_1 (^{1}H): 4.7	t_3 (^{1}H): 14 t_2 (^{13}C): 80 t_1 (^{1}H): 14	[31,39,48]
2D ^{13}C-^{1}H HSQC	t_2 (^{1}H): 1024 t_1 (^{13}C): 196	32	t_2 (^{1}H): 4.7 t_1 (^{13}C): 72	t_2 (^{1}H): 16 t1 (^{13}C): 165	[39]

4.3. Restrained Modeling

The set of the experimentally determined backbone and Cβ chemical shifts were input to run restrained MD modeling by means of the CS-ROSETTA server [19]. The chemical shifts represent experimental information that is employed to restrain the backbone dihedral angles φ and ψ by means of a pseudopotential term that introduces an energy penalty upon violation [19]. The same energy-penalty-driven approach was employed to calculate the structure based on the inter-proton distances obtained from the 2D and 3D NOESY spectra. The NOE-restrained structure determination was handled by means of the software suite PONDEROSA-C/S, using PONDEROSA-X refinement by which automated database-assisted NOE assignment is done (AUDANA algorithm) [50]. Experimentally determined chemical shift assignments for backbone, sidechain, and aromatic residues were input to automatically assign the 3D ^{15}N-^{1}H NOESY HSQC spectrum, and aliphatic and aromatic 3D ^{13}C-^{1}H NOESY HSQC spectra and calculate the structure as per the above procedure. Automated NOE-assignments were manually checked to remove ambiguous assignments and to add additional constraints. NOE intensities were considered only qualitatively as strong, medium and weak, corresponding to upper limit distances of 0.25, 0.35 and 0.5 nm, respectively.

4.4. Energy Minimization

The best 20 structures from the PONDEROSA C/S modeling were energy minimized first to remove the few (7.5 on average per each structure) bad contacts present, for 2000 minimization steps, using the steepest descent minimization algorithm. Since the solvent was not present at this stage, the GBSA implicit solvent model was adopted as implemented in the NAMD simulation software [51] according to the model by Onufriev, Bashford and Case [52]. Energy minimization resulted in structures devoid of bad contacts (according to the software Procheck [53]), except for two structures for which bad contacts persisted even after lengthening the minimization to 10,000 steps. The latter two structures were removed from the ensemble for MD simulations. At the same time, the ensemble of the ten most similar structures after energy minimization was retained as representing the NOE-restrained best cluster.

4.5. Molecular Dynamics Simulations

The best 18 structures resulting from energy minimization of the PONDEROSA C/S modeling were clustered by the PDB validation server (URL: www.wwpdb.org, accessed on 3 March 2021) into one 11-structure, two 2-structure and three 1-structure clusters. The best structure from each cluster was selected and subjected to MD simulations. Six MD simulations lasting 200 ns were performed using NAMD simulation software [51]. TIP3P water molecules (Jorgensen, 1983) and ions, to reach a 0.150 M ionic concentration, were added using the solvate module of the program VMD [54]. The simulation box was on average ca. 260,000 Å^3 and the average number of atoms was 25,554. Molecular interactions were described by amber99sb-ildn force field [55]. Protein atoms were placed at the center of a cubic box at a minimum distance of 12 Å from the edge of the box. We used Periodic Boundary Conditions set by the size of the box. The solvated systems were energy minimized by 2000 steepest descent minimization steps. The equilibration phase was performed by increasing gradually the temperature from 0 to 310 K in 100 ps followed by further 900 ps. At this stage temperature was controlled by a simple velocity rescaling procedure and pressure at 1 atm was controlled by a pressure Langevin piston [56,57], with the period of 200.0 fs and decay constant of 100 fs. The time step was 1 fs, bonded interactions were computed every 1 fs and non-bonded interactions every 2 fs. Finally, MD simulation lasted 200 ns at constant pressure and temperature, the latter controlled through Langevin dynamics with damping constant of 1 ps^{-1}. Snapshots were collected every 1 ns along the trajectory, giving a total of 200 snapshots which have been used in the analysis.

A total of 200 structures obtained from each MD simulation at 1ns time interval were analyzed as an ensemble of structures. The RMSD from the initial energy minimized structure was obtained by superimposing the backbone atoms of the residues structured in beta sheet based on multiple alignment of annotated sequences, i.e., residues 3–7, 10–12, 18–27, 34–39, 46–51, 55–60, 68–73, 78–83, 92–98. The time evolution of RMSD during the simulation was computed in the same way. From all pairwise snapshots superpositions, the root mean square fluctuations (RMSFs) for the backbone atoms of each residue were computed. The comparison between different simulations was performed by considering the ensemble of structures from each simulation and the ensemble obtained joining the two ensembles. A large increase in RMSF upon joining the two ensembles, compared to RMSFs observed in both ensembles, is indicative of local fluctuations about different conformations, i.e., the two simulations are sampling a different conformational space.

Supplementary Materials: The following supplementary material is available online. Supplementary information (Validation, Energy minimization, Assignment and Structure data); Table S1: H-bond list; Table S2: Pairwise RMSD; Figure S1: CD spectroscopy of Nb23 in H_2O; Figure S2: CD spectroscopy of Nb23 in H_2O with TCEP; Figure S3: ANSURR assessment for the CS-Rosetta and the NOE-restrained best cluster; Figure S4: RMSF at each residue of Nb23 in the MD simulation.

Author Contributions: Conceptualization: G.E., M.P., J.S. and F.F.; Methodology: G.E., M.P., Y.H. and F.F.; Formal analysis: M.P., F.F. and G.E.; Investigation: M.P., Y.H., F.F.and G.E.; Resources: G.E., M.P. and Y.H.; Writing—Original Draft Preparation: M.P.; Writing—Review & Editing: M.P., G.E., F.F., Y.H. and J.S.; Supervision: G.E., F.F.and Y.H.; Project Administration: M.P., G.E., Y.H. and F.F.; Funding Acquisition: G.E. All authors have read and agreed to the published version of the manuscript.

Funding: This research was funded by NYUAD (grant No. 76 71260 ADHPG VP046).

Institutional Review Board Statement: Not applicable.

Informed Consent Statement: Not applicable.

Data Availability Statement: The data presented in this study are openly available in the Biological Magnetic Resonance Bank (BMRB), accession number 50808, and in the Protein Data Bank (PDB), PDB ID 7EH3.

Acknowledgments: We thank the Core Technology Platform (CTP) of NYUAD for the access and the support with the instrumentation that was essential to achieve the reported result. To the best of our knowledge, this is the first protein structure determined in the UAE using only NMR spectroscopy. We also thank N.J. Fowler and Makek A. for the help with ANSURR software.

Conflicts of Interest: The authors declare no conflict of interest. The funder had no role in the design of the project, in the collection, analysis and interpretation of the data, in the writing of the paper, or in the decision to publish the results.

Sample Availability: Samples of the compounds are available from the authors.

References

1. Muyldermans, S. Nanobodies: Natural Single-Domain Antibodies. *Annu. Rev. Biochem.* **2013**, *82*, 775–797. [CrossRef] [PubMed]
2. Jovčevska, I.; Muyldermans, S. The Therapeutic Potential of Nanobodies. *BioDrugs* **2020**, *34*, 11–26. [CrossRef]
3. Boulenouar, H.; Amar, Y.; Bouchoutrouch, N.; Faouzi, M.E.A.; Cherrah, Y.; Sefrioui, H. Nanobodies and Their Medical Applications. *Genet. Mol. Res.* **2020**, *19*, 1–12. [CrossRef]
4. Hassanzadeh-Ghassabeh, G.; Devoogdt, N.; De Pauw, P.; Vincke, C.; Muyldermans, S. Nanobodies and their potential applications. *Nanomedicine* **2013**, *8*, 1013–1026. [CrossRef]
5. Bannas, P.; Hambach, J.; Koch-Nolte, F. Nanobodies and Nanobody-Based Human Heavy Chain Antibodies As Antitumor Therapeutics. *Front. Immunol.* **2017**, *8*, 1603. [CrossRef]
6. Lafaye, P.; Achour, I.; England, P.; Duyckaerts, C.; Rougeon, F. Single-domain antibodies recognize selectively small oligomeric forms of amyloid beta, prevent Abeta-induced neurotoxicity and inhibit fibril formation. *Mol. Immunol.* **2009**, *46*, 695–704. [CrossRef] [PubMed]
7. Gejyo, F.; Yamada, T.; Odani, S.; Nakagawa, Y.; Arakawa, M.; Kunitomo, T.; Kataoka, H.; Suzuki, M.; Hirasawa, Y.; Shirahama, T.; et al. A new form of amyloid protein associated with chronic hemodialysis was identified as beta 2-microglobulin. *Biochem. Biophys. Res. Commun.* **1985**, *129*, 701–706. [CrossRef]
8. Esposito, G.; Michelutti, R.; Verdone, G.; Viglino, P.; Hernández, H.; Robinson, C.V.; Amoresano, A.; Dal Piaz, F.; Monti, M.; Pucci, P.; et al. Removal of the N-terminal hexapeptide from human beta2-microglobulin facilitates protein aggregation and fibril formation. *Protein Sci.* **2000**, *9*, 831–845. [CrossRef] [PubMed]
9. Valleix, S.; Gillmore, J.D.; Bridoux, F.; Mangione, P.P.; Dogan, A.; Nedelec, B.; Boimard, M.; Touchard, G.; Goujon, J.M.; Lacombe, C.; et al. Hereditary systemic amyloidosis due to Asp76Asn variant β2-microglobulin. *N. Engl. J. Med.* **2012**, *366*, 2276–2283. [CrossRef] [PubMed]
10. Raimondi, S.; Porcari, R.; Mangione, P.P.; Verona, G.; Marcoux, J.; Giorgetti, S.; Taylor, G.W.; Ellmerich, S.; Ballico, M.; Zanini, S.; et al. A specific nanobody prevents amyloidogenesis of D76N β(2)-microglobulin in vitro and modifies its tissue distribution in vivo. *Sci. Rep.* **2017**, *7*, 46711. [CrossRef]

11. Domanska, K.; Vanderhaegen, S.; Srinivasan, V.; Pardon, E.; Dupeux, F.; Marquez, J.A.; Giorgetti, S.; Stoppini, M.; Wyns, L.; Bellotti, V.; et al. Atomic structure of a nanobody-trapped domain-swapped dimer of an amyloidogenic beta2-microglobulin variant. *Proc. Natl. Acad. Sci. USA* **2011**, *108*, 1314–1319. [CrossRef] [PubMed]
12. Vanderhaegen, S.; Fislage, M.; Domanska, K.; Versees, W.; Pardon, E.; Bellotti, V.; Steyaert, J. Structure of an early native-like intermediate of β2-microglobulin amyloidogenesis. *Protein Sci.* **2013**, *22*, 1349–1357. [CrossRef] [PubMed]
13. Kinjo, A.R.; Nishikawa, K. Eigenvalue analysis of amino acid substitution matrices reveals a sharp transition of the mode of sequence conservation in proteins. *Bioinformatics* **2004**, *20*, 2504–2508. [CrossRef] [PubMed]
14. Schulte-Herbrüggen, T.; Sørensen, O.W. Clean TROSY: Compensation for Relaxation-Induced Artifacts. *J. Magn. Reason.* **2000**, *144*, 123–128. [CrossRef]
15. Salzmann, M.; Pervushin, K.; Wider, G.; Senn, H.; Wüthrich, K. TROSY in triple-resonance experiments: New perspectives for sequential NMR assignment of large proteins. *Proc. Natl. Acad. Sci. USA* **1998**, *95*, 13585. [CrossRef]
16. Yamazaki, T.; Forman-Kay, J.D.; Kay, L.E. Two-dimensional NMR experiments for correlating carbon-13 beta. and proton.delta./.epsilon. chemical shifts of aromatic residues in 13C-labeled proteins via scalar couplings. *J. Am. Chem. Soc.* **1993**, *115*, 11054–11055. [CrossRef]
17. Hafsa, N.E.; Arndt, D.; Wishart, D.S. CSI 3.0: A web server for identifying secondary and super-secondary structure in proteins using NMR chemical shifts. *Nucleic Acids Res.* **2015**, *43*, W370–W377. [CrossRef] [PubMed]
18. Wishart, D.S.; Sykes, B.D. The 13C chemical-shift index: A simple method for the identification of protein secondary structure using 13C chemical-shift data. *J. Biomol. NMR* **1994**, *4*, 171–180. [CrossRef]
19. Shen, Y.; Bax, A. Protein backbone and sidechain torsion angles predicted from NMR chemical shifts using artificial neural networks. *J. Biomol. NMR* **2013**, *56*, 227–241. [CrossRef]
20. Micsonai, A.; Wien, F.; Bulyáki, É.; Kun, J.; Moussong, É.; Lee, Y.H.; Goto, Y.; Réfrégiers, M.; Kardos, J. BeStSel: A web server for accurate protein secondary structure prediction and fold recognition from the circular dichroism spectra. *Nucleic Acids Res.* **2018**, *46*, W315–W322. [CrossRef]
21. Shen, Y.; Lange, O.; Delaglio, F.; Rossi, P.; Aramini, J.M.; Liu, G.; Eletsky, A.; Wu, Y.; Singarapu, K.K.; Lemak, A.; et al. Consistent blind protein structure generation from NMR chemical shift data. *Proc. Natl. Acad. Sci. USA* **2008**, *105*, 4685–4690. [CrossRef] [PubMed]
22. Lee, W.; Kim, J.H.; Westler, W.M.; Markley, J.L. PONDEROSA, an automated 3D-NOESY peak picking program, enables automated protein structure determination. *Bioinformatics* **2011**, *27*, 1727–1728. [CrossRef] [PubMed]
23. Lee, W.; Stark, J.L.; Markley, J.L. PONDEROSA-C/S: Client-server based software package for automated protein 3D structure determination. *J. Biomol. NMR* **2014**, *60*, 73–75. [CrossRef]
24. Wüthrich, K.; Billeter, M.; Braun, W. Polypeptide secondary structure determination by nuclear magnetic resonance observation of short proton-proton distances. *J. Mol. Biol.* **1984**, *180*, 715–740. [CrossRef]
25. Englander, S.W.; Kallenbach, N.R. Hydrogen exchange and structural dynamics of proteins and nucleic acids. *Q. Rev. Biophys.* **1983**, *16*, 521–655. [CrossRef] [PubMed]
26. Gore, S.; Sanz García, E.; Hendrickx, P.M.S.; Gutmanas, A.; Westbrook, J.D.; Yang, H.; Feng, Z.; Baskaran, K.; Berrisford, J.M.; Hudson, B.P.; et al. Validation of Structures in the Protein Data Bank. *Structure* **2017**, *25*, 1916–1927. [CrossRef] [PubMed]
27. Davis, I.W.; Leaver-Fay, A.; Chen, V.B.; Block, J.N.; Kapral, G.J.; Wang, X.; Murray, L.W.; Arendall, W.B., 3rd; Snoeyink, J.; Richardson, J.S.; et al. MolProbity: All-atom contacts and structure validation for proteins and nucleic acids. *Nucleic Acids Res.* **2007**, *35*, W375–W383. [CrossRef]
28. Williams, C.J.; Headd, J.J.; Moriarty, N.W.; Prisant, M.G.; Videau, L.L.; Deis, L.N.; Verma, V.; Keedy, D.A.; Hintze, B.J.; Chen, V.B.; et al. MolProbity: More and better reference data for improved all-atom structure validation. *Protein Sci.* **2018**, *27*, 293–315. [CrossRef] [PubMed]
29. Fowler, N.J.; Sljoka, A.; Williamson, M.P. A method for validating the accuracy of NMR protein structures. *Nature Commun.* **2020**, *11*, 6321. [CrossRef]
30. Gallivan, J.P.; Dougherty, D.A. Cation-pi interactions in structural biology. *Proc. Natl. Acad. Sci. USA* **1999**, *96*, 9459–9464. [CrossRef]
31. Palmer, A.G.; Cavanagh, J.; Wright, P.E.; Rance, M. Sensitivity improvement in proton-detected two-dimensional heteronuclear correlation NMR spectroscopy. *J. Magn. Reason.* **1991**, *93*, 151–170. [CrossRef]
32. States, D.J.; Haberkorn, R.A.; Ruben, D.J. A two-dimensional nuclear overhauser experiment with pure absorption phase in four quadrants. *J. Magn. Reason.* **1982**, *48*, 286–292. [CrossRef]
33. Marion, D.; Wüthrich, K. Application of phase sensitive two-dimensional correlated spectroscopy (COSY) for measurements of 1H-1H spin-spin coupling constants in proteins. *Biochem. Biophys. Res. Commun.* **1983**, *113*, 967–974. [CrossRef]
34. Grzesiek, S.; Bax, A. The importance of not saturating water in protein NMR. Application to sensitivity enhancement and NOE measurements. *J. Am. Chem. Soc.* **1993**, *115*, 12593–12594. [CrossRef]
35. Piotto, M.; Saudek, V.; Sklenár, V. Gradient-tailored excitation for single-quantum NMR spectroscopy of aqueous solutions. *J. Biomol. NMR* **1992**, *2*, 661–665. [CrossRef] [PubMed]
36. Hwang, T.L.; Shaka, A.J. Water Suppression That Works. Excitation Sculpting Using Arbitrary Wave-Forms and Pulsed-Field Gradients. *J. Magn. Reason. Series A* **1995**, *112*, 275–279. [CrossRef]
37. Orekhov, V.Y.; Ibraghimov, I.; Billeter, M. Optimizing resolution in multidimensional NMR by three-way decomposition. *J. Biomol. NMR* **2003**, *27*, 165–173. [CrossRef]

38. Lee, W.; Tonelli, M.; Markley, J.L. NMRFAM-SPARKY: Enhanced software for biomolecular NMR spectroscopy. *Bioinformatics* **2015**, *31*, 1325–1327. [CrossRef]
39. Kay, L.; Keifer, P.; Saarinen, T. Pure absorption gradient enhanced heteronuclear single quantum correlation spectroscopy with improved sensitivity. *J. Am. Chem. Soc.* **1992**, *114*, 10663–10665. [CrossRef]
40. Nietlispach, D. Suppression of anti-TROSY lines in a sensitivity enhanced gradient selection TROSY scheme. *J. Biomol. NMR* **2005**, *31*, 161–166. [CrossRef]
41. Grzesiek, S.; Bax, A. An efficient experiment for sequential backbone assignment of medium-sized isotopically enriched proteins. *J. Magn. Reason.* **1992**, *99*, 201–207. [CrossRef]
42. Grzesiek, S.; Bax, A. Correlating backbone amide and side chain resonances in larger proteins by multiple relayed triple resonance NMR. *J. Am. Chem. Soc.* **1992**, *114*, 6291–6293. [CrossRef]
43. Grzesiek, S.; Anglister, J.; Bax, A. Correlation of Backbone Amide and Aliphatic Side-Chain Resonances in 13C/15N-Enriched Proteins by Isotropic Mixing of 13C Magnetization. *J. Magn. Reason. Ser. B* **1993**, *101*, 114–119. [CrossRef]
44. Grzesiek, S.; Bax, A. Amino acid type determination in the sequential assignment procedure of uniformly 13C/15N-enriched proteins. *J. Biomol. NMR* **1993**, *3*, 185–204. [CrossRef]
45. Salzmann, M.; Wider, G.; Pervushin, K.; Senn, H.; Wüthrich, K. TROSY-type Triple-Resonance Experiments for Sequential NMR Assignments of Large Proteins. *J. Am. Chem. Soc.* **1999**, *121*, 844–848. [CrossRef]
46. Shaka, A.J.; Lee, C.J.; Pines, A. Iterative schemes for bilinear operators; application to spin decoupling. *J. Magn. Reason.* **1988**, *77*, 274–293. [CrossRef]
47. Bax, A.; Clore, G.M.; Gronenborn, A.M. 1H-1H correlation via isotropic mixing of 13C magnetization, a new three-dimensional approach for assigning 1H and 13C spectra of 13C-enriched proteins. *J. Magn. Reason.* **1990**, *88*, 425–431. [CrossRef]
48. Schleucher, J.; Schwendinger, M.; Sattler, M.; Schmidt, P.; Schedletzky, O.; Glaser, S.J.; Sørensen, O.W.; Griesinger, C. A general enhancement scheme in heteronuclear multidimensional NMR employing pulsed field gradients. *J. Biomol. NMR* **1994**, *4*, 301–306. [CrossRef]
49. Kumar, A.; Ernst, R.R.; Wüthrich, K. A two-dimensional nuclear Overhauser enhancement (2D NOE) experiment for the elucidation of complete proton-proton cross-relaxation networks in biological macromolecules. *Biochem. Biophys. Res. Commun.* **1980**, *95*, 1–6. [CrossRef]
50. Lee, W.; Petit, C.M.; Cornilescu, G.; Stark, J.L.; Markley, J.L. The AUDANA algorithm for automated protein 3D structure determination from NMR NOE data. *J. Biomol. NMR* **2016**, *65*, 51–57. [CrossRef]
51. Kalé, L.; Skeel, R.; Bhdarkar, M.; Brunner, R.; Gursoy, A.; Krawetz, N.; Phillips, J.; Shinozaki, A.; Varadarajan, K.; Schulten, K. NAMD2: Greater Scalability for Parallel Molecular Dynamics. *J. Comput. Phys.* **1999**, *151*, 283–312. [CrossRef]
52. Onufriev, A.; Bashford, D.; Case, D.A. Exploring protein native states and large-scale conformational changes with a modified generalized born model. *Proteins* **2004**, *55*, 383–394. [CrossRef] [PubMed]
53. Laskowski, R.A.; MacArthur, M.W.; Moss, D.S.; Thornton, J.M. PROCHECK: A program to check the stereochemical quality of protein structures. *J. Appl. Crystallogr.* **1993**, *26*, 283–291. [CrossRef]
54. Humphrey, W.; Dalke, A.; Schulten, K. VMD: Visual molecular dynamics. *J. Mol. Graph.* **1996**, *14*, 33–38. [CrossRef]
55. Lindorff-Larsen, K.; Piana, S.; Palmo, K.; Maragakis, P.; Klepeis, J.L.; Dror, R.O.; Shaw, D.E. Improved side-chain torsion potentials for the Amber ff99SB protein force field. *Proteins* **2010**, *78*, 1950–1958. [CrossRef]
56. Martyna, G.J.; Tobias, D.J.; Klein, M.L. Constant pressure molecular dynamics algorithms. *J. Chem. Phys.* **1994**, *101*, 4177–4189. [CrossRef]
57. Feller, S.E.; Zhang, Y.; Pastor, R.W.; Brooks, B.R. Constant pressure molecular dynamics simulation: The Langevin piston method. *J. Chem. Phys.* **1995**, *103*, 4613–4621. [CrossRef]

Article

Patterns in Protein Flexibility: A Comparison of NMR "Ensembles", MD Trajectories, and Crystallographic B-Factors

Christopher Reinknecht , **Anthony Riga, Jasmin Rivera and David A. Snyder** *

Department of Chemistry, College of Science and Health, William Paterson University, 300 Pompton Rd, Wayne, NJ 07470, USA; reinknechtc1@student.wpunj.edu (C.R.); anthony_riga@aol.com (A.R.); jaz_x33@yahoo.com (J.R.)

* Correspondence: snyderd@wpunj.edu

Abstract: Proteins are molecular machines requiring flexibility to function. Crystallographic B-factors and Molecular Dynamics (MD) simulations both provide insights into protein flexibility on an atomic scale. Nuclear Magnetic Resonance (NMR) lacks a universally accepted analog of the B-factor. However, a lack of convergence in atomic coordinates in an NMR-based structure calculation also suggests atomic mobility. This paper describes a pattern in the coordinate uncertainties of backbone heavy atoms in NMR-derived structural "ensembles" first noted in the development of FindCore2 (previously called Expanded FindCore: DA Snyder, J Grullon, YJ Huang, R Tejero, GT Montelione, *Proteins: Structure, Function, and Bioinformatics* 82 (S2), 219–230) and demonstrates that this pattern exists in coordinate variances across MD trajectories but not in crystallographic B-factors. This either suggests that MD trajectories and NMR "ensembles" capture motional behavior of peptide bond units not captured by B-factors or indicates a deficiency common to force fields used in both NMR and MD calculations.

Keywords: Friedman's test; backbone atom coordinate variances and uncertainties; superimposition

Citation: Reinknecht, C.; Riga, A.; Rivera, J.; Snyder, D.A. Patterns in Protein Flexibility: A Comparison of NMR "Ensembles", MD Trajectories, and Crystallographic B-Factors. *Molecules* **2021**, *26*, 1484. https://doi.org/10.3390/molecules26051484

Academic Editor: Marilisa Leone

Received: 31 December 2020
Accepted: 28 February 2021
Published: 9 March 2021

Publisher's Note: MDPI stays neutral with regard to jurisdictional claims in published maps and institutional affiliations.

Copyright: © 2021 by the authors. Licensee MDPI, Basel, Switzerland. This article is an open access article distributed under the terms and conditions of the Creative Commons Attribution (CC BY) license (https://creativecommons.org/licenses/by/4.0/).

1. Introduction

Large molecules and biomolecules can have a high degree of motional flexibility, affecting their function [1]. Common sources of information about protein flexibility and modes of motion include crystallographic B-factors [2], molecular dynamics (MD) simulations [3], and Nuclear Magnetic Resonance (NMR) spectroscopy, including relaxation measurements [4–7] and even chemical shift data [8,9].

Each of the above techniques for evaluating protein flexibility yields an incomplete picture of protein dynamics in solution. Crystallographic B-factors are affected by packing and other special features of the crystalline state [10]. In addition, many factors may reduce the intensities of the "reflections" in a protein crystal's X-ray diffraction pattern, and, hence, elevated crystallographic B-factors that may not solely indicate macromolecular flexibility [11]. The quality of MD simulations is dependent on the quality of the seed structure and force field used, despite recent efforts applying MD simulations to NMR-derived structures [12]. NMR relaxation experiments provide a critical source of data for evaluating individual MD trajectories as well as the force fields and other methodological details of MD simulations [13,14]. The combination of multiple assessments of protein flexibility has proven particularly illuminating [15]. For example, the combination of NMR-relaxation data with MD simulations yields a detailed picture of protein dynamics and motional modes [16].

While lacking a universally accepted analog of the B-factor, the NMR-based structure determination process itself provides insight into protein flexibility. Atoms in loop residues and other flexible regions of a protein typically have fewer long-range "contacts" to atoms in other residues. This paucity of contacts leads to both increased flexibility of loop regions [17,18] as well as poor convergence for loop residue positions in NMR-based structure calculations [19–22], provided the structure refinement process does not lead to inaccurate rather than imprecise coordinates [23]. Moreover, the primary source of structural restraints in NMR-based structure calculations are NOESY (Nuclear Overhauser Effect Spectroscopy) experiments. Fast motions reduce NOEs while intermediate time scale motion causes line broadening that can interfere with the identification of NOESY cross-peaks. Thus, NMR yields a paucity of restraints for particularly flexible regions of a protein leading to poor convergence in NMR-based structure determination, and coordinate uncertainties in an NMR-derived "ensemble" of structures [24]. While, strictly speaking, such coordinate uncertainties measure the local reproducibility of the NMR-based structural determination process, coordinate uncertainties across NMR ensembles are highly correlated to coordinate variances across MD trajectories [25].

While they can provide key insights into protein flexibility and dynamics, evaluation of uncertainties in protein structure coordinates inferred from NMR data is a non-trivial and non-physical process. Typically, NMR-based structure calculations generate multiple (typically 10–40) models [22]. Such collections of structural models are called "ensembles". Although NMR ensemble generation can effect Boltzmann sampling [19–21], generally NMR ensembles, including those analyzed in this study, are not actually Boltzmann ensembles.

Calculation of coordinate variances requires the superimposition of NMR ensembles. However, inclusion of poorly converged coordinates can bias the superimposition process, reducing the applicability of the resulting coordinate variances [26,27]. Limiting the calculation of an optimal superimposition to a core atom set, determined in a superimposition independent manner using either circular variances of backbone dihedral angles [28] or an interatomic variance matrix [27,29], ensures calculation of optimal superimpositions and, hence, of appropriate coordinate uncertainties. Alternatively, assumptions concerning the distribution of coordinate variances can lead to model-based superimposition methods such as THESEUS, which assumes a multivariate Gaussian distribution of coordinate uncertainties [30,31].

Identification of a core atom set is a critical step in solving two distinct, albeit related, problems. Not only does identification of a core atom set an important step in calculating coordinate uncertainties via superimposition, but such a core atom or residue sets also convey in which regions the NMR-based structure calculation process has converged [22]. Since these two problems are different, their optimal solutions may differ slightly. For example, application of the FindCore method, which identifies core atom sets for use in assessing the precision of NMR ensembles, to the distinct, albeit related problem of identifying well-converged core atom sets for CASP10 [32,33], required extension of the FindCore method into an approach known as Expanded FindCore [22].

Software used in the CASP10 competition also required any residue with core atoms to have all backbone heavy atoms in the core. The process of modifying Expanded FindCore to meet this requirement revealed carbonyl oxygens from otherwise well-defined residues whose positions were poorly defined in NMR-based structural calculations. Given the relation between coordinate uncertainties in NMR-derived structures and physical flexibility as described above, this discovery raised questions about the high uncertainties (relative to other backbone heavy atoms in the same residue) of those carbonyl oxygens. How common are these relatively uncertain carbonyl oxygens and is this high relative uncertainty an artifact of the NMR-based structure determination process or is it indicative of a pattern in backbone atom flexibilities?

Addressing these questions requires a comparison of NMR ensembles with complementary structural information, such as that obtained from crystallographic data, as well as with MD trajectories that provide insight into protein flexibility. Protein structures obtained by the North East Structure Genomics (NESG; http://www.nesg.org/ accessed on 31 December 2020) consortium facilitated this analysis. The NESG performed crystallization and HSQC (Heteronuclear Single Quantum Coherence Spectroscopy) screening in parallel for robustly expressed protein targets resulted in more than 40 NMR/X-ray crystal structural pairs [34,35].

The analysis presented here demonstrates the persistence of a pattern in coordinate variances across structural "ensembles" obtained using multiple force fields, superimposition techniques, and sampling schemes (i.e., restrained, simulated annealing and similar schemes in NMR structural refinement vs. the unrestrained constant temperature approach used in MD). This persistent pattern does not necessarily occur in Crystallographic B-factors of backbone heavy atoms. That the relatively high uncertainty of carbonyl oxygens persists, in almost all MD trajectories simulated in this study, indicates that the relatively high uncertainty of carbonyl oxygens is not solely an artifact of NMR-based structural determination. The pattern in backbone heavy atom coordinate uncertainties reflects either a physical reality of peptide bond motion not evident in crystallographic data or a shortcoming common to multiple force fields. If the latter explanation is true, the analysis presented here underscores that further improvements in force field parameterization are necessary for better prediction and calculation of a protein structure and dynamics.

2. Results and Discussion

Figure 1 illustrates how coordinate uncertainty in NMR-derived "ensembles" (panel B) tracks coordinate variance in MD simulations (e.g., at 300 K in panel D). The position of the carbonyl oxygen atom in residue 42 varies both across structural models in the NMR ensemble and over MD trajectories (panels C and D), and this oxygen atom is splayed more than the carbonyl carbon to which it is attached in panels B–D. However, the crystallographic B-factor for this carbonyl oxygen (22.15) is not particularly high nor is it much larger than that of the carbonyl carbon (21.83). Meanwhile, on the opposite side of that peptide bond's plane, the amide nitrogen from residue 43 is relatively well superimposed in the NMR ensemble and MD trajectory. The motion of the peptide plane appears to pivot around the amide nitrogen and proton. However, in the crystallographic structure, the B-factor (21.47) is barely lower than that of the carbonyl atoms.

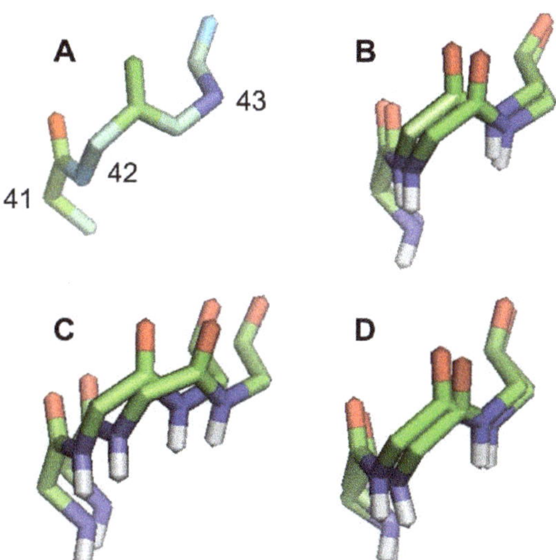

Figure 1. Backbone traces of residues 41–43 from Q8ZRJ2. (**A**) Crystallographic structure (PDB ID 2ES9) colored by a B-factor with blue being low, green being moderate, and red being high. Residue numbers shown in this panel reflect residue numbers in all panels. (**B**) FindCore superimposition of NMR ensemble (PDB ID 2JN8). This superimposition was calculated using a core atom set drawn from all heavy atoms (using all deposited models in the FindCore calculation) and not merely the residues shown. THESEUS superimposition, calculated from the entire MD trajectory using all heavy atoms, of MD trajectories simulated using the AMBER force field, showing snapshots 100 and 1000, at (**C**) 100 K and (**D**) 300 K. In panels (**B–D**), carbonyl oxygens are red, amide nitrogens are blue, carbons

are green, and amide hydrogens are white. Note the splaying in the carbonyl oxygens in panels (**B–D**) and the relatively well superimposed amide nitrogens in panels (**B**) and (**D**). Even in panel (**C**), amide nitrogens are better superimposed than carbonyl oxygens. In general, peptide planes appear to pivot with the amide protons and/or amide nitrogens being relatively immobile with the carbonyl oxygens at the opposite end of the peptide plane being relatively mobile. This pattern is not apparent in the B-factors depicted in panel (**A**).

Application of Friedman's test [36] to coordinate uncertainties (Figure 2, first two columns), ranked from lowest to highest on a per-residue basis, of NMR structures, yielded results that confirmed what was observed in the development of the Expanded Findcore method [22]. For almost all NMR ensembles considered, whether superimposed using FindCore or THESEUS, the average rank of the carbonyl oxygen (O) atoms was higher than the average ranks of the amide nitrogen (N), Cα, and carbonyl carbon (C′) atoms. In many structures, the average rank of C′ and N atoms was lower than the average rank of the Cα atoms. Average ranks (averaged on a per-structure basis) of backbone heavy atoms in THESEUS superimposed MD trajectories (Figure 2, third column) were also higher for O atoms and lower for C′ and N atoms. When analyzing crystallographic B-factors, however, average ranks did not generally vary much with the atom type (Figure 2, fourth column).

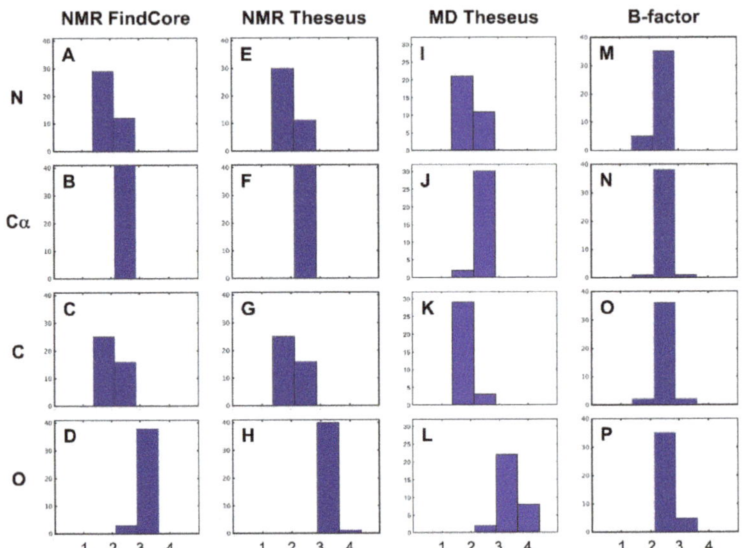

Figure 2. Distribution of average ranks of coordinate uncertainties, variances, and B-factors of backbone heavy atoms. As described in the main text, atoms in each residue are ranked by (**A–D**) coordinate uncertainty of FindCore superimposed NMR ensembles, (**E–H**) THESEUS superimposed NMR structures, coordinate variances of (**I–L**) THESEUS superimposed MD trajectories and (**M–P**) B-factors. For each structure, an average rank is calculated for each backbone heavy atom type: (first row) amide N, (second row) Cα, (third row) carbonyl C, and (fourth row) carbonyl O. For superimposed NMR ensembles (columns one and two) and MD trajectories (column three), a clear pattern is visible: average ranks for amide nitrogen atoms and carbonyl carbon atoms are often lower than average ranks for Cα atoms. The average ranks for carbonyl oxygen atoms are usually higher. When backbone heavy atoms are ranked by a B-factor, however, the average ranks for all backbone heavy atoms typically are between 2–3. The average ranks plotted in this figure are tabulated in Tables S2–S5, Supplementary Materials.

Multiple comparisons subsequent to Friedman's test (Figure 3) indicated that, for NMR ensembles and MD trajectories, the coordinate uncertainties and, respectively, variances (as ranked on a per-residue basis) for O atoms were significantly higher than the coordinate uncertainties or variances for N and C′ atoms in almost all ensembles or trajectories explored. In many superimposed NMR ensembles, coordinate uncertainties for O atoms were also significantly higher than coordinate uncertainties for Cα atoms, and, in a few superimposed NMR ensembles, coordinate uncertainties for Cα atoms were higher than those for N and C′ atoms. In most superimposed MD trajectories, coordinate uncertainties for O atoms were also significantly higher than coordinate uncertainties for Cα atoms, but coordinate uncertainties for Cα atoms were not significantly higher than those for N and C′ atoms. However, only a few crystal structures showed any significant differences in coordinate uncertainties between atom types.

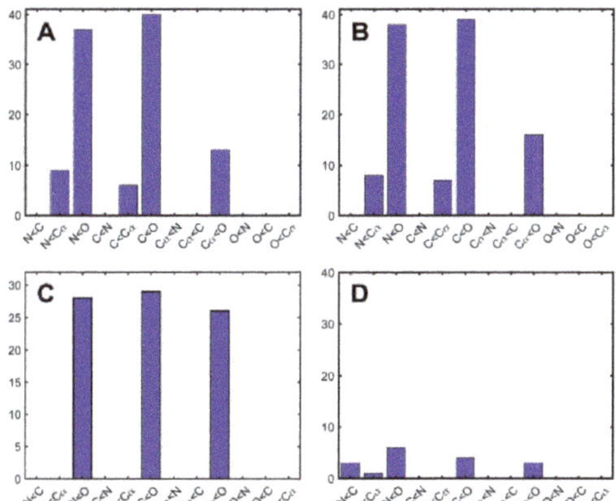

Figure 3. Results of Friedman's Test and subsequent multiple comparisons analysis. A bar, associated with a comparison X < Y, that is n units high, indicates that, in n structures, the assessed measure of coordinate variability is significantly lower for atom type X than for atom type Y. e.g., in panel A, the bar associated with C < O being 39 units high indicates that in 39 NMR ensembles, the coordinate uncertainties (calculated using FindCore superimpositions) for carbonyl carbons are significantly less (according to Friedman's test) than those for carbonyl oxygens. Mean ranks are considered significantly different if they differ by more than three standard deviations. Assessed measures of coordinate variability are (**A**) coordinate uncertainties in FindCore superimposed NMR ensembles, (**B**) coordinate uncertainties in THESEUS superimposed NMR ensembles, (**C**) coordinate uncertainties in THESEUS superimposed MD trajectories, and (**D**) crystallographic B-factors. Note that, in almost all superimposed NMR ensembles (independent of superimposition method), as well as in almost all THESEUS superimposed MD trajectories, amide nitrogen and carbonyl carbons have significantly lower coordinate uncertainties than carbonyl oxygens. However, only a small number of crystallographic structures have any significant results using the Friedman's test to compare B-factors of different atom types.

Unlike, in the case of superimposed NMR ensembles and MD trajectories, where the coordinate uncertainties or variances of backbone heavy atoms in a residue had a tendency to be lowest for N and C′ atoms and highest for O atoms, no such persistent pattern existed for crystallographic B factors. On the other hand, the pattern in coordinate variances in superimposed MD structures persisted across MD trajectories ran using different forcefields (AMBER99SB vs. OPLS) as well as temperatures (100 K vs. room temperature) and did not depend on whether the SeMET residues found in the crystal structures used to seed MD calculations were replaced with MET residues or not.

That carbonyl oxygens possess a significant tendency to have higher coordinate variances in THESEUS superimposed MD ensembles, as well as having higher coordinate uncertainties across FindCore superimposed NMR "ensembles" indicates the pattern of coordinate uncertainties observed in NMR-derived structures is not solely an artifact of the superimposition method (THESEUS vs. FindCore), not a particular force field used (AMBER and OPLS in MD simulations, CNS [37,38], and XPLOR-NIH [39] in NMR refinement), nor the particular characteristics of an NMR-based structural determination (e.g., a lack of experimentally derived restraints on carbonyl oxygen atoms). The persistence of the tendency for carbonyl oxygens to have higher coordinate variability between ensembles explored via MD simulation and NMR-derived "ensembles", which typically consist of models resulting from replicated, simulated, annealing calculations, indicates that this tendency is not solely an artifact of the structure sampling scheme used in NMR calculations. It may be the case that NMR structures not refined using CNS or XPLOR-NIH do not generally have carbonyl oxygens with high relative coordinate uncertainties. The one unrefined structure (1XPV) analyzed in this study did have carbonyl oxygens with high relative coordinate uncertainties.

One possible explanation of the high relative carbonyl oxygen uncertainties in superimposed NMR structures and variances in MD trajectories is that forcefields do not adequately restrain the positions of carbonyl oxygens. Carbonyl oxygen atoms are known to favorably interact with aromatic rings via n-π^* interactions [40,41] and also participate in hydrogen bonding, whose representation in classical forcefields is often deficient [42]. Hydrogen bonding is important in stabilizing the protein tertiary structure [43], and carbonyl oxygen atoms in regions of a secondary structure typically participate in hydrogen bonds.

Figure 4 shows that carbonyl oxygen atoms with relatively high coordinate uncertainties in NMR structures and with relatively high coordinate variances in MD trajectories occur in carbonyl oxygen atoms participating in intramolecular hydrogen bonding as well as those which are only hydrogen bonded to solvent. Nevertheless, some carbonyl oxygen atoms in a secondary structure have relatively greater coordinate variances across MD trajectories than in NMR structures. As NMR-based structure calculations typically involve additional restraints on hydrogen bonding atoms (based on H/D exchange data and/or secondary structure as established based on resonance assignments), it may be the case that MD simulations could benefit from better representation of hydrogen bonding [42] and other non-covalent interactions [41] in MD forcefields.

In addition to potentially inadequately representing quantum mechanical phenomena such as hydrogen bonding and n-π^* interactions, many force fields strongly penalize any deviation of a peptide bond from planarity. In particular, requiring peptide bonds to remain planar may cause more complex motions of the amide backbone to be represented by simple rocking motions along an axis near the N–C bond axis but angled slightly toward the Cα. This motional model, by placing carbonyl oxygens furthest from the axis of motion (and Cα atoms second furthest), inappropriately represents them as being most mobile. Deficiencies in representing hydrogen bonding in force fields [42] may also be problematic when such deficiencies result in insufficient restraints on carbonyl oxygen positions. Hydrogen bonds that are important in stabilizing protein tertiary structure [43], may represent important restraints in a carbonyl oxygen position across MD trajectories just as they are in NMR-based structural determination.

It is possible that the pattern of coordinate uncertainties and variances observed, respectively, in superimposed NMR and MD ensembles actually represents internal motions of peptide bond units in proteins in the solution state. Carbonyl oxygen atoms, branching off from the main polypeptide chain, may have enhanced thermal motion relative to backbone atoms on the main chain. In fact, other atoms branching off from the main chain, including Cβ atoms and even amide protons, tend to have significantly more coordinate uncertainty in superimposed NMR ensembles and coordinate variance across superimposed MD trajectories than amide nitrogen or carbonyl carbon atoms (Figures S1–S4, Supplementary Materials). However, more crystallographic structures have significantly higher Cβ B-factors, as compared to amide nitrogen B-factors by Friedman's test, than higher carbonyl oxygen B-factors as compared to amide nitrogen B-factors.

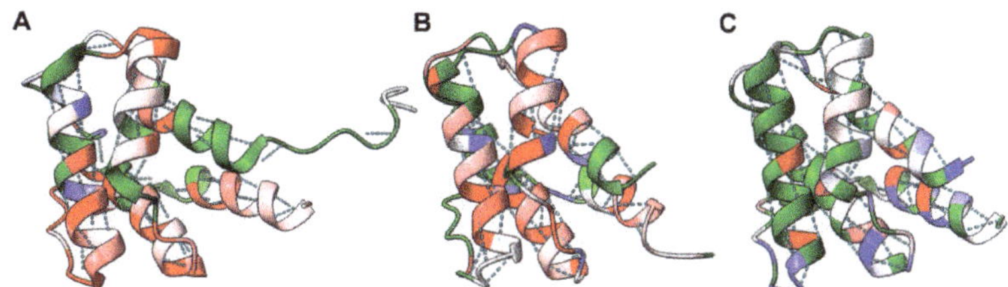

Figure 4. F-scores comparing UniProt ID Q8ZRJ2 backbone heavy atom coordinate uncertainties and variances. (**A**) The first model in the NMR "ensemble" 2JN8, (**B**) final snapshot of the MD trajectory seeded with 2ES9 (replacing Se-MET residues with MET residues, ran at 300 K with the AMBER 99SB forcefield), and (**C**) crystallographic structure, PDB ID 2ES9, each colored on a per residue bases by the F-score described in the Materials and Methods section (Equation (1)). Red indicates an F-score greater than 10 (relative uncertainty, variance, or B-factor of carbonyl oxygen coordinates quite high), white an F-score equal to 1 and blue and F-score less than 0.1. Green indicates residues for which the carbonyl oxygen coordinate uncertainty, variance, or B-factor for the carbonyl oxygen was actually less than the uncertainty, variance, or B-factor for the corresponding amide nitrogen. Dotted lines indicate hydrogen bonds: carbonyl oxygen atoms with high relative coordinate uncertainties and variances occur in both a hydrogen-bonded secondary structure as well as in loop regions. Some helical regions, likely endowed with extra restraints in the NMR-based structure determination process, do have slightly fewer carbonyl oxygens with high relative coordinate uncertainty as compared with the MD trajectory, illustrating the potential importance of hydrogen bonding in "fixing" the position of carbonyl oxygen atoms with high relative coordinate variances. By comparison, the crystallographic structure, PDB ID 2ES9, has relatively few carbonyl oxygens with high relative B-factors as indicated by the relative dearth of red in panel (**C**).

It is often assumed that crystallographic B-factors correlate well with internal flexibility. The absence of a persistent pattern in the B-factors for backbone atoms suggests that any such pattern observed in MD trajectories and NMR "ensembles" is an artifact. However, even in an ideal case where Crystallographic B-factors arise entirely from static and dynamic disorder, these B-factors reflect protein dynamics in the crystalline state and not in the solution state [44]. Moreover, previous studies have not only shown that NMR coordinate uncertainties correlate well to coordination variances in MD trajectories but also have demonstrated that crystallization has a "flattening" effect on protein flexibility [25]. Additionally, since Debye-Waller theory attributes any reduction in diffraction pattern intensities relative to those expected given a static protein structure to localize harmonic motion, other processes that reduce diffraction pattern intensities may result in over-estimation or even under-estimation of protein flexibility [45]. Relatedly, values obtained for B-factors are dependent on the refinement techniques used in interpreting X-ray data [25].

Nevertheless, the patterns described in this paper as well as the relatively high correlations between the statistical coordinate uncertainties derived from NMR and the putatively physical coordinate variances across MD ensembles may very well indicate deficiencies common to all force fields. Fully exploring the pervasiveness of the patterns described in this paper necessitates MD simulations and analysis of NMR structures beyond the systems studied here. However, the analysis presented in this paper identifies that coordinate variances/uncertainties from at least some MD trajectories and NMR ensembles have properties not found in B-factors. This divergence between B-factors and coordinate variances potentially indicates that there remain critical concerns in force field development. Future studies of MD trajectories will hopefully reveal which potentially inaccurate aspects of force fields, such as the requirement that peptide bonds remain planar and inadequacies in the representation of non-covalent interactions, such as hydrogen bonding as well as solvent/protein interactions, need the most adjustment. Addressing such deficiencies in force field construction can result in better descriptions of protein structure and, hence, facilitate the accurate prediction of protein dynamics, structure, and folding pathways.

3. Materials and Methods

The NMR and crystallographic structures analyzed in this study consisted of all (41) NMR structures and all but one (40) crystallographic structure listed in the "community resource" described by Everett et al. [35], which also outlines standardized methods used by the NESG for solving crystallographic and NMR structures. All but one of the NMR structures (1XPV) analyzed here were refined using CNS [37,38] and/or XPLOR-NIH [39]. MD simulations were performed on a randomly selected set of 12 targets from the community resource, using the conditions indicated in Table S1, Supplementary Materials. Most simulations used the OPLS [46] forcefield, but several simulations were performed with the AMBER99SB [47,48] forcefield as well.

MD simulations were initiated using crystallographic structures retrieved from the Protein Data Bank (PDB, [49]) with the identifications (IDs) listed in Table S1, Supplementary Materials. Simulations were prepared with Schrodinger's Maestro GUI made available as part of the Desmond [50] software package (which also ran MD simulations), using Na^{+} or Cl^{-} ions to achieve electrical neutrality and the TIP4PEW water model. In order to avoid artifacts due to truncation of the simulated constructs and facilitate parameterization in AMBER99SB, the terminal amino acid residues present in the coordinate sets obtained from the PDB were capped. Simulations ran for up to 36 ns (following default relaxation/minimization protocols), with snapshots recorded every 14.4 ps (up to 2500 snapshots). Re-parameterization of each simulation to use the AMBER99SB force field was performed using Desmond's Viparr utility. Most simulations were run at room temperature (generally defined for each protein by the temperature at which NMR experiments used to solve the protein's structure were performed. For all proteins in this study the temperature was very nearly 300 K in order to mimic the conditions in both the NMR tube and during (room temperature) crystallization. Some simulations were also performed at 100 K to mimic conditions obtained during cryo-cooled x-ray diffraction experiments. Simulations were ran both with and without substituting methionine (MET) for the seleno-methionine (SeMET) residues found in crystallographic structures. Dangling ends of protein chains absent from the crystallographic coordinates deposited in the PDB were not filled in computationally but rather were omitted from each simulation.

Initial parsing and visualization of each trajectory were performed using VMD [51]. A simple trajectory rescuer was used prior to initial parsing in VMD for simulations that turned into hung processes. Reformatting was completed for the multi-structural PDB file output from VMD into a multi-model format suitable for further analysis. THESEUS [30] superimposed MD trajectories prior to a coordinate variance calculation and the MATLAB [52] implementation of the FindCore Toolbox superimposed NMR ensembles. Calculation of coordinate uncertainties (calculated as coordinate variances) from FindCore superimposed NMR ensembles used the FindCore Toolbox and calculation of coordinate uncertainties and variances from THESEUS superimposed NMR ensembles and MD trajectories was also performed in MATLAB.

Friedman's test [36] is a non-parametric analog of ANOVA with repeated measures used here to compare whether coordinate uncertainties, variances, and B-factors are significantly different for different atom types. Application of Friedman's test proceeded as follows. For each residue in each structure, backbone heavy atom coordinate uncertainties, coordinate variances, or B-factors (depending on the analysis performed) were ranked (from 1–4). For each structure, the resulting ranks were tabulated with columns (treatments) corresponding to a heavy atom type (amide N, Cα, carbonyl carbon, and carbonyl oxygen) and one row (block) for each residue, and the resulting table was subjected to Friedman's test, which compared column averages (average rank by heavy atom type, averaged on a per-structure basis). MATLAB scripts tabulated B-factor and coordinate variance/uncertainty data for analysis via Friedman's test and subsequent multiple comparisons, which were also performed in MATLAB. MATLAB was also used to calculate an F-score measuring the relative uncertainties, variances, or B-factors of carbonyl oxygen atoms in a given residue (Equation (1)):

$$F = (u(O) - u(N))^2 / (u(C\prime) - u(N))^2, \quad (1)$$

where $u(.)$ denotes the coordinate uncertainty, variance, or B-factor of the given atom and O, N, and $C\prime$ are the carbonyl oxygen, amide nitrogen, and carbonyl carbon atoms, respectively.'

Supplementary Materials: The following are available online: Table S1: Parameters/Input for MD Simulations. Table S2: Average Ranks of Backbone Atom B-Factors for Crystallographic Structures. Table S3: Average Ranks of Backbone Atom Coordinate Uncertainties for Theseus Superimposed NMR "Ensembles". Table S4: Average Ranks of Backbone Atom Coordinate Uncertainties for FindCore Superimposed NMR "Ensembles". Table S5: MD Simulation Results. Table S6: Average Ranks of N, C′, Cα, and Cβ B-Factors for Crystallographic Structures. Table S7: Average Ranks of N, C′, Cα, and Cβ Coordinate Uncertainties for Theseus Superimposed NMR "Ensembles". Table S8: Average Ranks of N, C′, Cα, and Cβ Coordinate Uncertainties for FindCore Superimposed NMR "Ensembles". Table S9: Average Ranks of N, C′, Cα, and Cβ Coordinate Variances in Theseus Superimposed MD Trajectories. Table S10: Average Ranks of N, C′, Cα, and H Coordinate Uncertainties for Theseus Superimposed NMR "Ensembles". Table S11: Average Ranks of N, C′, Cα, and H Coordinate Uncertainties for FindCore Superimposed NMR "Ensembles". Table S12: Average Ranks of N, C′, Cα, and H Coordinate Variances in Theseus Superimposed MD Trajectories. Figure S1: Distribution of average ranks of coordinate uncertainties, variances, and B-factors of N, Cα, carbonyl C, and Cβ atoms. Figure S2: Results of Friedman's Test and subsequent multiple comparisons analysis with Cβ atoms. Figure S3: Distribution of average ranks of coordinate uncertainties, variances, and B-factors of N, Cα, carbonyl C, and Cβ atoms. Figure S4: Results of Friedman's Test and subsequent multiple comparisons analysis with amide H atoms.

Author Contributions: Conceptualization: D.A.S.; Data curation: D.A.S., C.R., A.R. and J.R. Formal analysis: D.A.S., C.R., A.R. and J.R. Investigation: D.A.S., C.R., A.R. and J.R. Methodology: D.A.S. Software: D.A.S., C.R., A.R. and J.R. Writing—original draft: D.A.S., C.R., A.R. and J.R. Writing—review and editing: D.A.S., C.R., A.R. and J.R. All authors have read and agreed to the published version of the manuscript.

Funding: While working on this project during the Summer of 2019, Christopher Reinknecht received a stipend from the Garden State—Louis Stokes Alliance for Minority Participation (GS-LSAMP), an NSF funded program.

Data Availability Statement: Input files used to run the MD simulations analyzed in this paper, the resulting (superimposed) MD trajectories, and all the scripts used to perform the analyses reported here are all archived on Zenodo, doi:10.5281/zenodo.4323630.

Acknowledgments: The authors graciously acknowledge Adrian Roitberg for his insights into the results presented by DAS at the ACS Spring 2014 Meeting in Dallas. The authors also thank Gaetano Montelione for his insights into NMR-based structural determination as well as for his helpful comments and constructive discussion, and John Chodera for his suggestion to publish a previous iteration of this study on bioRxiv. An award of Assigned Release Time for research from the Office of the Provost of William Paterson University of NJ facilitated completion of this work.

Conflicts of Interest: The authors have no conflicts of interest to report.

Sample Availability: Samples of the compounds are not available from the authors.

References

1. Rashin, A.A.; Rashin, A.H.; Jernigan, R.L. Protein flexibility: Coordinate uncertainties and interpretation of structural differences. *Acta Crystallogr. D Biol. Crystallogr.* **2009**, *65*, 1140–1161. [CrossRef]
2. Fenwick, R.B.; Bedem, H.V.D.; Fraser, J.S.; Wright, P.E. Integrated description of protein dynamics from room-temperature X-ray crystallography and NMR. *Proc. Natl. Acad. Sci. USA* **2014**, *111*, E445–E454. [CrossRef] [PubMed]
3. Karplus, M.; McCammon, J.A. Molecular dynamics simulations of biomolecules. *Nat. Genet.* **2002**, *9*, 646–652. [CrossRef]

4. Sapienza, P.J.; Lee, A.L. Using NMR to study fast dynamics in proteins: Methods and applications. *Curr. Opin. Pharmacol.* **2010**, *10*, 723–730. [CrossRef] [PubMed]
5. Wand, A.J. The dark energy of proteins comes to light: Conformational entropy and its role in protein function revealed by NMR relaxation. *Curr. Opin. Struct. Biol.* **2013**, *23*, 75–81. [CrossRef]
6. Lipari, G.; Szabo, A. Model-free approach to the interpretation of nuclear magnetic resonance relaxation in macromolecules. 1. Theory and range of validity. *J. Am. Chem. Soc.* **1982**, *104*, 4546–4559. [CrossRef]
7. Lipari, G.; Szabo, A. Model-free approach to the interpretation of nuclear magnetic resonance relaxation in macromolecules. 2. Analysis of experimental results. *J. Am. Chem. Soc.* **1982**, *104*, 4559–4570. [CrossRef]
8. Robustelli, P.; Stafford, K.A.; Palmer, A.G. Interpreting Protein Structural Dynamics from NMR Chemical Shifts. *J. Am. Chem. Soc.* **2012**, *134*, 6365–6374. [CrossRef]
9. Berjanskii, M.V.; Wishart, D.S. A Simple Method to Measure Protein Side-Chain Mobility Using NMR Chemical Shifts. *J. Am. Chem. Soc.* **2013**, *135*, 14536–14539. [CrossRef]
10. Eyal, E.; Gerzon, S.; Potapov, V.; Edelman, M.; Sobolev, V. The Limit of Accuracy of Protein Modeling: Influence of Crystal Packing on Protein Structure. *J. Mol. Biol.* **2005**, *351*, 431–442. [CrossRef]
11. Read, R.J. Structure-factor probabilities for related structures. *Acta Crystallogr. Sect. A Found. Crystallogr.* **1990**, *46*, 900–912. [CrossRef]
12. Li, D.-W.; Brüschweiler, R. Protocol to Make Protein NMR Structures Amenable to Stable Long Time Scale Molecular Dynamics Simulations. *J. Chem. Theory Comput.* **2014**, *10*, 1781–1787. [CrossRef]
13. Showalter, S.A.; Brüschweiler, R. Validation of molecular dynamics simulations of biomolecules using NMR spin relaxation as benchmarks: Application to the AMBER99SB force field. *J. Chem. Theory Comput.* **2007**, *3*, 961–975. [CrossRef] [PubMed]
14. Brüschweiler, R. Certification of Molecular Dynamics Trajectories with NMR Chemical Shifts. *J. Phys. Chem. Lett.* **2009**, *1*, 246–248. [CrossRef]
15. Rueda, M.; Ferrer-Costa, C.; Meyer, T.; Pérez, A.; Camps, J.; Gelpí, J.L.; Orozco, M. A consensus view of protein dynamics. *Proc. Natl. Acad. Sci. USA* **2007**, *104*, 796–801. [CrossRef]
16. Showalter, S.A.; Johnson, E.; Rance, M.; Brüschweiler, R. Toward Quantitative Interpretation of Methyl Side-Chain Dynamics from NMR by Molecular Dynamics Simulations. *J. Am. Chem. Soc.* **2007**, *129*, 14146–14147. [CrossRef] [PubMed]
17. Zhang, F.; Brüschweiler, R. Contact Model for the Prediction of NMR N−H Order Parameters in Globular Proteins. *J. Am. Chem. Soc.* **2002**, *124*, 12654–12655. [CrossRef]
18. Ming, D.; Brüschweiler, R. Reorientational Contact-Weighted Elastic Network Model for the Prediction of Protein Dynamics: Comparison with NMR Relaxation. *Biophys. J.* **2006**, *90*, 3382–3388. [CrossRef]
19. Rieping, W.; Nilges, M.; Habeck, M. ISD: A software package for Bayesian NMR structure calculation. *Bioinformatics* **2008**, *24*, 1104–1105. [CrossRef]
20. Carstens, S.; Nilges, M.; Habeck, M. Inferential Structure Determination of Chromosomes from Single-Cell Hi-C Data. *PLoS Comput. Biol.* **2016**, *12*, e1005292. [CrossRef] [PubMed]
21. Richter, B.; Gsponer, J.; Várnai, P.; Salvatella, X.; Vendruscolo, M. The MUMO (minimal under-restraining minimal over-restraining) method for the determination of native state ensembles of proteins. *J. Biomol. NMR* **2007**, *37*, 117–135. [CrossRef] [PubMed]
22. Snyder, D.A.; Grullon, J.; Huang, Y.J.; Tejero, R.; Montelione, G.T. The expanded FindCore method for identification of a core atom set for assessment of protein structure prediction. *Proteins Struct. Funct. Bioinform.* **2013**, *82*, 219–230. [CrossRef] [PubMed]
23. Tejero, R.; Bassolino-Klimas, D.; Bruccoleri, R.E.; Montelione, G.T. Simulated annealing with restrained molecular dynamics using CONGEN: Energy refinement of the NMR solution structures of epidermal and type-α transforming growth factors. *Protein Sci.* **1996**, *5*, 578–592. [CrossRef] [PubMed]
24. Salvatella, X. *Understanding Protein Dynamics Using Conformational Ensembles*; Springer International Publishing: New York, NY, USA, 2013; Volume 805, pp. 67–85.
25. Jamroz, M.; Kolinski, A.; Kmiecik, S. CABS-flex predictions of protein flexibility compared with NMR ensembles. *Bioinformatics* **2014**, *30*, 2150–2154. [CrossRef] [PubMed]
26. Kirchner, D.K.; Güntert, P. Objective identification of residue ranges for the superposition of protein structures. *BMC Bioinform.* **2011**, *12*, 170. [CrossRef] [PubMed]
27. Snyder, D.A.; Montelione, G.T. Clustering algorithms for identifying core atom sets and for assessing the precision of protein structure ensembles. *Proteins Struct. Funct. Bioinform.* **2005**, *59*, 673–686. [CrossRef]
28. Hyberts, S.G.; Goldberg, M.S.; Havel, T.F.; Wagner, G. The solution structure of eglin c based on measurements of many NOEs and coupling constants and its comparison with X-ray structures. *Protein Sci.* **1992**, *1*, 736–751. [CrossRef]
29. Kelley, L.A.; Gardner, S.P.; Sutcliffe, M.J. An automated approach for defining core atoms and domains in an ensemble of NMR-derived protein structures. *Protein Eng.* **1997**, *10*, 737–741. [CrossRef] [PubMed]
30. Theobald, D.L.; Wuttke, D.S. THESEUS: Maximum likelihood superpositioning and analysis of macromolecular structures. *Bioinformatics* **2006**, *22*, 2171–2172. [CrossRef]
31. Theobald, D.L.; Wuttke, D.S. Accurate structural correlations from maximum likelihood superpositions. *PLoS Comput. Biol.* **2008**, *4*, e43. [CrossRef] [PubMed]

32. Moult, J.; Fidelis, K.; Kryshtafovych, A.; Tramontano, A. Critical assessment of methods of protein structure prediction (CASP)-round IX. *Proteins Struct. Funct. Bioinform.* **2011**, *79*, 1–5. [CrossRef]
33. Kryshtafovych, A.; Moult, J.; Bales, P.; Bazan, J.F.; Biasini, M.; Burgin, A.; Chen, C.; Cochran, F.V.; Craig, T.K.; Das, R.; et al. Challenging the state of the art in protein structure prediction: Highlights of experimental target structures for the 10th Critical Assessment of Techniques for Protein Structure Prediction Experiment CASP10. *Proteins Struct. Funct. Bioinform.* **2013**, *82*, 26–42. [CrossRef] [PubMed]
34. Mao, B.; Tejero, R.; Baker, D.; Montelione, G.T. Protein NMR Structures Refined with Rosetta Have Higher Accuracy Relative to Corresponding X-ray Crystal Structures. *J. Am. Chem. Soc.* **2014**, *136*, 1893–1906. [CrossRef]
35. Everett, J.K.; Tejero, R.; Murthy, S.B.K.; Acton, T.B.; Aramini, J.M.; Baran, M.C.; Benach, J.; Cort, J.R.; Eletsky, A.; Forouhar, F.; et al. A community resource of experimental data for NMR/X-ray crystal structure pairs. *Protein Sci.* **2015**, *25*, 30–45. [CrossRef] [PubMed]
36. Hollander, M.; Wolfe, D.A.; Chicken, E. *Nonparametric Statistical Methods*, 3rd ed.; Wiley: Hoboken, NY, USA, 2015; pp. 1–10.
37. Brunger, A.T. Version 1.2 of the Crystallography and NMR system. *Nat. Protoc.* **2007**, *2*, 2728–2733. [CrossRef] [PubMed]
38. Brunger, A.T.; Adams, P.D.; Clore, G.M.; Delano, W.L.; Gros, P.; Grosse-Kunstleve, R.W.; Jiang, J.-S.; Kuszewski, J.; Nilges, M.; Pannu, N.S.; et al. Crystallography & NMR System: A New Software Suite for Macromolecular Structure Determination. *Acta Crystallogr. Sect. D Biol. Crystallogr.* **1998**, *54*, 905–921. [CrossRef]
39. Schwieters, C.D.; Kuszewski, J.J.; Tjandra, N.; Clore, G.M. The Xplor-NIH NMR molecular structure determination package. *J. Magn. Reson.* **2003**, *160*, 65–73. [CrossRef]
40. Jain, A.; Purohit, C.S.; Verma, S.; Sankararamakrishnan, R. Close Contacts between Carbonyl Oxygen Atoms and Aromatic Centers in Protein Structures: $\pi \cdots \pi$ or Lone-Pair$\cdots\pi$ Interactions? *J. Phys. Chem. B* **2007**, *111*, 8680–8683. [CrossRef] [PubMed]
41. Singh, S.K.; Das, A. The n $\rightarrow \pi^*$ interaction: A rapidly emerging non-covalent interaction. *Phys. Chem. Chem. Phys.* **2015**, *17*, 9596–9612. [CrossRef]
42. Lange, O.F.; Van Der Spoel, D.; De Groot, B.L. Scrutinizing Molecular Mechanics Force Fields on the Submicrosecond Timescale with NMR Data. *Biophys. J.* **2010**, *99*, 647–655. [CrossRef]
43. Pace, C.N.; Fu, H.; Lee Fryar, K.; Landua, J.; Trevino, S.R.; Schell, D.; Thurlkill, R.L.; Imura, S.; Scholtz, J.M.; Gajiwala, K.; et al. Contribution of hydrogen bonds to protein stability. *Protein Sci.* **2014**, *23*, 652–661. [CrossRef]
44. Reichert, D.; Zinkevich, T.; Saalwächter, K.; Krushelnitsky, A. The relation of the X-ray B-factor to protein dynamics: Insights from recent dynamic solid-state NMR data. *J. Biomol. Struct. Dyn.* **2012**, *30*, 617–627. [CrossRef]
45. Kuzmanic, A.; Pannu, N.S.; Zagrovic, B. X-ray refinement significantly underestimates the level of microscopic heterogeneity in biomolecular crystals. *Nat. Commun.* **2014**, *5*, 3220. [CrossRef] [PubMed]
46. Shivakumar, D.; Williams, J.; Wu, Y.; Damm, W.; Shelley, J.; Sherman, W. Prediction of Absolute Solvation Free Energies using Molecular Dynamics Free Energy Perturbation and the OPLS Force Field. *J. Chem. Theory Comput.* **2010**, *6*, 1509–1519. [CrossRef] [PubMed]
47. Salomon-Ferrer, R.; Case, D.A.; Walker, R.C. An overview of the Amber biomolecular simulation package. *Wiley Interdiscip. Rev. Comput. Mol. Sci.* **2013**, *3*, 198–210. [CrossRef]
48. Hornak, V.; Abel, R.; Okur, A.; Strockbine, B.; Roitberg, A.; Simmerling, C. Comparison of multiple Amber force fields and development of improved protein backbone parameters. *Proteins Struct. Funct. Bioinform.* **2006**, *65*, 712–725. [CrossRef] [PubMed]
49. Berman, H.M.; Westbrook, J.; Feng, Z.; Gilliland, G.; Bhat, T.N.; Weissig, H.; Shindyalov, I.N.; Bourne, P.E. The Protein Data Bank. *Nucleic Acids Res.* **2000**, *28*, 235–242. [CrossRef] [PubMed]
50. Bowers, K.J.; Chow, D.E.; Xu, H.; Dror, R.O.; Eastwood, M.P.; Gregersen, B.A.; Klepeis, J.L.; Kolossvary, I.; Moraes, M.A.; Sacerdoti, F.D.; et al. Scalable algorithms for molecular dynamics simulations on commodity clusters. In Proceedings of the 2006 ACM/IEEE Conference on Supercomputing, New York, NY, USA, 11 November 2006; p. 43.
51. Humphrey, W.; Dalke, A.; Schulten, K. VMD: Visual molecular dynamics. *J. Mol. Graph.* **1996**, *14*, 33–38. [CrossRef]
52. *MATLAB*; version 9.2.0.538062 (R2017a); The MathWorks Inc.: Natick, MA, USA, 2017.

MDPI

Article

Swapping the Positions in a Cross-Strand Lateral Ion-Pairing Interaction between Ammonium- and Carboxylate-Containing Residues in a β-Hairpin

Cheng-Hsin Huang [1,†], Tong Wai Wong [1,†], Chen-Hsu Yu [1], Jing-Yuan Chang [1], Shing-Jong Huang [2], Shou-Ling Huang [2] and Richard P. Cheng [1,*]

1 Department of Chemistry, National Taiwan University, Taipei 10617, Taiwan; R07223208@ntu.edu.tw (C.-H.H.); R02223133@ntu.edu.tw (T.W.W.); R05223205@ntu.edu.tw (C.-H.Y.); R08223149@ntu.edu.tw (J.-Y.C.)

2 Instrumentation Center, National Taiwan University, Taipei 10617, Taiwan; shingjonghuang@ntu.edu.tw (S.-J.H.); shouling@ntu.edu.tw (S.-L.H.)

* Correspondence: rpcheng@ntu.edu.tw

† These authors contributed equally to this study.

Abstract: Cross-strand lateral ion-pairing interactions are important for antiparallel β-sheet stability. Statistical studies suggested that swapping the position of cross-strand lateral residues should not significantly affect the interaction. Herein, we swapped the position of ammonium- and carboxylate-containing residues with different side-chain lengths in a cross-strand lateral ion-pairing interaction in a β-hairpin. The peptides were analyzed by 2D-NMR. The fraction folded population and folding free energy were derived from the chemical shift data. The ion-pairing interaction energy was derived using double mutant cycle analysis. The general trends for the fraction folded population and interaction energetics remained similar upon swapping the position of the interacting charged residues. The most stabilizing cross-strand interactions were between short residues, similar to the unswapped study. However, the fraction folded populations for most of the swapped peptides were higher compared to the corresponding unswapped peptides. Furthermore, subtle differences in the ion-pairing interaction energy upon swapping were observed, most likely due to the "unleveled" relative positioning of the interacting residues created by the inherent right-handed twist of the structure. These results should be useful for developing functional peptides that rely on lateral ion-pairing interactions across antiparallel β-strands.

Keywords: ion-pairing interaction; side-chain length; charged amino acids; β-hairpin; peptide

Citation: Huang, C.-H.; Wong, T.W.; Yu, C.-H.; Chang, J.-Y.; Huang, S.-J.; Huang, S.-L.; Cheng, R.P. Swapping the Positions in a Cross-Strand Lateral Ion-Pairing Interaction between Ammonium- and Carboxylate-Containing Residues in a β-Hairpin. *Molecules* **2021**, *26*, 1346. https://doi.org/10.3390/molecules26051346

Academic Editor: Marilisa Leone

Received: 25 December 2020
Accepted: 11 February 2021
Published: 3 March 2021

Publisher's Note: MDPI stays neutral with regard to jurisdictional claims in published maps and institutional affiliations.

Copyright: © 2021 by the authors. Licensee MDPI, Basel, Switzerland. This article is an open access article distributed under the terms and conditions of the Creative Commons Attribution (CC BY) license (https://creativecommons.org/licenses/by/4.0/).

1. Introduction

The β-sheet is an important protein secondary structure. About one-fourth of protein residues adopt a β-sheet conformation in protein structures [1–3]. Furthermore, β-sheets are also formed in amyloid fibrils involved in various diseases, including Alzheimer's disease [4,5], Huntington's disease [6], and Parkinson's disease [7,8]. Therefore, understanding the folding energetics of β-sheets is scientifically important with potential therapeutic applications [9,10].

The side-chains of the closest residues on adjacent strands are on the same face of a β-sheet. This would enable cross strand lateral side-chain-side-chain interactions. Statistical analysis showed that oppositely charged residues are frequently observed across antiparallel β-sheets [11–13], suggesting that cross-strand interactions between oppositely charged residues may be important for β-sheet stability. Accordingly, the energetics of cross strand ion pairs have been measured in sheet-containing host systems, including the protein G B1 domain [14,15], the zinc finger domain [16], and β-hairpins [11,17–23]. For the protein G B1 domain, a cross strand lateral Glu44-Lys53 ion-pairing interaction increased the protein stability by 1.0 kcal/mol based on thermal denaturation studies [14].

For the zinc finger domain, cross strand ion-pairing interactions involving Asp were more stabilizing compared to those involving Glu based on competitive metal ion binding studies [16]. In particular, cross strand Lys3-Asp10 and Arg3-Asp10 interactions stabilized the system by 0.48 and 0.26 kcal/mol, respectively [16].

The effect of charged amino acid side-chain length on cross strand lateral ion-pairing interaction was investigated in hairpin peptides [22,23]. The negatively charged carboxylate-containing amino acids with different side-chain lengths were incorporated at the N-terminal strand guest site (position 4), whereas the ammonium-containing amino acids with different side-chain lengths were incorporated at the C-terminal strand guest site (position 9) [22]. The results showed that length matching was necessary to form a stabilizing interaction, i.e., the side-chain length of the carboxylate- and ammonium-containing residues were either both long or both short [22]. The long side-chains provided large hydrophobic surfaces to interact with one another. Alternatively, the short side-chains paid less side-chain entropic penalties to interact with one another.

Statistical analysis showed that cross strand lateral residue pairs in antiparallel β-sheets are symmetric [24], meaning that swapping the position of a pair of cross strand lateral residues (i.e., orientation) should not significantly affect the interaction. However, two different experimental studies showed that swapping the positions of an amino acid pair in antiparallel β-sheets changed the stability of the system [14,19]. For the protein G B1 domain, the cross strand Phe44-Thr53 interaction stabilized the protein by 0.19 kcal/mol, but the Thr44-Phe53 interaction destabilized the protein by 0.36 kcal/mol based on thermal denaturation studies [14]. In addition, the Ile44-Phe53 and Ile44-Thr53 interactions were non-identical compared to the corresponding swapped interactions, with a change in overall thermal stability of the system [14]. Similarly, swapping the oppositely charged residues in the cross strand Lys3-Glu12 ion pair in a hairpin peptide altered the fraction folded population of the system based on NMR data [19]. As such, it appears that the statistical studies and the experimental studies contradict one another. Herein, we report the effect of lateral ion-pair interactions in a β-hairpin with the positively charged ammonium-containing residue at the N-terminal strand guest site (position 4) and negatively charged carboxylate-containing residue at the C-terminal strand guest site (position 9), effectively swapping the positions of the oppositely charged residues in a previous study [22].

2. Results

2.1. Peptide Design and Synthesis

The experimental HPTXaaZbb peptides were designed based on Gellman's YKL peptide [11,25,26] and hairpin peptides in our previous studies [3,22,23] (Figure 1a). The Tyr2 (in peptide YKL) was replaced with Thr [22] because the aromatic side-chain of Tyr may interact diagonally with the residue at position 9 due to the right-handed twist [11,27,28]. An acetyl group and a carboxamide group was incorporated at the N- and C-termini, respectively, to remove the terminal charges, preventing unintended electrostatic interactions [17]. Non-hydrogen-bonded sites at positions 4 and 9 were chosen as guest sites [3,22,23], which were near the center of the strands to avoid end fraying near the termini and excessive folding near the turn [20,21]. Our previous study placed the negatively charged carboxylate-containing residues at position 4 and the positively charged ammonium- (or guanidinium-)-containing residues at position 9 [22,23]. To investigate the effect of charged amino acid side-chain length upon swapping the position of the charged residues in the lateral cross strand ion-pairing interaction, the positively charged ammonium-containing residues (Xaa = Lys, Orn, Dab, Dap) and negatively charged carboxylate-containing residues (Zbb = Aad, Glu, Asp) were incorporated at positions 4 and 9, respectively, to give the experimental HPTXaaZbb peptides (Figure 1b). The peptides were named with an "HPT" prefix, representing hairpin peptide with Thr at position 2, followed by the positively charged Xaa residue at position 4 and the negatively charged Zbb residue at position 9.

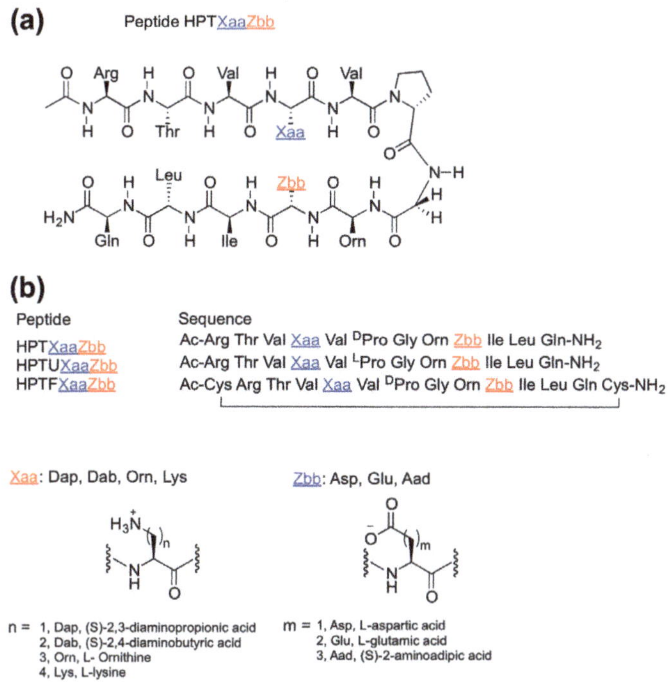

Figure 1. Design of peptides to study the effect of charged amino acid side-chain length upon swapping the charged amino acid positions in lateral ion-pairing interactions. (**a**) The chemical structure of the experimental HPTXaaZbb peptides; (**b**) The sequences of the experimental HPTXaaZbb peptides, the unfolded reference HPTUXaaZbb peptides, and the folded reference HPTFXaaZbb peptides.

The fully folded reference peptides and the fully unfolded reference peptides were necessary to determine the fraction folded population of the experimental HPTXaaZbb peptides [26]. For the fully folded reference peptides, cysteine residues were added to both termini of the experimental HPTXaaZbb peptides to form intramolecular disulfide bonds to give macrocyclic peptides to serve as the fully folded reference peptides HPTFXaaZbb [3,11,22,23,25,26]. For the fully unfolded reference peptides, the DPro6 in the experimental HPTXaaZbb peptides was replaced with Pro to give the fully unfolded reference peptides HPTUXaaZbb [3,11,22,23,25,26], because Pro does not favor β-hairpin structures [11,26].

The peptides were synthesized by solid-phase peptide synthesis using Fmoc-based chemistry [29,30]. The disulfide bond in the folded reference HPTFXaaZbb peptides was formed via charcoal mediated air oxidation [31]. All peptides were purified by reverse-phase high-performance liquid chromatography (RP-HPLC) to higher than 95% purity and confirmed by matrix-assisted laser desorption ionization time-of-flight mass spectrometry (MALDI-TOF). Since the nuclear magnetic resonance (NMR) spectra (chemical shift and line width) of analogous hairpin peptides did not change with concentration (20 μM to 10 mM) [3,11,12,32], the peptides in this study (2.0–15.4 mM) should not aggregate in solution. Accordingly, the experimental data should reflect the intramolecular interactions with minimal interference from intermolecular interactions.

2.2. *β-Hairpin Structure Characterization by NMR*

The peptides were analyzed by ^{1}H–^{1}H homonuclear two-dimensional solution NMR spectroscopy, including double-quantum filtered-correlated spectroscopy (DQF-COSY) [33], total correlation spectroscopy (TOCSY) [34], and rotating-frame nuclear Overhauser effect spectroscopy (ROESY) [35] at 298 K. Sequence-specific assignment of all peptides was performed based on the TOCSY and ROESY spectra (Tables S1–S36) [36]. For a given Xaa4-Zbb9 pair, the chemical shift dispersion of the peptides followed the trend HPTFXaaZbb > HPTXaaZbb > HPTUXaaZbb (Tables S1–S36). Since the higher the fraction folded population, the higher the chemical shift dispersion [37], this trend is consistent with the intended designs of the peptides.

The β structure of the experimental and fully folded reference peptides was confirmed by the chemical shift deviations of the Hα signals, the $^3J_{HN\alpha}$ spin–spin coupling constants, and the NOE cross-peaks. The Hα chemical shift deviation (ΔδHα) is the difference between the Hα signal for the residue of interest and the corresponding random coil Hα signal [38]. In this study, the fully unfolded reference peptides were considered to be random coil [11,22,23,26]. A positive ΔδHα value suggests an extended β-sheet conformation [38,39]. The ΔδHα values of the residues Thr2 through Val5 and Orn8 through Leu11 for the experimental HPTXaaZbb peptides and the fully folded reference HPTFXaaZbb peptides were positive (Figure 2, Figures S1 and S2), suggesting an extended β-strand conformation for these residues. This is consistent with the intended design. In general, the ΔδHα values for the residues in the strand regions (residues 2–5 and residues 8–11) of the fully folded reference peptides were more positive compared to those for the corresponding experimental peptides (Figure 2, Figures S1 and S2), suggesting that the fully folded reference peptides were more well folded than the corresponding experimental peptides. The ΔδHα values of the terminal residues Arg1 and Gln12 for the experimental peptides were near zero (Figure 2, Figures S1 and S2), most likely due to end fraying effects [21]. The ΔδHα values for Gly7 were negative or mostly close to zero (Figure 2, Figures S1 and S2), consistent with turn formation [12].

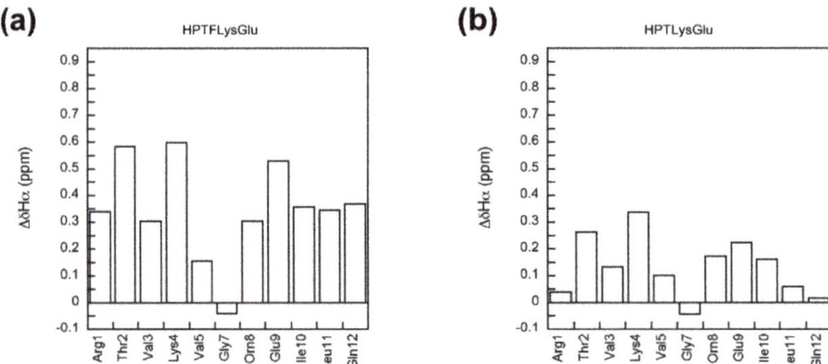

Figure 2. The chemical shift deviation (ΔδHα) for the residues in peptides HPTFLysGlu (**a**) and HPTLysGlu (**b**).

The DQF-COSY spectra were used to determine the $^3J_{HN\alpha}$ spin–spin coupling constants for each residue in the peptides (Tables S37–S45) [33,40]. The $^3J_{HN\alpha}$ coupling constants of the residues in the fully folded reference HPTFXaaZbb peptides showed values higher than 7 Hz (Tables S43–S45), consistent with a β-hairpin structure [36,41]. The experimental HPTXaaZbb peptides also exhibited $^3J_{HN\alpha}$ coupling constants higher than 7 Hz, but slightly lower $^3J_{HN\alpha}$ values compared to those for the fully folded reference HPTFXaaZbb peptides (Tables S37–S39 and S43–S45). This suggested that the experimental HPTXaaZbb peptides may not be as well folded as the fully folded reference HPTFXaaZbb peptides. For the unfolded reference HPTUXaaZbb peptides, some residues exhibited $^3J_{HN\alpha}$ values near or less than 7 Hz (Tables S40–S42), suggesting that these peptides may not be as well folded as the experimental HPTXaaZbb peptides or the fully folded reference HPTFXaaZbb peptides.

The NOE cross-peaks in the ROESY spectra included sequential, intra-residues, medium-range, and long-range NOEs with a number of cross strand Hα–Hα, Hα–HN,

HN–HN correlations (Figures S3–S50). All sequential Hα–HN NOE correlations in every strand for all peptides were observed (Figures S39–S50), consistent with β-strand formation [42,43]. In addition, the lack of $d_{\alpha N}(i, i + n)$ (n = 2,3,4) and $d_{NN}(i, i + n)$ (n = 1, 2) patterns rules out the formation of other secondary structures (Figures S39–S50) [42,43]. A network of cross strand side-chain-side-chain NOEs between residues on the two β-strands was observed for the experimental peptides HPTXaaZbb and fully folded reference peptides HPTFXaaZbb (Figures S3−S38), consistent with β-hairpin formation for these peptides. Long-range NOE cross-peaks between Thr2 and Xaa9 were observed for most of the experimental HPTXaaZbb peptides and fully folded reference HPTFXaaZbb peptides (Figures S3−S38), consistent with a right-handed twist [11,27,28]. The number of cross-peaks in the ROESY spectra followed the general trend HPTFXaaZbb > HPTXaaZbb > HPTUXaaZbb (Figures S3−S38), consistent with the intended fraction folded population for our designs [3,22,23].

2.3. Fraction Folded Population and ΔG_{fold}

The fraction folded population and folding free energy (ΔG_{fold}) of each residue on the experimental peptides were derived from the Hα chemical shift deviation data (Figures S51 and S52). The residues close to the termini suffered from the end fraying effects [3,21–23]. The residues next to the turn were intrinsically highly folded due to proximity to the turn residues. Therefore, the residues near the center of the strands (positions 2, 3, 9, 10) were used to derive the fraction folded population and ΔG_{fold} for each peptide (Tables 1 and 2) [3,11,20,22,23]. Both hydrogen-bonded sites (positions 3 and 10) and non-hydrogen-bonded sites (positions 2 and 9) were included [3,11,20,22,23]. Since the fraction folded population and the folding free energy showed the same trends (i.e., the more negative the folding free energy, the higher the fraction folded population), further discussion will only focus on the fraction folded data.

Table 1. The fraction folded population (%) for the HPTXaaZbb peptides [1].

Xaa4	Zbb9		
	Asp	Glu	Aad
Dap	63 ± 1	72 ± 3	35 ± 2
Dab	47 ± 5	45 ± 2	50 ± 1
Orn	46 ± 2	47 ± 2	57 ± 1
Lys	41 ± 2	44 ± 1	52 ± 2

[1] Average value for residues 2, 3, 9, and 10.

Table 2. The folding free energy (ΔG_{fold}, kcal/mol) for the HPTXaaZbb peptides [1].

Xaa4	Zbb9		
	Asp	Glu	Aad
Dap	−0.30 ± 0.03	−0.57 ± 0.08	0.36 ± 0.04
Dab	0.08 ± 0.03	0.12 ± 0.04	0.00 ± 0.03
Orn	0.10 ± 0.04	0.08 ± 0.06	−0.17 ± 0.02
Lys	0.23 ± 0.04	0.14 ± 0.03	−0.05 ± 0.04

[1] Average value for residues 2, 3, 9, and 10.

The fraction folded populations for the peptides were between 35% and 72%, and the standard deviations were within 5% (Table 1). Peptides HPTDapAsp and HPTDapGlu, containing the shortest positively charged residue Dap, exhibited exceptionally high fraction folded populations. In particular, HPTDapGlu exhibited the highest fraction folded population among all the HPTXaaZbb peptides. In contrast, HPTDapAad exhibited the least fraction folded population.

The fraction folded population of the HPTXaaAsp peptides followed the trend HPTDapAsp > HPTDabAsp ~ HPTOrnAsp > HPTLysAsp. Similarly, the fraction folded population of the HPTXaaGlu peptides followed the trend HPTDapGlu > HPTDabGlu ~ HPTOrnGlu ~ HPTLysGlu. However, the fraction folded population of the HPTXaaAad peptides followed the trend HPTDapAad < HPTDabAad < HPTOrnAad > HPTLysAad. If one disregards HPTDapAad and HPTDabAad, the fraction folded population of the HPTXaaZbb peptides for a given negatively charged residue Zbb9 generally decreased upon increasing the side-chain length of the positively charged residue Xaa4.

The fraction folded population of the HPTDapZbb peptides followed the trend HPT-DapAsp < HPTDapGlu > HPTDapAad. The fraction folded population of the HPTDabZbb peptides followed the trend HPTDabAsp ~ HPTDabGlu < HPTDabAad. Similarly, the fraction folded population of the HPTOrnZbb peptides followed the trend HPTOrnAsp ~ HPTOrnGlu < HPTOrnAad. The fraction folded population of the HPTLysZbb peptides followed the trend HPTLysAsp ~ HPTLysGlu < HPTLysAad. Again, if one disregards HPTDapAad, the fraction folded population of the HPTXaaZbb peptides for a given positively charged residue Xaa4 generally increased with increasing side-chain length of the negatively charged residue Zbb9.

2.4. Lateral Cross Strand Xaa-Zbb Interactions

Double mutant cycle analysis was performed to derive the interaction free energy (ΔG_{int}) for each lateral Xaa4-Zbb9 interaction (Table 3) [44,45]. For the reference peptides with minimal cross strand interaction, Ala was incorporated at position 4, position 9, or both positions 4 and 9 simultaneously because of the small side-chain of Ala [22,23,44]. The difference in folding energetics between peptides HPTXaaZbb and HPTAlaAla [22] would reflect the effect of simultaneously incorporating Xaa at position 4 and Zbb at position 9. This energy difference would include the effect of incorporating the Xaa residue and Zbb residue individually at positions 4 and 9, respectively, and the interaction between Xaa4 and Zbb9. Therefore, the effect of individually incorporating Xaa and Zbb would need to be considered to derive the Xaa4-Zbb9 interaction energy. The difference in folding energetics between peptides HPTXaaAla [3] and HPTAlaAla [22] would represent the effect of only incorporating Xaa at position 4. Similarly, the difference in folding energetics between peptides HPTAlaZbb [3] and HPTAlaAla [22] would represent the effect of only incorporating Zbb at position 9. The Xaa4-Zbb9 interaction energy (ΔG_{int}) was determined from the folding energetics for the peptides HPTXaaZbb, HPTXaaAla [3], HPTAlaZbb [3], and HPTAlaAla [22] using Equation (7) (Table 3).

Table 3. The Xaa4-Zbb9 ion-pairing interaction energy (ΔG_{int}, kcal/mol) [1].

Xaa4	Zbb9		
	Asp	**Glu**	**Aad**
Dap	-1.09 ± 0.05	-1.10 ± 0.07	-0.07 ± 0.06
Dab	-0.54 ± 0.05	-0.23 ± 0.08	-0.27 ± 0.05
Orn	-0.40 ± 0.05	-0.17 ± 0.12	-0.32 ± 0.05
Lys	-0.24 ± 0.06	-0.13 ± 0.12 [2]	-0.17 ± 0.11

[1] Average value for residues 2, 3, 9, and 10. [2] Average value for residues 2, 3, and 10.

All of the cross strand lateral Xaa4-Zbb9 ion-pairing interactions were apparently stabilizing (Table 3). For the HPTXaaAsp peptides, the Xaa4-Asp9 interaction energy followed the trend Dap < Dab < Orn < Lys. Similarly, for the HPTXaaGlu peptides, the Xaa4-Glu9 interaction energy followed the trend Dap < Dab ~ Orn ~ Lys. For the HPTXaaAad peptides, the Xaa4-Aad9 interaction energy followed the trend Dap > Dab ~ Orn < Lys. If one disregards the Dap4-Aad9 interaction, the Xaa4-Zbb9 interaction generally becomes more stabilizing with decreasing Xaa4 side-chain length for a given Zbb9. Interestingly, the Dap4-Asp9 and Dap4-Glu9 interactions were the most stabilizing, providing more than 1 kcal/mol stabilization (Table 3). In contrast, the Dap4-Aad9 interaction provided the least stabilization, being essentially nonexistent. This showed that interaction between oppositely charged residues with short side-chains form stabilizing lateral cross strand ion-pairing interactions.

3. Discussion

The effect of side-chain length on lateral cross strand ion-pairing interactions between ammonium- and carboxylate-containing amino acids upon swapping the position of the charged amino acids was investigated. The fraction folded population for the HPTXaaZbb peptides was between 35% and 72% (Table 1). The extensive range of fraction folded population of the HPTXaaZbb peptides can be rationalized by the individual effects of the side-chain length of the ammonium- and carboxylate-containing at positions 4 and 9 on the hairpin formation, respectively, and the lateral cross strand Xaa4-Zbb9 interaction. In general, the fraction folded population of the HPTXaaZbb peptides for a given positively charged residue Xaa4 increased with increasing side-chain length of the negatively charged residue Zbb9 (except for peptide HPTDapAad; vide supra) (Table 1). This is consistent with the increased fraction folded hairpin population upon increasing the negatively charged residue side-chain length at position 9 for the HPTAlaZbb peptides [3]. In general, the fraction folded population of the HPTXaaZbb peptides for a given negatively charged residue Zbb9 decreased upon increasing the side-chain length of the positively charged residue Xaa4 (except for peptides HPTDapAad and HPTDabAad; vide supra) (Table 1). However, the fraction folded hairpin population for the HPTXaaAla peptides increased upon increasing the positively charged residue side-chain length at position 4 [3], suggesting the presence of cross strand Xaa4-Zbb9 interactions.

The two peptides with the highest fraction folded populations were HPTDapGlu ($72 \pm 3\%$) and HPTDapAsp ($63 \pm 1\%$). Similarly, the peptides with the same interacting residues, but the positions unswapped also exhibited the highest fraction folded populations in our previous study on HPTZbbXaa peptides (HPTGluDap: $63 \pm 2\%$; HPTAspDap: $55 \pm 3\%$) [22]. Nonetheless, the fraction folded populations of the HPTXaaZbb peptides with the charged residues swapped in this study (Table 1) were consistently higher compared to the corresponding unswapped HPTZbbXaa peptides in our previous study [22]. This is consistent with the higher fraction folded hairpin population for the HPTXaaAla peptides with the positively charged residue (Xaa) at position 4 compared to the corresponding HPTAlaXaa peptides with the positively charged residues (Xaa) at position 9 [3]. The change in fraction folded population upon swapping the residues in an interacting pair was consistent with studies on the protein G B1 domain [14] and a different hairpin system [19], which both showed a change in the stability of the host system upon swapping the position of interacting residues.

The largest difference in fraction folded population upon swapping was between the unswapped peptide HPTAadDab ($26 \pm 2\%$) [22] and the corresponding swapped peptide HPTDabAad ($50 \pm 1\%$). To gain further insight into this difference in the fraction folded population upon swapping, side-chain conformational analysis was performed on these two peptides by molecular mechanics calculations. The initial model was generated based on the solution structure of an analog of the parent YKL peptide (pdb code 1JY9 [46]). All possible combinations of low-energy side-chain dihedral angles (χ) for Aad and Dab were investigated. A combined total of 2916 conformations were minimized. The lowest energy conformation for the unswapped HPTAadDab exhibited higher energy (less negative energy, i.e., less stable) compared to that for the swapped HPTDabAad (Table 4), consistent with the fraction folded population for the two peptides. Conformations within 4 kcal/mol of the lowest energy conformer for each peptide were then examined (i.e., low-energy conformations, Table 4) because room temperature can provide up to 4 kcal/mol of thermal energy. All but one low-energy conformation exhibited salt bridges between the charged residues at positions 4 and 9. There were more low-energy conformations for the unswapped HPTAadDab compared to the swapped HPTDabAad. The energy reflects the enthalpic component of the conformation, whereas the number of low-energy conformations reflects the entropic component of the folded form of the peptide. The side-chain conformational entropy contribution of the residues at positions 4 and 9 to the free energy of the folded form for the two peptides was calculated based on the Boltzmann distribution of the various low-energy conformations (Table 4). The more negative −TS reflected the higher side-chain conformational entropy in the folded form for the unswapped HPTAadDab compared to the swapped HPTDabAad, despite involving the same two potentially interacting residues. The conformation of the low-energy conformers was examined in detail. Each χ_1 dihedral was divided into three categories: gauche− (60°, g−), trans (180°, t), and gauche+ (300°, g+) [47,48]. The combination of the χ_1 dihedrals was represented in parentheses (Table 4, Figures S53 and S54), showing the conformation for the residue at position 4 followed by the conformation for the residue at position 9. For example, a conformation with t at position 4 and g+ at position 9 would be designated (t, g+). For the unswapped HPTAadDab, 8 of the 9 possible combinations were present (Table 4 and Figure S53), whereas only 4 of the 9 possible combinations were observed for the swapped HPTDabAad (Table 4 and Figure S54). Importantly, the majority of the low-energy conformations did not involve g− (for χ_1) at either position in either peptide (68% for HPTAadDab, and 94% for HPTDabAad). This is most likely because the g− conformation is higher in energy compared to t and g+ [47,48], and the g− conformation would inherently point the side-chain away from the neighboring strand.

Table 4. Summary of the low-energy conformations from the side-chain conformational analysis of peptides HPTAadDab and HPTDabAad by molecular mechanics calculations.

Peptide	**Lowest Energy Conformation Energy**	**Side-Chain Conformational Entropy Contribution**	**Conformations within 4 kcal/mol of the Lowest Energy Conformer**		
	(kcal/mol)	**−TS (kcal/mol)**	**No** [1]	**Salt Bridge** [2] **(%)**	**Conformations, Number (Residue 4 χ_1, Residue 9 χ_1)**
HPTAadDab	−384.4	−1.67	65	98	16 (g+, g+), 14 (t, g+) 13 (g+, t), 8 (g−, g+) 6 (g−, t), 5 (g+, g−) 2 (g+, g−), 1 (t, t)
HPTDabAad	−387.8	−1.05	16	100	7 (t, g+), 5 (g+, t), 3 (g+, g+), 1 (g−, g+)

[1] The number of conformations within 4 kcal/mol of the lowest energy conformer for each peptide. [2] The percentage of conformations within 4 kcal/mol of the lowest energy conformer with an Aad–Dab salt bridge, which is a hydrogen-bonded ion pair [49].

More low-energy conformations were observed for the unswapped Aad4-Dab9 interaction compared to the swapped Dab9-Aad9 interaction (Table 4 and Figure 3). Apparently, the right-handed twist of the hairpin structure [11] raised the residue at position 4 and lowered the residue at position 9 (Figures 3 and 4). This "unleveled" relative positioning of the interacting residues resulted in more proper length matching for the Aad4-Dab9 interaction, enabling more low-energy conformations with the Aad4-Dab9 interaction and a higher proportion of the g− conformation in χ_1 (21 in 65, or 32%). For the swapped Dab4-Aad9 interaction, the unleveled relative positioning exacerbated the length difference between Dab4 and Aad9, leading to less low-energy conformations with the Dab4-Aad9 interaction and a relatively low proportion of the g− conformation in χ_1 (1 in 16, or 6%). This unleveled positioning created by the right-handed twist appeared to be one of the factors giving rise to the difference between the unswapped and swapped peptides.

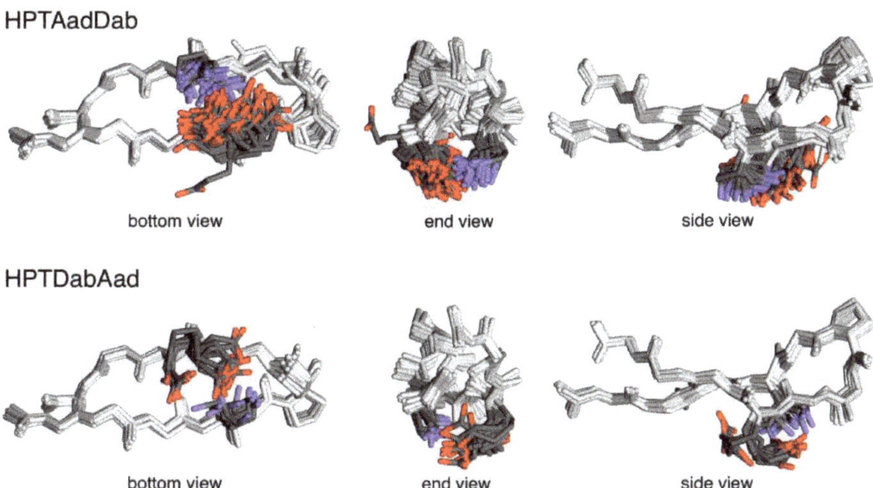

Figure 3. The low-energy conformations from molecular mechanics calculations for peptides HPTAadDab and HPTDabAad. The backbone and DPro side-chain are shown in white. The residues at positions 4 and 9 are colored according to element: carbon in gray, oxygen in red, and nitrogen in blue. The other side-chains and all hydrogen atoms are omitted for clarity.

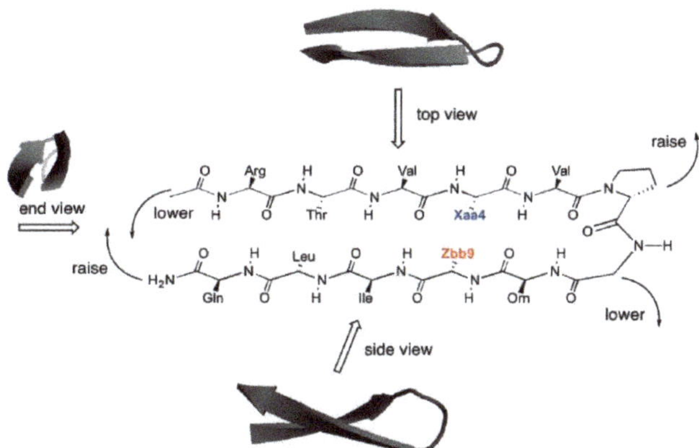

Figure 4. The chemical structure of HPTXaaZbb shown in perspective view, showing the consequence of the right-handed twist on a flat structure. The cartoon ribbon representation was generated from the solution structure of an analog of the parent YKL peptide using PyMOL (pdb code 1JY9 [46]).

The energetic contribution of lateral cross strand Xaa4-Zbb9 interactions to hairpin formation was determined by double mutant cycle analysis (Table 3). For the HPTXaaZbb swapped peptides in this study, the Dap4-Glu9 and Dap4-Asp9 interactions were most stabilizing (Table 3). Similarly, the Glu4-Dap9 and Asp4-Dap9 interactions were the most stabilizing in our previous study on the original unswapped HPTZbbXaa peptides [22]. This may be because shorter amino acids such as Dap and Asp are conformationally less flexible compared to amino acids with longer side-chains, leading to higher stabilizing lateral cross strand ion-pairing interactions due to less entropic penalty [22]. The mean conformational entropic penalty of one side-chain rotatable bond upon folding is 0.5 kcal/mol [50].

There is a dramatic decrease in fraction folded population and stabilizing side-chain interaction for peptides HPTDapAsp and HPTDapGlu upon increasing the Dap side-chain length by just one methylene to Dab (Tables 1 and 3). Increasing the side-chain length of Dap by one methylene to Dab would increase the electron-donating characteristics, decrease the electron-withdrawing characteristics from the backbone functionality, and thus decrease the cationic charge density on the ammonium group. This decrease in cationic charge density would decrease the electrostatic interaction with the negatively charged Asp and Glu, resulting in the decreased fraction folded population for peptides HPTDabGlu and HPTDabAsp compared to peptides HPTDapGlu and HPTDapAsp, respectively (Table 1 and Figure 3).

Further lengthening the positively charged residue Dab in HPTDabGlu and HPTDabAsp did not alter the fraction folded population or the Xaa4-Zbb9 interaction as drastically as the change upon lengthening Dap to Dab. The longer side-chains are more flexible, and therefore, more energy would be needed to confine the side-chain conformation to enable the cross strand ion-pairing interaction. This would decrease the overall energetic contribution of the cross strand ion-pairing interaction for residues with longer side-chains. However, as the side-chain length of the positively charged residue Xaa4 increases, the fraction folded population of HPTXaaAla peptide increases [3]. As such, this increase in hairpin formation (due to the positively charged residue Xaa) compensates for the increase in the side-chain entropic penalty for the longer side-chains to form a cross strand ion-pairing interaction, leading to less drastic changes in the fraction folded population.

In general, the interaction free energy became less stabilizing with increasing side-chain length of the positively charged residue Xaa4 for a given negatively charged residue Zbb9 (except for Dap4-Aad9) in the swapped HPTXaaZbb peptides in this study (Table 3). The same general trend was also observed in our previous study on the original unswapped HPTZbbXaa peptides, but with less stabilizing lateral cross strand interactions between residues at positions 4 and 9 [22]. This difference in interaction energy could be due to the difference in the relative placement of the residues at position 9 on the C-terminal strand and position 4 on the N-terminal strand (Figure 4, vide infra), stemming from the inherent right-handed twist of sheet structures [28].

The Xaa4-Asp9 interactions were more stabilizing compared to the corresponding Xaa4-Glu9 and Xaa4-Aad9 interactions for a given Xaa4 (Table 3). This is perhaps the result of the relative positioning of Xaa4 and Zbb9. The right-handed twist of the hairpin structure [11] lowers Zbb9 and raises Xaa4 (Figure 4). For Xaa4 to interact with Zbb9, the ammonium group on Xaa4 and the carboxylate group on Zbb9 need to be close to one another. Apparently, length matching is critical for lateral cross strand interactions [22,23]. Since the carboxylate group on Zbb9 is inherently longer than the ammonium group on Xaa4, the shorter Asp9 would be more well suited to interact with Xaa4 (especially shorter residues) compared to the longer Glu9 and Aad9 due to the "unleveled" relative positioning of the residues created by the right-handed twist (Figure 4). In comparison, interactions between the longer Glu9 (and Aad9) and Xaa4 would be weaker compared to the corresponding Xaa4-Asp9 interactions. Furthermore, the Xaa4-Asp9 interaction would be more stabilizing because of the need to pay less of an entropic penalty to confine the short Asp9 side-chain conformation to enable the Xaa4-Asp9 interaction compared to the longer Glu and Aad. For the original unswapped HPTZbbXaa peptides [22], there is no general trend among the Asp4-Xaa9, Glu4-Xaa9, and Aad4-Xaa9 interactions for a given Xaa9 residue. This may be because the unleveled relative positioning created by the right-handed twist facilitates the length matching, bringing the inherently shorter ammonium functionality closer to the inherently longer carboxylate functionality. These results are consistent with the studies on cross-strand interactions in the protein G B1 domain [14], showing that swapping the amino acid positions in lateral cross-strand interactions changed the interaction energy. Overall, our results suggest that there is an orientation preference for lateral cross strand interactions to stabilize sheet systems, despite the apparent symmetry of lateral cross strand interactions based on statistical studies [24].

The Dap4-Aad9 and Lys4-Aad9 interactions were less stabilizing compared to the Dab4-Aad9 and Orn4-Aad9 interactions if one disregards the error bars. The low stabilization of the Lys4-Aad9 interaction may be due to the conformationally more flexible

Lys side-chain, leading to the need to pay a higher entropic penalty to confine the long Lys and Aad side-chains to enable the Lys4-Aad9 interaction. The low stabilization of the Dap4-Aad9 interaction may be due to the length discrepancy between Dap and Aad, which is further magnified by the unleveled relative placement of residues at positions 4 and 9 resulting from the right-handed twist. Importantly, the combination of unleveled relative placement of the interacting functional groups and difference in entropic penalty necessary to form the lateral cross-strand interaction resulted in the observed trends and the effects upon swapping interacting residues.

4. Materials and Methods

4.1. Peptide Synthesis

Peptides were synthesized by solid-phase peptide synthesis using Fmoc-based chemistry [29,30]. The disulfide bond in the Cys-containing HPTFXaaZbb peptides was formed via charcoal-mediated air oxidation [31]. All peptides were purified by reverse-phase high-performance liquid chromatography (RP-HPLC) (Waters, Milford, MA, USA) to higher than 95% purity. The identity of the peptides was confirmed by matrix-assisted laser desorption ionization time-of-flight mass spectrometry (MALDI-TOF) (Bruker, Billerica, MA, USA). More detailed procedures and peptide characterization data are provided in the Supplementary Materials.

4.2. Nuclear Magnetic Resonance Spectroscopy

Purified peptides were dissolved in H_2O/D_2O (9:1 ratio by volume) in the presence of 50 mM sodium deuterioacetate buffer (pH 5.5 uncorrected). Peptide concentrations were 2.0–15.4 mM. 2-Dimethyl-2-silapentane-5-sulfonate (DSS) was added to the sample as an internal reference. All NMR experiments were performed on a Brüker AVIII 800 MHz spectrometer (Bruker, Billerica, MA, USA). ^{1}H-^{1}H homonuclear phase-sensitive double-quantum filtered-correlated spectroscopy (DQF-COSY) [33], total correlation spectroscopy (TOCSY) [34], and rotating-frame nuclear Overhauser effect spectroscopy (ROESY) [35] experiments were performed by collecting 2048 point in f_2 with 4–8 scans and 256–512 points in f_1 at 298 K. Solvent suppression was achieved by the WATERGATE solvent suppression sequence [51]. TOCSY and ROESY experiments employed a spin locking field of 10 kHz. Mixing times of 60 and 200 ms were used for the TOCSY and ROESY experiment, respectively.

4.3. Chemical Shift Deviation

Sequence-specific assignments for all peptides were completed by using the 2D-NMR spectra (TOCSY and ROESY). The chemical shift deviation (ΔδHα) for each residue of the experimental peptide ΔδHα(exp)) and the folded reference peptide (ΔδHα(F)) was derived using Equations (1) and (2), respectively [38]. δHα(exp) is the chemical shift for the residue of interest on the experimental peptide, and δHα(U) is the chemical shift for the corresponding residue of interest on the fully unfolded reference peptide. δHα(F) is the chemical shift for the residue of interest on the fully folded reference peptide.

$$\Delta\delta H\alpha(\text{exp}) = \delta H\alpha(\text{exp}) - \delta H\alpha(\text{U}) \quad (1)$$

$$\Delta\delta H\alpha(\text{F}) = \delta H\alpha(\text{F}) - \delta H\alpha(\text{U}) \quad (2)$$

4.4. $J_{HN\alpha}$ Spin–Spin Coupling Constant

The peak-to-peak separation in the absorptive (v_a) and the dispersive (v_d) spectra were measured to derive the *J* coupling constant. The v_a and v_d values were obtained using values on the f_2 axis. Equation (3) was used to derive the coupling constants [40].

$$J^6 - \upsilon_d{}^2 J^4 + \left(-\frac{9}{4}\upsilon_a{}^4 + \frac{3}{2}\upsilon_a{}^2\upsilon_d{}^2 + \frac{3}{4}\upsilon_d{}^4\right)J^2 + \frac{81}{64}\upsilon_a{}^6 - \frac{9}{16}\upsilon_a{}^4\upsilon_d{}^2 - \frac{21}{32}\upsilon_a{}^2\upsilon_d{}^4 - \frac{1}{16}\upsilon_d{}^6 - \frac{\upsilon_d{}^8}{64\upsilon_a{}^2} = 0 \quad (3)$$

4.5. Interproton Distance Determination via NOE Integration

The NOE cross-peaks for all peptides were assigned from the corresponding ROESY spectra. Integration was performed based on Gaussian peak modeling to obtain the intensity of cross-peaks (I). The distance between the β-hydrogen atoms on the proline side-chain (regardless of stereochemistry) was set as the standard (1.77 Å) to derive the interproton distance for the cross-peak of interest using Equation (4). The distances (r) were grouped into short (≤2.5 Å), medium (2.5~3.5 Å), and long (>3.5 Å) for the depictions in the Wüthrich diagrams (Figures S39–S50).

$$r = 1.77 \times 10^{-10} \times \left(\frac{I_{standard}}{I_{NOE}}\right)^{\frac{1}{6}} \quad (4)$$

4.6. Fraction Folded Population and Folding Free Energy (ΔG_{fold})

The equilibrium constant between the unfolded and folded states of an experimental peptide is the ratio of the folded and unfolded populations. The fraction folded population for each residue was derived from the chemical shift data according to Equation (5). The folding free energy ΔG_{fold} for each residue was derived using Equation (6). The fraction folded population and folding free energy (ΔG_{fold}) of the peptide was obtained by averaging the fraction folded population and ΔG_{fold}, respectively, for the residues 2, 3, 9, and 10 [11,20,22,23].

$$\text{Fraction Folded Population} = \frac{\delta H\alpha(exp) - \delta H\alpha(U)}{\delta H\alpha(F) - \delta H\alpha(U)} \times 100\% \quad (5)$$

$$\Delta G_{fold} = -RT\ln\frac{\delta H\alpha(exp) - \delta H\alpha(U)}{\delta H\alpha(F) - \delta H\alpha(exp)} \quad (6)$$

4.7. Double Mutant Cycle Analysis

Double mutant cycle analysis [44] was performed to determine the interaction free energy (ΔG_{int}) between charged residues Xaa4 and Zbb9 in the HPTXaaZbb peptides using Equation (7) [44,45]. This analysis accounted for the effect of each charged residue (individually) on strand stability using data from the corresponding Ala-containing peptides HPTXaaAla [3] and HPTAlaZbb [3] to determine the Xaa4-Zbb9 ion-pairing interaction exclusively (Table 3). The peptide with Ala incorporated at positions 4 and 9, HPTAlaAla [22], was used as the reference peptide.

$$\Delta G_{int} = (\Delta G_{HPTXaaZbb} - \Delta G_{HPTAlaAla}) - (\Delta G_{HPTXaaAla} - \Delta G_{HPTAlaAla}) - (\Delta G_{HPTAlaZbb} - \Delta G_{HPTAlaAla}) \quad (7)$$

4.8. Side-Chain Conformational Analysis by Molecular Mechanics Calculations

The conformational analysis was performed using the program Discovery Studio 2.1 (Accelrys, CA, USA) on an IBM x3550M2 workstation (CPU: Dual Xeon E5530 2.4 GHz with Quad cores; RAM: 48 G) running the operating system CentOS 5.3. The models were created based on the solution structure of an analogous peptide of the parent YKL peptide [46] with various combinations of potential low-energy side-chain dihedrals. For each side-chain dihedral angle (χ) involving sp^3 carbons, three possible low-energy staggered conformations were considered: gauche− (60°, g−), trans (180°, t), and gauche+ (300°, g+) [47,48]. For the dihedral angle involving the sp^2 carboxylate carbon of Aad, six conformations were considered: 0°, 30°, 60°, 90°, 120°, and 150°. For each peptide (HPTAadDab and HPTDabAad), 1458 conformations were evaluated. Each conformation was minimized using the CFF forcefield. The nonbond radius of 99 Å, nonbond higher cutoff distance of 98 Å and nonbond lower cutoff distance of 97 Å were employed to perform the calculations with effectively no cutoffs. Distance dependent dielectric constant of 2 was used as the implicit solvent model. Minimization was performed by steepest descent and conjugate gradient protocols until convergence (converging slope was set to 0.1 kcal/(mol × Å). After minimization, each conformation was reexamined to remove duplicating conformations because minimization with different starting conformations occasionally resulted in the same final conformation. When the same conformation was obtained more than once, only the lowest energy conformation was considered in further analyses. The probability of conformation i at 298 K (p_i) was calculated based on Boltzmann distribution using Equation (8), in which ε_i is the energy of conformation i, k_B is the Boltzmann constant, and T is the temperature (298 K). The entropic contribution to the folded form at 298 K was calculated using Equation (9).

$$p_i = \frac{e^{\frac{-\varepsilon_i}{k_B T}}}{\sum_j e^{\frac{-\varepsilon_j}{k_B T}}} \tag{8}$$

$$-TS = -T\cdot(-k_B)\sum_i p_i \cdot \ln(p_i) \tag{9}$$

5. Conclusions

We investigated the effect of swapping the cross strand interacting charged amino acid positions in a β-hairpin from the original Zbb4-Xaa9 in a previous study (HPTZbbXaa peptides) to the swapped Xaa4-Zbb9 in this study (HPTXaaZbb peptides). The general trends for the fraction folded population, and side-chain interaction energetics remained similar upon swapping the position of potentially interacting charged residues. Nonetheless, the fraction folded populations for most of the swapped HPTXaaZbb peptides were higher compared to the corresponding original HPTZbbXaa peptides, consistent with the inherent effect of the positively charged Xaa residue on hairpin formation at the two different positions. The most stabilizing cross strand interactions were between short residues (Dap4-Asp9 and Dap4-Glu9) even after swapping the position of the charged residues. However, subtle differences were present, most likely due to the unleveled relative placement of the residues at positions 4 and 9 created by the right-handed twist of the sheet structure. These results should be useful for developing functional peptides that rely on lateral ion-pairing interactions across antiparallel β-strands.

Supplementary Materials: The following are available online. Tables S1–S36: The 1H chemical shift assignments for the peptides. Tables S37–S45: The 3JNHα values of the peptides. Figure S1: The Hα chemical shift deviations for the residues in the experimental HPTXaaZbb peptides. Figure S2: The Hα chemical shift deviations for the residues in the fully folded reference HPTFXaaZbb peptides. Figures S3–S38: The NOEs observed involving the side-chains of the peptides. Figures S39–S50: Wüthrich diagrams of the backbone NOE connectivities involving the α-protons and amide protons for the peptides. Figure S51: The fraction folded of the residues in the peptides. Figure S52: The ΔGfold of the residues in the peptides. Figure S53: The low-energy conformations for peptide HPTAadDab. Figure S54: The low-energy conformations for peptide HPTDabAad. Detailed peptide synthesis procedures and peptide characterization data.

Author Contributions: Conceptualization, R.P.C.; methodology, R.P.C.; software, C.-H.H., T.W.W., and C.-H.Y.; validation, C.-H.H. and C.-H.Y.; formal analysis, C.-H.H., T.W.W., and C.-H.Y.; investigation, C.-H.H., S.-J.H., and S.-L.H.; resources, C.-H.H., T.W.W., and C.-H.Y.; data curation, C.-H.H., T.W.W., C.-H.Y., and J.-Y.C.; writing—original draft preparation, C.-H.H. and T.W.W.; writing—review and editing, J.-Y.C. and R.P.C.; visualization, C.-H.H., T.W.W., C.-H.Y., and R.P.C.; supervision, R.P.C.; project administration, R.P.C.; funding acquisition, R.P.C. All authors have read and agreed to the published version of the manuscript.

Funding: This research was funded by the Ministry of Science and Technology in Taiwan, MOST-101-2113-M-002-006-MY2,MOST-103-2113-M-002-018-MY3,and MOST-109-2113-M-002-011.

Institutional Review Board Statement: Not applicable.

Informed Consent Statement: Not applicable.

Data Availability Statement: The data presented in this study are available in the Supplementary Materials. The raw data are available on request from the corresponding author.

Acknowledgments: The authors would like to thank Hsiou-Ting Kuo, Jhe-Hao Li, Po-Yi Wu, and Chien-Hsiang Liu for their assistance in acquiring the NMR data. The authors would like to thank the Computer and Information Networking Center at National Taiwan University for the support of the high-performance computing facilities.

Conflicts of Interest: The authors declare no conflict of interest.

Sample Availability: Samples of the compounds are not available from the authors.

References

1. Chou, P.Y.; Fasman, G.D. Conformational parameters for amino acids in helical, β-sheet, and random coil regions calculated from proteins. *Biochemistry* **1974**, *13*, 211–222. [CrossRef] [PubMed]
2. Muñoz, V.; Serrano, L. Intrinsic secondary structure propensities of the amino acids, using statistical φ-ψ matrices: Comparison with experimental scales. *Proteins* **1994**, *20*, 301–311. [CrossRef] [PubMed]
3. Kuo, L.-H.; Li, J.-H.; Kuo, H.-T.; Hung, C.-Y.; Tsai, H.-Y.; Chiu, W.-C.; Wu, C.-H.; Wang, W.-R.; Yang, P.-A.; Yao, Y.-C.; et al. Effect of charged amino acid side chain length at non-hydrogen bonded strand positions on β-hairpin stability. *Biochemistry* **2013**, *52*, 7785–7797. [CrossRef]
4. Hardy, J.; Allsop, D. Amyloid deposition as the central event in the aetiology of Alzheimer's disease. *Trends Pharmacol. Sci.* **1991**, *12*, 383–388. [CrossRef]
5. Bartzokis, G.; Lu, P.H.; Mintz, J. Human brain myelination and amyloid β deposition in Alzheimer's disease. *Alzheimers Dement.* **2007**, *3*, 122–125. [CrossRef]
6. Scherzinger, E.; Lurz, R.; Turmaine, M.; Mangiarini, L.; Hollenbach, B.; Hasenbank, R.; Bates, G.P.; Davies, S.W.; Lehrach, H.; Wanker, E.E. Huntingtin-encoded polyglutamine expansions form amyloid-like protein aggregates in vitro and in vivo. *Cell* **1997**, *90*, 549–558. [CrossRef]
7. Mastaglia, F.L.; Johnsen, R.D.; Byrnes, M.L.; Kakulas, B.A. Prevalence of amyloid-β deposition in the cerebral cortex in Parkinson's disease. *Mov. Disord.* **2003**, *18*, 81–86. [CrossRef] [PubMed]
8. Irvine, G.B.; El-Agnaf, O.M.; Shankar, G.M.; Walsh, D.M. Protein aggregation in the brain: The molecular basis for Alzheimer's and Parkinson's diseases. *Mol. Med.* **2008**, *14*, 451–464. [CrossRef]
9. Cheng, Y.-S.; Chen, Z.-t.; Liao, T.-Y.; Lin, C.; Shen, H.C.-H.; Wang, Y.-H.; Chang, C.-W.; Liu, R.-S.; Chen, R.P.-Y.; Tu, P.-h. An intranasally delivered peptide drug ameliorates cognitive decline in Alzheimer transgenic mice. *EMBO Mol. Med.* **2017**, *9*, 703–715. [CrossRef]
10. Chen, R.P.-Y. From nose to brain: The promise of peptide therapy for Alzheimer's disease and other neurodegenerative diseases. *J. Alzheimers Dis. Parkinsonism* **2017**, *7*, 1000314. [CrossRef]
11. Syud, F.A.; Stanger, H.E.; Gellman, S.H. Interstrand side chain-side chain interactions in a designed β-hairpin: Significance of both lateral and diagonal pairings. *J. Am. Chem. Soc.* **2001**, *123*, 8667–8677. [CrossRef] [PubMed]
12. Ramirez-Alvarado, M.; Blanco, F.J.; Serrano, L. De novo design and structural analysis of a model β-hairpin peptide system. *Nat. Struct. Mol. Biol.* **1996**, *3*, 604–612. [CrossRef] [PubMed]
13. Gellman, S.H. Minimal model systems for β-sheet secondary structure in proteins. *Curr. Opin. Chem. Biol.* **1998**, *2*, 717–725. [CrossRef]
14. Smith, C.K.; Regan, L. Guidelines for protein design: The energetics of β-sheet side chain interactions. *Science* **1995**, *270*, 980–982. [CrossRef] [PubMed]
15. Merkel, J.S.; Sturtevant, J.M.; Regan, L. Sidechain interactions in parallel β sheets: The energetics of cross-strand pairings. *Structure* **1999**, *7*, 1333–1343. [CrossRef]
16. Blasie, C.A.; Berg, J.M. Electrostatic interactions across a β-sheet. *Biochemistry* **1997**, *36*, 6218–6222. [CrossRef] [PubMed]
17. Searle, M.S.; Griffiths-Jones, S.R.; Skinner-Smith, H. Energetics of weak interactions in a β-hairpin peptide: Electrostatic and hydrophobic contributions to stability from lysine salt bridges. *J. Am. Chem. Soc.* **1999**, *121*, 11615–11620. [CrossRef]
18. Russell, S.J.; Cochran, A.G. Designing Stable β-Hairpins: Energetic Contributions from Cross-Strand Residues. *J. Am. Chem. Soc.* **2000**, *122*, 12600–12601. [CrossRef]
19. Ramírez-Alvarado, M.; Blanco, F.J.; Serrano, L. Elongation of the BH8 β-hairpin peptide: Electrostatic interactions in β-hairpin formation and stability. *Protein Sci.* **2001**, *10*, 1381–1392. [CrossRef] [PubMed]
20. Kiehna, S.E.; Waters, M.L. Sequence dependence of β-hairpin structure: Comparison of a salt bridge and an aromatic interaction. *Protein Sci.* **2003**, *12*, 2657–2667. [CrossRef]
21. Ciani, B.; Jourdan, M.; Searle, M.S. Stabilization of β-hairpin peptides by salt bridges: Role of preorganization in the energetic contribution of weak interactions. *J. Am. Chem. Soc.* **2003**, *125*, 9038–9047. [CrossRef] [PubMed]
22. Kuo, H.-T.; Fang, C.-J.; Tsai, H.-Y.; Yang, M.-F.; Chang, H.-C.; Liu, S.-L.; Kuo, L.-H.; Wang, W.-R.; Yang, P.-A.; Huang, S.-J.; et al. Effect of charged amino acid side chain length on lateral cross-strand interactions between carboxylate-containing residues and lysine analogues in a β-hairpin. *Biochemistry* **2013**, *52*, 9212–9222. [CrossRef] [PubMed]
23. Kuo, H.-T.; Liu, S.-L.; Chiu, W.-C.; Fang, C.-J.; Chang, H.-C.; Wang, W.-R.; Yang, P.-A.; Li, J.-H.; Huang, S.-J.; Huang, S.-L.; et al. Effect of charged amino acid side chain length on lateral cross-strand interactions between carboxylate- and guanidinium-containing residues in a β-hairpin. *Amino Acids* **2015**, *47*, 885–898. [CrossRef] [PubMed]

24. Cootes, A.P.; Curmi, P.M.G.; Cunningham, R.; Donnelly, C.; Torda, A.E. The dependence of amino acid pair correlations on structural environment. *Proteins* **1998**, *32*, 175–189. [CrossRef]
25. Stanger, H.E.; Gellman, S.H. Rules for antiparallel β-sheet design: D-Pro-Gly is superior to L-Asn-Gly for β-hairpin nucleation. *J. Am. Chem. Soc.* **1998**, *120*, 4236–4237. [CrossRef]
26. Syud, F.A.; Espinosa, J.F.; Gellman, S.H. NMR-based quantification of β-sheet populations in aqueous solution through use of reference peptides for the folded and unfolded states. *J. Am. Chem. Soc.* **1999**, *121*, 11577–11578. [CrossRef]
27. Weatherford, D.W.; Salemme, F.R. Conformations of twisted parallel β-sheets and the origin of chirality in protein structures. *Proc. Natl. Acad. Sci. USA* **1979**, *76*, 19–23. [CrossRef] [PubMed]
28. Yang, A.-S.; Honig, B. Free energy determinants of secondary structure formation: II. Antiparallel β-sheets. *J. Mol. Biol.* **1995**, *252*, 366–376. [CrossRef] [PubMed]
29. Fields, G.B.; Noble, R.L. Solid phase peptide synthesis utilizing 9-fluorenylmethoxycarbonyl amino acids. *Int. J. Pept. Protein Res.* **1990**, *35*, 161–214. [CrossRef]
30. Atherton, E.; Fox, H.; Harkiss, D.; Logan, C.J.; Sheppard, R.C.; Williams, B.J. A mild procedure for solid phase peptide synthesis: Use of fluorenylmethoxycarbonylamino-acids. *J. Chem. Soc. Chem. Commun.* **1978**, 537–539. [CrossRef]
31. Volkmer-Engert, R.; Landgraf, C.; Schneider-Mergener, J. Charcoal surface-assisted catalysis of intramolecular disulfide bond formation in peptides. *J. Pept. Res.* **1998**, *51*, 365–369. [CrossRef]
32. Russell, S.J.; Blandl, T.; Skelton, N.J.; Cochran, A.G. Stability of cyclic β-hairpins: Asymmetric contributions from side chains of a hydrogen-bonded cross-strand residue pair. *J. Am. Chem. Soc.* **2003**, *125*, 388–395. [CrossRef]
33. Aue, W.P.; Bartholdi, E.; Ernst, R.R. Two-dimensional spectroscopy. Application to nuclear magnetic resonance. *J. Chem. Phys.* **1976**, *64*, 2229–2246. [CrossRef]
34. Bax, A.; Davis, D.G. MLEV-17-based two-dimensional homonuclear magnetization transfer spectroscopy. *J. Magn. Reson.* **1985**, *65*, 355–360. [CrossRef]
35. Bothner-By, A.A.; Stephens, R.L.; Lee, J.M.; Warren, C.D.; Jeanloz, R.W. Structure determination of a tetrasaccharide—Transient nuclear overhauser effects in the rotating frame. *J. Am. Chem. Soc.* **1984**, *106*, 811–813. [CrossRef]
36. Wüthrich, K. *NMR of Proteins and Nucleic Acids*; John Wiley & Sons: New York, NY, USA, 1986.
37. Yao, J.; Dyson, H.J.; Wright, P.E. Chemical shift dispersion and secondary structure prediction in unfolded and partly folded proteins. *FEBS Lett.* **1997**, *419*, 285–289. [CrossRef]
38. Dalgarno, D.C.; Levine, B.A.; Williams, R.J.P. Structural information from NMR secondary chemical-shifts of peptide α-C-H protons in proteins. *Biosci. Rep.* **1983**, *3*, 443–452. [CrossRef] [PubMed]
39. Wishart, D.S.; Sykes, B.D.; Richards, F.M. Relationship between nuclear magnetic resonance chemical shift and protein secondary structure. *J. Mol. Biol.* **1991**, *222*, 311–333. [CrossRef]
40. Kim, Y.; Prestegard, J.H. Measurement of vicinal couplings from cross peaks in COSY spectra. *J. Magn. Reson.* **1989**, *84*, 9–13. [CrossRef]
41. Pardi, A.; Billeter, M.; Wüthrich, K. Calibration of the angular dependence of the amide proton-C^{α} proton coupling constants, $^{3}J_{HN\alpha}$, in a globular protein. Use of $^{3}J_{HN\alpha}$ for identification of helical secondary structure. *J. Mol. Biol.* **1984**, *180*, 741–751. [CrossRef]
42. Wüthrich, K.; Billeter, M.; Braun, W. Polypeptide secondary structure determination by nuclear magnetic resonance observation of short proton-proton distances. *J. Mol. Biol.* **1984**, *180*, 715–740. [CrossRef]
43. Wagner, G.; Neuhaus, D.; Wörgötter, E.; Vasák, M.; Kägi, J.H.R.; Wüthrich, K. Nuclear magnetic resonance identification of "half-turn" and 3_{10}-helix secondary structure in rabbit liver metallothionein-2. *J. Mol. Biol.* **1986**, *187*, 131–135. [CrossRef]
44. Horovitz, A. Double-mutant cycles: A powerful tool for analyzing protein structure and function. *Folding Des.* **1996**, *1*, R121–R126. [CrossRef]
45. Cockroft, S.L.; Hunter, C.A. Chemical double-mutant cycles: Dissecting non-covalent interactions. *Chem. Soc. Rev.* **2007**, *36*, 172–188. [CrossRef] [PubMed]
46. Stanger, H.E.; Syud, F.A.; Espinosa, J.F.; Giriat, I.; Muir, T.; Gellman, S.H. Length-dependent stability and strand length limits in antiparallel β-sheet secondary structure. *Proc. Natl. Acad. Sci. USA* **2001**, *98*, 12015–12020. [CrossRef] [PubMed]
47. McGregor, M.J.; Islam, S.A.; Sternberg, M.J.E. Analysis of the relationship between side-chain conformation and secondary structure in globular proteins. *J. Mol. Biol.* **1987**, *198*, 295–310. [CrossRef]
48. Dunbrack, R.L.; Karplus, M. Backbone-dependent rotamer library for proteins—Application to side-chain prediction. *J. Mol. Biol.* **1993**, *230*, 543–574. [CrossRef]
49. Marqusee, S.; Baldwin, R.L. Helix stabilization by Glu-... Lys+ salt bridges in short peptides of de novo design. *Proc. Natl. Acad. Sci. USA* **1987**, *84*, 8898–8902. [CrossRef]
50. Doig, A.J.; Sternberg, M.J.E. Side-chain conformational entropy in protein folding. *Protein Sci.* **1995**, *4*, 2247–2251. [CrossRef] [PubMed]
51. Piotto, M.; Saudek, V.; Sklenár, V. Gradient-tailored excitation for single-quantum NMR-spectroscopy of aqueous-solutions. *J. Biomol. NMR* **1992**, *2*, 661–665. [CrossRef]

Review

Structural and Energetic Characterization of the Denatured State from the Perspectives of Peptides, the Coil Library, and Intrinsically Disordered Proteins

Elisia A. Paiz, Karen A. Lewis and Steven T. Whitten *

Department of Chemistry and Biochemistry, Texas State University, San Marcos, TX 78666, USA; elisia.paiz@utsouthwestern.edu (E.A.P.); karen.lewis@txstate.edu (K.A.L.)
* Correspondence: sw50@txstate.edu; Tel.: +1-512-245-7893; Fax: +1-512-245-2374

Abstract: The α and polyproline II (PPII) basins are the two most populated regions of the Ramachandran map when constructed from the protein coil library, a widely used denatured state model built from the segments of irregular structure found in the Protein Data Bank. This indicates the α and PPII conformations are dominant components of the ensembles of denatured structures that exist in solution for biological proteins, an observation supported in part by structural studies of short, and thus unfolded, peptides. Although intrinsic conformational propensities have been determined experimentally for the common amino acids in short peptides, and estimated from surveys of the protein coil library, the ability of these intrinsic conformational propensities to quantitatively reproduce structural behavior in intrinsically disordered proteins (IDPs), an increasingly important class of proteins in cell function, has thus far proven elusive to establish. Recently, we demonstrated that the sequence dependence of the mean hydrodynamic size of IDPs in water and the impact of heat on the coil dimensions, provide access to both the sequence dependence and thermodynamic energies that are associated with biases for the α and PPII backbone conformations. Here, we compare results from peptide-based studies of intrinsic conformational propensities and surveys of the protein coil library to those of the sequence-based analysis of heat effects on IDP hydrodynamic size, showing that a common structural and thermodynamic description of the protein denatured state is obtained.

Keywords: denatured state ensemble; protein coil library; peptides; intrinsically disordered proteins

Citation: Paiz, E.A.; Lewis, K.A.; Whitten, S.T. Structural and Energetic Characterization of the Denatured State from the Perspectives of Peptides, the Coil Library, and Intrinsically Disordered Proteins. *Molecules* **2021**, *26*, 634. https://doi.org/10.3390/molecules26030634

Academic Editor: Marilisa Leone
Received: 1 January 2021
Accepted: 23 January 2021
Published: 26 January 2021

Publisher's Note: MDPI stays neutral with regard to jurisdictional claims in published maps and institutional affiliations.

Copyright: © 2021 by the authors. Licensee MDPI, Basel, Switzerland. This article is an open access article distributed under the terms and conditions of the Creative Commons Attribution (CC BY) license (https://creativecommons.org/licenses/by/4.0/).

1. Introduction

Proteins under biological conditions exhibit marginal structural stability [1], and they unfold and refold repeatably in vivo [2]. Consequently, many of the biological processes that are facilitated by protein macromolecules are modulated by the properties and energetic character of the denatured state. Indeed, numerous efforts have shown that denatured state effects, such as residual structure [3], excluded volume [4], and intrinsic conformational propensities [5], have key roles in molecular recognition [6], allosteric signaling [7], folding [8,9], and stability [10]. A molecular-level understanding of how proteins are utilized for biological work thus requires characterization of the native, as well as the myriad of non-native, conformational states that exist in solution for a protein, the latter of which is referred to as its denatured state ensemble (DSE).

Despite its importance in understanding protein function, the probability and structural character of the full spectrum of states sampled by proteins are not known. Numerous studies have used short peptides as experimental models from which to probe the characteristics of the DSE [11–13]. The use of short peptides is advantageous because, being too short to fold, they offer access to unfolded states under otherwise folding conditions. Moreover, in the absence of folding, conformational preferences are simplified and locally driven by factors such as hydration [14] and steric hindrance [15]. These studies find that peptides have strong preferences for the polyproline II (PPII) backbone conformation, even at non-proline positions [12,16,17], suggesting that PPII structures are dominant components of the DSE. The PPII conformation is characterized by an extended left-handed helical turn with the amide hydrogen and the carboxyl oxygen of each peptide backbone projecting into solution, presumably making favorable contact with the solvent [18–20]. In addition, the PPII conformation appears to facilitate favorable intrachain $n \rightarrow \pi^*$ interactions, which should be a stabilizing factor [21]. Short peptides also exhibit conformational preferences for other backbone structures. At cold temperatures, alanine residues have intrinsic α-helix-forming tendencies (i.e., even in the absence of favorable side chain interactions) that are stabilized predominantly by peptide hydrogen bonds [22]. Elevated temperatures have been observed to promote low levels of β-strand [16] or β-turn [23], though the amino acid preferences for forming strand [24] or reverse turn structures [25,26] are thought to be highly context-dependent.

The protein coil library [27] also has been used as a structural model for the DSE [28–30]. These libraries are constructed from the segments of protein structure in the Protein Data Bank (PDB) that are found outside the α-helix and β-strand domains. Some libraries further omit additional conformationally restricted positions, such as those in reverse turns, or preceding prolines, or immediately flanking a region of secondary structure [29]. The underlying assumption when using a coil library as a DSE model is that site-specific effects on the intrinsic conformational preferences of the amino acids are minimized by averaging over many environments, and also by removing the regular and repetitive interactions associated with folded structures. Overall, coil libraries exhibit structural trends that are in good agreement with the results from peptide structural studies [29,31]. For example, chemical shifts and three-bond J couplings ($^3J_{HN\alpha}$) measured in peptides by NMR spectroscopy can be reproduced from structural models made from the protein coil library [32–34]. Notably, and similar to the results obtained from peptides, strong preferences for PPII that vary by amino acid type are found in structural surveys of the protein coil library [28–30].

Intrinsically disordered proteins (IDPs) offer another experimental system from which to assess structural preferences in unfolded states under nondenaturing conditions [35]. While chemically denatured proteins are known to adopt macromolecular sizes that depend weakly on sequence details other than chain length [36], IDPs in water exhibit strong sequence-dependent influences on structural size [37]. Computer simulations show that steric effects on disordered structure cannot account for the hydrodynamic size dependence on sequence observed in IDPs [38]. Additionally, temperature changes are found to induce large shifts in the hydrodynamic size for disordered proteins in water [39–41] that can exceed the change in size associated with the heat denaturation of folded proteins of the same chain length [42]. The implication of these findings, albeit expected, is that monomeric disordered protein structure is both under thermodynamic control and highly sensitive to the primary sequence.

In this review, we show that the sequence dependence of IDP hydrodynamic size can be described from the amino acid-specific biases for PPII in the denatured state. Because PPII-rich structures are extended [43], the magnitude of a PPII preference in the denatured state can affect its mean hydrodynamic size [44,45]. Specifically, experiments that evaluate how IDP hydrodynamic size changes with compositional changes in the protein give an independent measure of PPII bias, and further reveal amino acid-specific preferences for PPII that are in good quantitative agreement with PPII bias determined experimentally in peptides [37]. Good agreement is also found when the IDP results are compared to PPII bias in the protein coil library. Moreover, the analysis of heat effects on IDP hydrodynamic size indicates the PPII bias is driven by a significant and favorable enthalpy, and is partially offset by an unfavorable entropy [37], which, again, agrees quantitatively with the peptide results [46]. Across these three models (i.e., peptides, the coil library, and IDPs), the data indicate that the structural and energetic character of the DSE at normal temperatures follows the predictions of a PPII-dominant ensemble. At cold temperatures, both peptides and IDPs reveal the DSE can shift in population toward the α-helix backbone conformation. To demonstrate these conclusions, the following sections review results obtained from numerous spectroscopic and calorimetric studies on short peptides [11–13,16,17,46], surveys of structures in the protein coil library [28–30], and the more recently acquired sequence- and temperature-based analysis of IDP hydrodynamic sizes [37], showing that these three experimental systems used for characterizing unfolded proteins under folding conditions convey a surprisingly consistent structural and energetic view of the DSE.

2. Peptide Models of the DSE

The structural preferences associated with unfolded proteins are often described in terms of a predisposition for specific pairs of backbone dihedral angles, phi (Φ) and psi (Ψ). Visually, this is demonstrated with a Ramachandran plot, shown in Figure 1, where pairs of Φ, Ψ angles that are sterically accessible to a polypeptide chain are mapped [47]. For example, using a representative plot computed for the central residue in a poly-alanine tripeptide, it shows that (Φ, Ψ) = (0°, 0°) is found in a disallowed region of the plot because these angles for the central residue place the backbone carbonyl oxygen and backbone nitrogen from the first and third residues, respectively, inside the normal contact limits, creating a steric conflict. In contrast, (Φ, Ψ) = (−90°, 90°) for the central alanine has no such contact violations for any of the tripeptide atoms, and thus this angle pair is physically allowed. When an unfolded protein shows preferences for some allowed Φ, Ψ pairs at the expense of others, specifically during the rapid interconversion between states of its conformational ensemble, it is said that the unfolded protein exhibits a conformational bias.

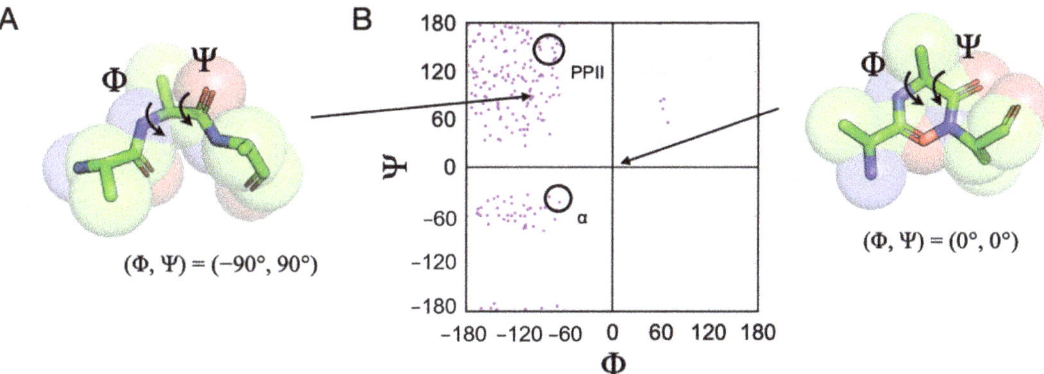

Figure 1. Sterically allowed backbone conformations in polypeptides. (**A**) Peptide backbone dihedral angles, Φ and Ψ. (**B**) Ramachandran plot of allowed Φ, Ψ for the central residue in a poly-alanine tripeptide, calculated from structures generated computationally using a hard sphere collision (HSC) model [48,49] and the "normal" atom pair distances from Ramachandran et al. [47]. Approximately 9000 random structures were generated to find 200 sterically allowed configurations. Highlighted by the circled areas are Φ, Ψ regions corresponding to the PPII and α-helix backbone conformations, as indicated.

The idea that unfolded proteins and polypeptides in water may exhibit intrinsic biases for some backbone conformations at the expense of others began to receive widespread consideration when the observation was made that, for a protein chain to achieve its unique structure in a biologically relevant time frame, a random search of all accessible conformations is not possible [50]. The unfolded chain, accordingly, must search a smaller conformational space to what would be predicted from steric considerations alone. This observation predicted that folding is guided by the structural characteristics of the DSE, and experiments to identify folding intermediates, both kinetic [51,52] and equilibrium [53,54], and measure the intrinsic conformational propensities of the amino acids [5] have been extensively pursued over the many decades since.

Early experimental evidence indicating structural preferences in the DSE was provided by Tiffany and Krimm from studies on short poly-proline and poly-lysine peptides using circular dichroism (CD) and optical rotatory dispersion (ORD) spectroscopies [55–57]. Though these short peptides were unfolded, owing to insufficient chain length for forming compact, globular structures, Tiffany and Krimm found strong preferences for PPII structures. This structural motif at the residue level corresponds to the *trans* isomer of the peptide bond and (Φ, Ψ) of approximately ($-75°$, $+145°$) [43,55]. Its presence in a polypeptide can be established from positive and negative bands in the spectroscopic readings at ~220 nm and ~200 nm, respectively [55,56]. The predisposition for adopting PPII was linked to a variety of factors, such as low temperatures, steric hindrance between side chains, a lack of internal hydrogen bonding, and protonation [57]. Short peptides of poly-glutamic acid also were observed to transition from α-helix at low pH to PPII at neutral pH and higher, identified from CD and ORD spectroscopies [56], indicating that structural transitions between one region of the Ramachandran plot to others could occur for some sequences owing to simple changes in the peptide charge state. These results, Tiffany and Krimm hypothesized, predict a DSE dominated by backbone interconversions between three main structural states: PPII, α-helix, and unordered, where unordered is represented by the random chain [57]. They also speculated, to some resistance [58–60], that solvation effects may contribute to the observed PPII preferences, since the PPII configuration places the backbone amide and backbone carbonyl oxygen polar groups in favorable positions for contact with water. Intrinsic PPII propensities thus could be helpful for keeping unfolded proteins solvated. Overall, their findings from these peptide-based studies supported the idea that unfolded proteins, though highly dynamic and exhibiting broad structural heterogeneity, nonetheless can show backbone conformational biases that are determined locally by sequence details.

Peptide studies have also made extensive use of poly-alanine, because of the natural abundance of alanine in proteins and its chemically simple side chain (i.e., a methyl group). Using a peptide called XAO, where A is an alanine heptamer and X and O are flanking diaminobutyric acid and ornithine, respectively, Kallenbach and coworkers found strong, temperature-dependent preferences for the PPII conformation [11]. $^3J_{HN\alpha}$ coupling constants measured by NMR techniques were used to estimate the Φ angle at each alanine position from the Karplus relationship [61], and it was found that Φ was approximately $-70°$ at low temperatures. Because both PPII and α-helix can have Φ angles near this value (Figure 1), the presence of the α-helix was ruled out by a lack of measurable NOEs between successive amides in the peptide chain, which is an indicator for α-helix formation. The CD spectrum of XAO also confirmed PPII content. Increasing temperatures caused gradual reductions in populating the PPII state that coincided with an increasing population of β-strand conformations to approximately 10% at 55 °C. The reduction in PPII content at high temperatures implied a favorable enthalpy of PPII formation that was also observed by Tiffany and Krimm [57]. Further studies of XAO by Asher et al. using UV Raman spectroscopy established that XAO is structurally similar to a 21-residue alanine-peptide, AP, that forms α-helix under cold conditions [62]. AP transitions to PPII at higher temperatures, and demonstrates that AP, similarly to XAO, shows temperature-dependent conformational preferences.

Additional studies that examined a single alanine flanked on both sides by two glycines (i.e., Ac-$(Gly)_2$-Ala-$(Gly)_2$-NH_2) found intrinsic preferences for PPII and heat-induced shifts toward β-strand backbone conformations [63]. Temperature-dependent transitions that exhibit similar structural characteristics have also been seen in alanine tripeptides, tetrapeptides, and octapeptides [18,64,65].

To explore the determinants of the PPII bias in greater detail, quantitative studies designed to measure its dependence on amino acid type were initially conducted by Creamer and coworkers [12]. Host–guest substitutions at an internal position in a proline-rich peptide (Ac-$(Pro)_3$-X-$(Pro)_3$-Gly-Tyr-NH_2, where X is the substitution site) were used to analyze substitution-induced effects on the CD spectrum and measure a scale of relative

PPII propensities for 18 of the 20 common amino acids. Bias estimates for tryptophan and tyrosine were not measured, because the aromatic contribution to the CD spectrum from their side chains overlaps with the region where signal height was used to determine PPII content [66,67], impeding their analysis. These experiments found that amino acids with charged side chains, except for histidine, had relatively high preferences for the PPII conformation in this peptide. The observed biases, measured at 5 °C, were mostly insensitive to changes in solution pH from 2 to 12. Residues with small, non-polar side chains, such as alanine and glycine, reported somewhat higher propensities for PPII that, in general, exceeded the biases observed from residues with non-polar and bulky side chains, such as isoleucine and valine. The list of amino acid-specific intrinsic propensities for PPII determined in these studies is given in Table 1.

Table 1. Experimental intrinsic propensity for the PPII backbone conformation measured in short peptides.

Amino Acid	PPII Propensity [a]	PPII Propensity [b]	PPII Propensity [c]
ALA (A)	0.61	0.818	0.37
CYS (C)	0.55	0.557	0.25
ASP (D)	0.63	0.552	0.30
GLU (E)	0.61	0.684	0.42
PHE (F)	0.58	0.639	0.17
GLY (G)	0.58	-	0.13
HIS (H)	0.55	0.428	0.20
ILE (I)	0.50	0.519	0.39
LYS (K)	0.59	0.581	0.56
LEU (L)	0.58	0.574	0.24
MET (M)	0.55	0.498	0.36
ASN (N)	0.55	0.667	0.27
PRO (P)	0.67	-	1.00
GLN (Q)	0.66	0.654	0.53
ARG (R)	0.61	0.638	0.38
SER (S)	0.58	0.774	0.24
THR (T)	0.53	0.553	0.32
VAL (V)	0.49	0.743	0.39
TRP (W)	-	0.764	0.25
TYR (Y)	-	0.630	0.25
average	0.58	0.626	0.35

[a] Measured at the X position in Ac-$(Pro)_3$-X-$(Pro)_3$-Gly-Tyr-NH_2 by Creamer and coworkers, at 5 °C, and excluding Tyr and Trp [12]. [b] Measured at the X position in Ac-$(Gly)_2$-X-$(Gly)_2$-NH_2 by Kallenbach and coworkers, at 20 °C, and excluding Gly and Pro [16]. [c] Measured at the X position in Ac-Val-$(Pro)_2$-X-Val-$(Pro)_2$-$(Arg)_3$-Tyr-NH_2 by Hilser and coworkers, at 25 °C [17].

Similarly, Kallenbach and coworkers extended their NMR- and CD-based structural studies of the short peptides mentioned above to include other amino acid types at the central residue position in Ac-$(Gly)_2$-X-$(Gly)_2$-NH_2, where X was the substitution site. Substitution-induced effects on peptide structure were then used to establish a scale of PPII bias in this glycine-rich host [16]. Substantial intrinsic PPII propensities were found, giving additional support to the idea that unfolded states are predisposed to PPII (see Table 1). The magnitude of the PPII bias at the peptide guest position, surrounded by glycine, however, was noticeably different (and typically larger) when compared to the amino acid-specific biases that were measured in the proline-based host by Creamer. This predicts position-specific PPII bias in an unfolded chain that is modulated by the amino acid identity at neighboring sites, which has been subsequently verified [68]. Moreover, the glycine-rich peptides exhibited a heat-induced shift in structure from PPII to nonPPII with a slight bias at high temperatures for strand-like conformations. The intrinsic PPII propensities reported in Table 1 from Kallenbach were measured at 20 °C.

A third experimental scale of PPII propensity in peptides was measured calorimetrically by Hilser and coworkers [13,17,69]. Their experiments utilized a peptide host–guest system in which the *Caenorhabditis elegans* Sem-5 SH3 domain binds a peptide in the PPII conformation [70]. This peptide (Ac-Val-$(Pro)_3$-Val-$(Pro)_2$-$(Arg)_3$-Tyr-NH_2) is derived from the recognition sequence of a SH3 binding partner, Sos (Son of Sevenless). A non-interacting residue of this peptide corresponding to its fourth position [13] was substituted for each amino acid before binding was measured by isothermal titration calorimetry. The observed change in binding affinity reflects a change in the conformational equilibrium between binding-incompetent and binding-competent (i.e., PPII) states of the peptide ligand, which can be interpreted as a PPII propensity [13,69]. Once again, a substantial intrinsic bias for PPII was observed, albeit at magnitudes and rank orders that were different when compared to the scales determined by either Creamer or Kallenbach. Elam et al. conclude that there is a general consensus regarding amino acids that are high in PPII propensity (proline, lysine, glutamine, and glutamic acid) and low in PPII propensity (histidine, tryptophan, tyrosine, and phenylalanine), with the other amino acids falling in between [17]. The intrinsic PPII propensities in Table 1 from Hilser's group were measured at 25 °C.

There are a number of other studies beyond the few described above, each of which uses their own system to examine the structural propensities of the different amino acids in peptides (reviewed in ref. [71]). While the ranks of relative PPII propensities are often both quantitatively and qualitatively different when compared between studies, possibly owing to the use of different host models, all studies have indicated the same general conclusions that (1) unfolded peptides have structural preferences that are predominantly locally determined [72]; (2) nevertheless, these preferences at individual positions can be modulated by the structural features of neighboring residues [68], and (3) importantly, the unfolded chain does not evenly sample the sterically allowed regions of Ramachandran space [71].

In addition to PPII propensities, alanine-based peptides have been utilized to measure intrinsic α-helix-forming tendencies in a host–guest model that was designed to avoid stabilizing side chain–side chain and side chain–macrodipole interactions [22]. Though cold temperatures were required for this peptide to populate helix at appropriate levels for study, Baldwin and coworkers measured amino acid substitution effects on the CD signal at 222 nm and determined an experimental scale of α-helix intrinsic propensities for each of the 20 common amino acids. At 0 °C, most of the amino acids disfavored forming helix at guest positions in the alanine-based host, while leucine and arginine were indifferent to helix-formation. Alanine, however, had a preference for forming helix in this host. The intrinsic propensity for forming α-helix determined by Baldwin and coworkers for each of the common amino acids is provided in Table 2.

Table 2. Experimental intrinsic propensity for the α-helix measured in short peptides.

Amino Acid	ΔG (kcal mol^{-1}) [a]	α-Helix Propensity [b]
ALA (A)	−0.258	0.62
CYS (C)	0.570	0.26
ASP (D)	0.635	0.24
GLU (E)	0.433	0.31
PHE (F)	0.672	0.22
GLY (G)	1.62	0.05
HIS (H)	0.525	0.28
ILE (I)	0.445	0.31
LYS (K)	0.108	0.45
LEU (L)	0.022	0.49
MET (M)	0.251	0.39
ASN (N)	0.635	0.24
PRO (P)	4	0.001
GLN (Q)	0.314	0.36
ARG (R)	−0.047	0.52
SER (S)	0.525	0.28

Table 2. *Cont.*

Amino Acid	ΔG (kcal mol^{-1}) [a]	α-Helix Propensity [b]
THR (T)	1.07	0.12
VAL (V)	0.797	0.19
TRP (W)	~0.6	0.25
TYR (Y)	~0.4	0.32
average		0.29

[a] Measured in an alanine-rich host at 0 °C by Baldwin and coworkers [22]. In the original report, bias for the α-helix was given as a free energy (ΔG) of helix formation. The values for ASP, GLU, LYS, and ARG represent the charged species; His value is for the neutral species. [b] α-helix propensities were estimated from the free energies as $K_\alpha/(1 + K_\alpha)$, where $K_\alpha = e^{-\Delta G/RT}$, ΔG is from column 2, R is the gas constant, and T is temperature.

3. Protein Coil Library Model of the DSE

The PDB [73] provides an ever-increasing number of high-resolution protein structures, which include both regularly ordered secondary structures (helices, sheets, and turns) and irregularly ordered structures (coils and loops). While any individual coil or loop was sufficiently ordered for structural determination, the assumption is that in aggregate, a large set of irregularly ordered structures would provide information on the conformational tendencies and properties of the polypeptide chain in the denatured state. Collectively, these models of the denatured state are constructed by examining the regions of resolved protein structures that are outside the α-helix and β-strand domains. Indeed, analyses of "protein coil libraries" generally support the structural preferences that have been observed in peptide-based models. As these libraries of coil structures have evolved, the field has gained valuable insights into the roles of sequence context, intramolecular interactions, and protein hydration in determining the intrinsic structural tendencies of the amino acids.

In 1995, Swindells and Thornton generated one of the first iterations of a protein coil library based on high-resolution protein structures [27]. Four basins were defined on the Ramachandran plot, corresponding to a (α-helix), b (β-sheet), p (PPII), and L (left-handed helix). Using 85 structures obtained from the PDB, they removed residues that were assigned helix or sheet conformation, retaining all coils, loops, and turns in the analyzed set. Within this set, residues Glu, Gln, Ser, Asp, and Thr demonstrated strong propensities for the "a" region, as their side chains have both the hydrogen bonding capacity and rotational flexibility to form hydrogen bonds to backbone groups. The "b" propensities appeared to be less sensitive to the chemistry and rotamer of the side chain, consistent with the location of the side chain relative to the backbone when in the β-sheet conformation. While the authors did not explicitly discuss the "p" region (PPII), their data show a significant redistribution of the population between the four basins when the "whole" and "coil library" sets are compared. When the entire polypeptide chain was considered, the a and b basins were the two most highly populated. In the coil library, with helices and sheets removed, the a and p basins exhibited the highest populations. This demonstrated that in the structures of intact proteins, PPII conformations are well represented in the non-alpha and non-beta regions.

This work was followed by an analysis of the PPII content in 274 high-resolution structures conducted by Stapley and Creamer [74]. In their analysis, they found the PPII conformation was common, with more than half of the proteins containing at least one PPII helix longer than three residues, despite PPII residues comprising just 2% of all residues in the dataset. This study was the first to detail the PPII propensities of each side chain. Predictably, Gly was disfavored, while Pro had a strong PPII propensity. Additionally, they observed that Gln, Arg, Lys, and Thr had generally strong propensities for adopting PPII conformations. Moreover, a positional dependence of PPII propensity within the PPII helix was also found. The ability of polar side chains, such as Gln, Lys, and Arg, to form hydrogen bonds with the backbone between i and $i + 1$ positions stabilizes the PPII helix. This is consistent with the overrepresentation of Gln, Arg, Lys, and Thr in the first PPII helix position. These data also supported the idea that PPII helices have extensive solvent exposure, as there was a significant negative correlation between nonpolar solvent accessibility and PPII propensity. Taken together, their work demonstrated that both solvent accessibility and the ability to form hydrogen bonds with the backbone were important elements of PPII propensity, consistent with prior work in peptides.

In 2005, Rose and coworkers developed a protein coil library (PCL) that is web-accessible [28]. The PCL becomes updated as the PDB is also updated. This repository of structure elements uses the regular expressions for α-helices and β-sheet and then extracts all non-helix and non-sheet residues from deposited structures that share <90% identity. Note that, as a result, the PCL contains both turns and homologous sequences. Additionally, for structure classification purposes, the PCL divides the Ramachandran plot into $30° \times 30°$ bins, whereby each bin refers to one of 144 different "mesostates".

An analysis in 2008 by Perskie et al. identified seven naturally clustering basins in a Ramachandran plot of PCL structures [30]. These basins represent the familiar α, β, PPII, αL, and τ (type II′ β-turn) structural motifs, and also a γ basin, for inverse γ turns, and a δ basin that captures residues preceding a proline in proline-terminated helices. This allowed amino acid preferences for the different basins (see Table 2 in ref. [30]) to be determined and studied. For example, solvent–backbone hydrogen bonding, which can favor PPII [14], and side chain–side chain sterics, which for branched amino acids adjacent to proline can favor δ at the expense of β, were found to be crucial determinants of the basin preferences.

To better understand how the conformational preferences of a residue in the denatured state depend on the identity and state of its adjacent (nearest) neighbor, Freed and coworkers constructed an increasingly stringent set of coil libraries [29]. Using 2020 nonhomologous polypeptide chains, the "full" set was defined as the entire polypeptide chain, sans the terminal residues. The first cull of the full set ($C_{\alpha\beta}$) removed the α-helix and β-sheet identified residues, similar to the original coil libraries and the PCL described above. This had the effect of reducing the number of residues to 40% of the original. The next restriction additionally removed hydrogen-bonded turns from the set ($C_{\alpha\beta t}$), slimming the library to 28% of the original. Finally, to produce the most restricted coil library, the authors retained only those residue positions located within contiguous stretches four residues or longer, and which were "internal" to coils. This had the effect of reducing "end bias" from structured regions, which is known to favor PPII at the expense of α and β.

The sequential removal of ordered residues had the overall effect of increasing PPII content and decreasing α populations in the coil library. Specifically, when all structured positions were included, α-helical conformations were the predominant state. Upon removing the α-helix and β-sheet residues—as Swindells and Thornton did a decade prior—the PPII conformation emerged as a major subpopulation. With turns also removed ($C_{\alpha\beta t}$), the most populated conformation was PPII, and there was a significant reduction in the α population. The dominance of the PPII conformation is not restricted to a particular subset of amino acids, as all 20 amino acids show a considerable propensity to adopt the PPII configuration (Table 3). The most restricted set (with only residues that are well within coil regions) showed little change in the population distribution, with the PPII population continuing to be dominant.

Using the most restricted set, the authors also found that the size of the PPII subpopulation is constant regardless of solvent accessibility [29]. Moreover, PPII is the dominant conformation in all but the 10% most surface-exposed residues. The α-helix dominates in the surface residues, due to the propensity of the polypeptide backbone at the surface to preferentially turn back toward the folded core of the protein. The independence of PPII content and solvent accessibility initially appears to contrast with earlier work with both peptides and earlier versions of PCLs. However, these results can be reconciled by understanding the PPII conformation as a mechanism for maximizing backbone hydrogen bonding. In the PPII conformation, the backbone amides and carbonyls are in positions that both minimize steric hindrance and enable both functional groups to form hydrogen bonds, either with solvent molecules or within the protein [29]. Therefore, the PPII propensity likely reflects the intrinsic hydrogen bond capacity of a polypeptide, not merely solvation.

Table 3. Amino acid specific propensity for the PPII backbone conformation in the protein coil library.

Amino Acid	PPII Propensity [a]
ALA (A)	0.48
CYS (C)	0.38
ASP (D)	0.34
GLU (E)	0.38
PHE (F)	0.36
GLY (G)	0.21
HIS (H)	0.28
ILE (I)	0.39
LYS (K)	0.35
LEU (L)	0.47
MET (M)	0.38
ASN (N)	0.31
PRO (P)	0.81
GLN (Q)	0.35
ARG (R)	0.32
SER (S)	0.31
THR (T)	0.24
VAL (V)	0.34
TRP (W)	0.35
TYR (Y)	0.37
average	0.37

[a] Calculated by Freed and coworkers using a restricted coil library that removed α-helices, β-sheets, turns, and residues flanking secondary structures from a set of protein structures [29].

These general results can be replicated using almost any set of nonhomologous protein structures. Figure 2 shows results from a curated set of 122 human protein structures, sharing less than 50% sequence identity and with structural resolution < 2.0 Å [75]. In the full set, containing 15,958 residue positions, the α conformation is the most populated (Figure 2A). When α-helices and β-strands are removed, PPII is the most favored conformation for the remaining 6418 residue positions (Figure 2B).

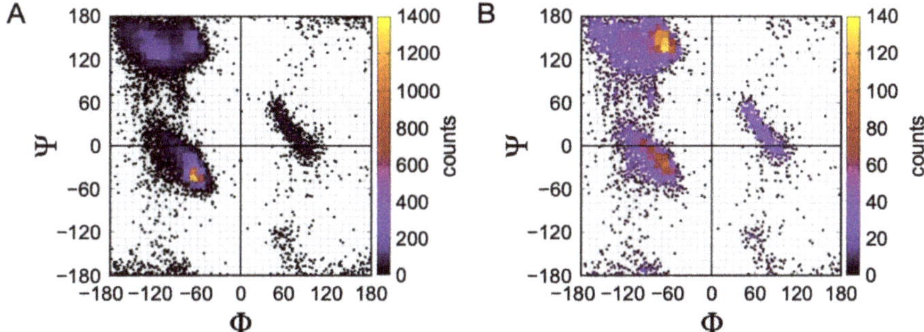

Figure 2. Protein coil libraries are dominated by PPII conformations. In total, 122 non-homologous human structures were analyzed for individual residue conformations (including Gly). (**A**) Ramachandran plot for every residue in the set (15,598 residues). The major population is in the α region, centered at (−65°, −45°). (**B**) Ramachandran plot of the same set after removing all identified α-helix and β-sheet residues (identified using the information provided in the PDB structure file header), yielding 6418 remaining. The major population has shifted to the PPII region, and peaks at (−65°, 135°). For both plots, color represents the count in 10° × 10° bins.

The consistency of PPII propensity in protein coil libraries, especially when viewed in light of hydrogen bonding capacity, therefore predicts that a bias toward PPII conformations is an inherent characteristic of the polypeptide backbone.

4. IDP Model of the DSE

The results of many studies (reviewed above) revealed a significant bias toward PPII in the unstructured states of proteins, even when no prolines are present in the sequence. This indicates that the PPII conformation is a dominant component of the DSE, and potentially an important structural descriptor for understanding the properties associated with IDPs and intrinsically disordered regions (IDRs). Although intrinsic PPII propensities have been determined for the common amino acids (see Tables 1 and 3), the ability of these experimentally determined propensities to quantitatively reproduce ID structural behavior in biological proteins has been difficult to establish.

An experimental system was designed to address this issue and provide an independent measure of the amino acid-specific bias for PPII in IDPs. Based on the hypothesis that the magnitude of a PPII preference in the disordered conformational ensemble can affect its population-weighted hydrodynamic size [41,44,45], it has been shown that intrinsic PPII propensities can be obtained by analyzing the sequence dependence of the mean hydrodynamic radius, R_h, of IDPs [37]. This method relies on two assumptions we demonstrate are reasonable. First, that PPII effects on mean R_h follow a simple power law scaling relationship [41,44,45], and second, that the protein net charge also can influence the hydrodynamic size [38,76].

To establish the relationship linking mean R_h to chain bias for PPII in an ensemble, a computer algorithm based on the hard sphere collision (HSC) model was used to generate polypeptide structures through a random search of conformational space [48,49]. The HSC model has no intrinsic bias for PPII, which was demonstrated previously [49], and thus a PPII sampling bias could be added to the algorithm as a user-defined parameter [41].

Briefly, in this model, individual conformers are generated by using the standard bond angles and bond lengths [77], and a random sampling of the backbone dihedral angles Φ, Ψ, and Ω. (Φ, Ψ) is restricted to the allowed Ramachandran regions [78]; the peptide bond dihedral angle, Ω, is given 100% the *trans* form for nonproline amino acids, while prolines sample the *cis* form at a rate of 6–10%, depending on the identity of the preceding amino acid [79]. The positions of side chain atoms are determined from sampling rotamer libraries [80]. Van der Waals atomic radii [47,81] are used as the only scoring function to eliminate grossly improbable conformations. To calculate state distributions typical of protein ensembles, a structure-based energy function parameterized to solvent-accessible surface areas is used to determine the population weight of the generated structures [82–90]. Random structures are generated until the population-weighted structural size, $<L>$, becomes stable [41]. L is the maximum C_α–C_α distance in a state, and $<L>$ is considered stable when its value changes by less than 1% upon a 10-fold increase in the number of ensemble states. $<L>/2$ is used to approximate the mean R_h of an ensemble.

Figure 3A shows the effect on simulated mean R_h (i.e., $<L>/2$) from increasing the applied PPII sampling bias, S_{PPII}, which is obtained by weighting the random selection of Φ and Ψ. For example, a 30% sampling bias for PPII had 30% of the paired (Φ, Ψ) values for any residue randomly distributed in the region of ($-75° \pm 10°$, $+145° \pm 10°$). The remaining 70% of paired (Φ, Ψ) were distributed in the allowed Ramachandran regions outside of ($-75° \pm 10°$, $+145° \pm 10°$). In this figure, each data point represents a computer-generated poly-alanine conformational ensemble (typically $>10^8$ states). These results are mostly insensitive to steric effects originating from the side chain atoms when biological sequences are used instead of poly-alanine [38]. Unusual sequences, such as all proline or all glycine, cause deviations from the poly-alanine trend.

The simulations revealed that increasing chain propensity for PPII gives rise to increased mean R_h, which is expected because PPII is an extended structure [43]. The dependence of mean R_h on chain length at each sampling bias was fit to the power law scaling relationship, $R_h = R_o \cdot N^v$, where N is chain length in number of residues, R_o the pre-factor, and v the polymer scaling exponent. Individual fits at a given S_{PPII} are shown by lines in Figure 3A, obtained by nonlinear least squares methods. R_o, on average, was 2.16 Å, except when the sampling bias was 100% PPII (Figure 3B). When R_o is held at 2.16 Å, the resulting v shows a logarithmic dependence on S_{PPII} (Figure 3C).

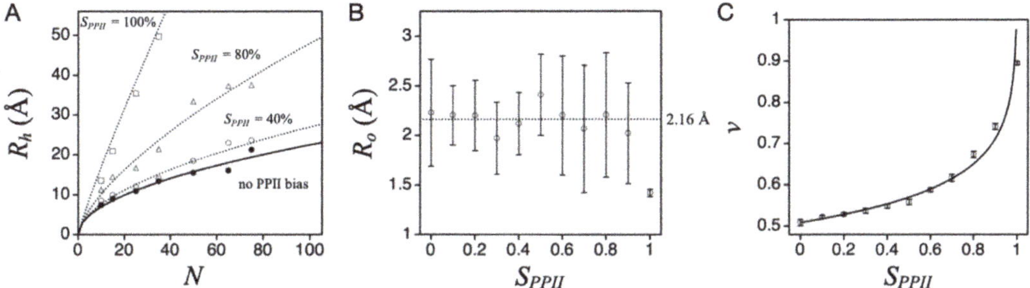

Figure 3. PPII bias expands the structural dimensions of the DSE. (**A**) The effect of an applied PPII sampling bias (S_{PPII}) on mean R_h (i.e., <L>/2) for poly-alanine at different N. Filled circles represent no preferential bias, while open circles, triangles, and squares show when S_{PPII} is 40%, 80%, and 100%, respectively. (**B**) S_{PPII} effects on fit parameter R_o. Average R_o for S_{PPII} range of 0–90% is 2.16 Å, indicated by the stippled line. (**C**) S_{PPII} effects on fit parameter v when R_o is held constant at 2.16 Å. Line is from nonlinear least squares fit of these data to the logarithmic equation, $v = v_o + a \cdot ln(1 - S_{PPII})$.

Because most computer-generated random structures have steric conflicts, and thus are removed by the hard sphere filter, the applied PPII bias, S_{PPII}, does not necessarily equal the population-weighted fractional number of residues in the PPII conformation in an ensemble of allowed states. By using $f_{PPII} = \langle N_{PPII} \rangle / N$ to account for this difference, where N_{PPII} is the number of residues in the PPII conformation in a state, and $\langle N_{PPII} \rangle$ is the population-weighted value for the ensemble (i.e., $\langle N_{PPII} \rangle = \sum N_{PPII,i} \cdot P_i$ with P_i the Boltzmann probability of state i), the simulation trends in Figure 3 can be combined into a simple relationship,

$$R_h = (2.16\ A) \cdot N^{0.503 - 0.11 \cdot ln(1 - f_{PPII})} \tag{1}$$

Additional simulations found that Equation (1) is independent of the specific pattern of PPII propensities in the polypeptide chain [45].

To test Equation (1) directly, mutational effects on experimental R_h were measured for an IDP [44]. Apparent changes in f_{PPII} were determined from amino acid substitutions, following the strategy shown in Figure 4. These experiments used the N-terminal end of the p53 tumor suppressor protein, a prototypical IDP consisting of 93 residues, p53(1-93). The apparent net charge, Q_{net}, calculated from sequence for p53(1-93), is −17. Thus, this test was conducted in the background of potentially strong intramolecular charge–charge interactions that were unaccounted for. Nonetheless, experiments with P→G and A→G substitutions applied to p53(1-93) gave reasonable results, indicating a per-position average PPII bias change of 0.76 at each proline site (i.e., relative to the intrinsic PPII bias of glycine) and 0.48 at each alanine site. These results are evidence of significant conformational bias for PPII in IDPs, even at nonproline positions.

Equation (1) was also used to predict R_h from sequence for a database of IDPs, using the experimental PPII propensities in Table 1 [45]. For each IDP, f_{PPII} was calculated by $\sum PPII_i/N$, where $PPII_i$ is the PPII propensity of amino acid type i, and the summation is over the protein sequence containing N number of amino acids. Figure 5A shows R_h predicted when using PPII propensities from Hilser and coworkers (column 4, Table 1). Compared to the null model where PPII is not strongly preferred and the chain is an unbiased statistical coil, Equation (1) indeed captures the overall experimental trend. Repeating these predictions using the PPII scales measured by Creamer or Kallenbach (columns 2 and 3, Table 1), both yield R_h values that are consistently larger than in the experiment [45], indicating these two scales may be overestimated, at least for describing structural preferences in prototypical IDPs. Moreover, the error from predicting R_h by Equation (1) when using the Hilser-measured PPII scale was found to trend strongly with Q_{net} when Q_{net} was normalized to chain length (Figure 5B), more so than >500 other physicochemical properties that can be calculated from the primary sequence [38]. The linear trend in prediction error to Q_{net} (determined from sequence as number of K and R minus number of D and E) was used to modify Equation (1), yielding

$$R_h = (2.16\ A)\cdot N^{0.503-0.11\cdot ln(1-f_{PPII})} + 0.26\cdot|Q_{net}| - 0.29\cdot N^{0.5} \quad (2)$$

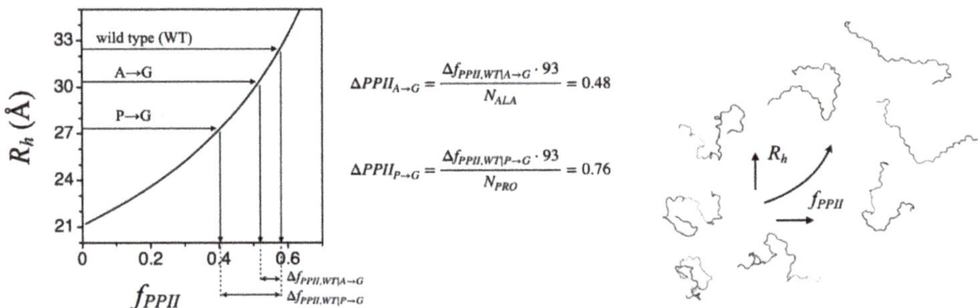

Figure 4. Using mutational effects on IDP R_h to estimate changes in chain bias for PPII. Computed R_h dependence on f_{PPII} for a 93-residue polypeptide, using Equation (1). Arrows show results from experimental R_h measured by both dynamic light scattering (DLS) and size exclusion chromatography (SEC) methods for wild type p53(1-93) and the P→G and A→G substitution mutants. In total, 22 proline (N_{PRO}) and 12 alanine residues (N_{ALA}) in the wild type sequence were substituted to glycine in the P→G and A→G mutants, respectively.

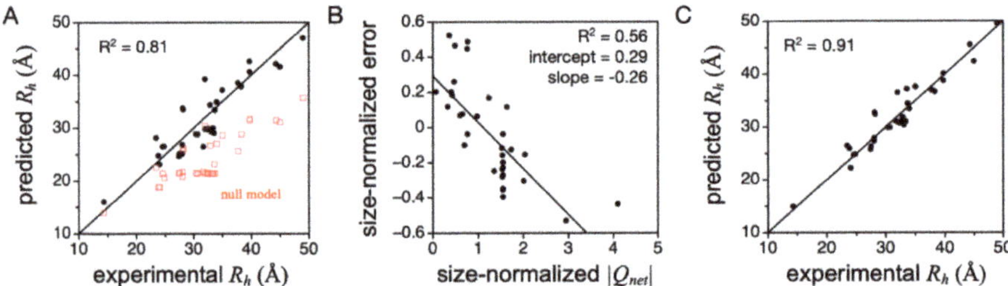

Figure 5. Predicting IDP R_h from sequence using experimental PPII propensities. (**A**) R_h predicted by Equation (1) compared to experimental R_h for 34 IDPs. Predicted values (black circles) were determined from sequence using experimental PPII propensities measured in peptides by Hilser and coworkers (column 4, Table 1). Red squares show R_h predictions when using a null model where PPII is not preferentially populated [45]. (**B**) Size-normalized error, (predicted—experimental R_h)/$N^{0.5}$, compared to size-normalized Q_{net} (i.e., $|Q_{net}|/N^{0.5}$) for each IDP in panel A. (**C**) Equation (2) predicted R_h compared to experimental R_h for 34 IDPs. The identity, primary sequence, and experimental R_h for the IDPs used to generate data in this Figure are provided in ref. [37]. In each plot, R^2 is the coefficient of determination.

Equation (2), which amends Equation (1) for Q_{net} effects on the hydrodynamic size, is highly accurate in predicting R_h from sequence for many IDPs (Figure 5C). Further, in this set of IDPs, mean R_h did not trend with κ [38], which is a measure of the mixing of positive and negative charges in the primary sequence [91]. This justified using Q_{net} to modify Equation (1) and obtain Equation (2), because mean R_h was independent of sequence organization of the charged side chains.

To further test Equation (2) and its ability to describe PPII effects on IDP R_h, random PPII scales were generated and tested for accuracy at predicting experimental R_h [37]—thus establishing the sensitivity of Equation (2) to scale variations. Briefly, each random scale, where the 20 common amino acids were individually assigned random values between 0 and 1, was used to predict R_h by Equation (1), and was then compared to experimental

R_h, an example of which is shown Figure 5A for the peptide-based PPII scale measured by Hilser and coworkers. Next, the linear trend in prediction error to size-normalized Q_{net} was determined, as in Figure 5B. These two steps generate two correlations (R^2), which were used to evaluate each random scale (Figure 6A). Because the slope and intercept from the error trend with size-normalized Q_{net} provides the coefficients preceding $|Q_{net}|$ and $N^{0.5}$ in Equation (2), each scale yields a unique empirical modification to Equation (1) that corrects for net charge effects on mean R_h. The results from analyzing 10^6 randomly generated scales in this manner are given in Figure 6A. Each data point represents a PPII scale. The color, from black to purple, red, and through yellow, is the average error in predicting R_h from sequence after correcting for net charge effects on hydrodynamic size (i.e., after using scale-specific Equation (2) to predict R_h). The abscissa is the correlation (R^2) of Equation (1)-predicted R_h with the experiment for a scale. The ordinate is the correlation (R^2) of size-normalized Equation (1) error with size-normalized Q_{net}.

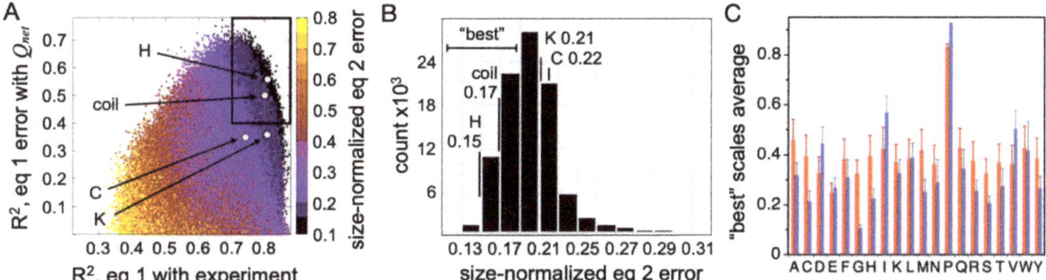

Figure 6. Using experimental R_h from IDPs to determine amino acid-specific intrinsic PPII propensities. (**A**) Ability of experimental PPII propensity scales (from Table 1) to describe the sequence dependence on IDP R_h compared to 10^6 random PPII propensity scales. Missing amino acids from scales measured by Kallenbach (column 3, Table 1) and Creamer (column 2, Table 1) were given the scale average (bottom value, Table 1). Compared as well is the result from using a coil library scale (Table 3). In panels A and B, results from using scales from Hilser and coworkers, Kallenbach and coworkers, Creamer and coworkers, and the coil library are labeled "H", "K", "C", and "coil", respectively. (**B**) Histogram of error distribution in the boxed region of panel A. Small errors are better. (**C**) Average scale value calculated for each of the 20 common amino acids using the "best" performing random PPII propensity scales (red bars). Average scale value using the "best" performing random scales that also maintain correct rank order for the nonpolar amino acids (blue bars), yielding an experimental PPII propensity scale based on IDPs. Error bars report standard deviations.

Two key observations are immediately apparent in the data given in Figure 6A. First, there is a set of random PPII propensity scales that are better than typical at predicting mean R_h from sequence when using f_{PPII}, Q_{net}, and N. These scales, highlighted by the boxed area, predict IDP R_h with good correlation with experimental R_h ($R^2 > 0.7$; x-axis) and a prediction error that also trends with Q_{net} ($R^2 > 0.4$; y-axis). Second, the experimental PPII propensities determined calorimetrically from host–guest analysis of the binding energetics of the Sos peptide (i.e., the peptide-based scale measured by Hilser and coworkers) outperform almost all random scales in their ability to describe sequence effects on mean hydrodynamic size when using only conformational bias and net charge considerations. This is particularly evident when comparing error magnitudes (Figure 6B).

To determine if Equation (2) is sufficiently sensitive to discern the differences in PPII bias of the amino acids, the average scale value for each amino acid type was computed from the "best" performing random scales. The "best" scales were defined as those in the boxed area of Figure 6A with the smallest error (i.e., less than the distribution mode; see Figure 6B). The computed averages, unfortunately, report a somewhat trivial specificity except for distinguishing proline and nonproline types (red bars, Figure 6C), most likely owing to the low representation of some amino acid types in the IDP dataset, specifically the nonpolar amino acids [92]. When substitution effects on mean R_h were measured experimentally in p53(1-93) to determine rank order in PPII propensities among the nonpolar amino acid types [37], and then used to restrict the "best" random scales to those that also maintain this rank order, the average scale value by amino acid type (blue bars, Figure 6C) exhibited strong correlations with the other experimental PPII scales (Figure 7). These amino acid-specific average scale values (blue bars, Figure 6C), which were obtained solely from analyzing sequence effects on IDP R_h, represent an independent measure of the intrinsic PPII bias in the ID states of biological proteins.

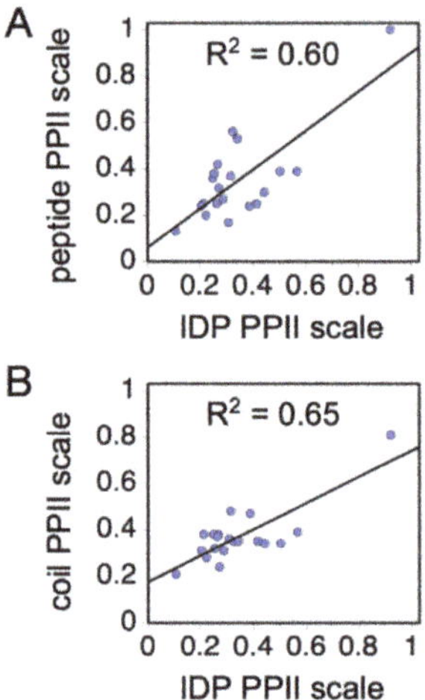

Figure 7. Comparison of experimental PPII propensities. (**A**) Correlation of the peptide-measured PPII scale from Hilser and coworkers (column 4, Table 1) with the IDP-measured PPII propensities (blue bars, Figure 6C). (**B**) Correlation of the coil library (Table 3) and IDP scales. In both plots, each blue circle represents an amino acid type.

Because ID has sequence characteristics that show fundamental disparities when compared to nonID sequences, using IDPs as a DSE model for folded protein is not fully supported. For example, unlike the heterogeneous composition of amino acids and the weak repetition found in the sequences of folded proteins [93,94], IDPs and IDRs have a lower sequence complexity [95] with strong preferences for hydrophilic and charged amino acid side chains over aromatic and hydrophobic side chains [92,96]. These disparate properties of the primary sequence suggest potentially disparate structural behavior. To investigate this issue, protein sequence reversal was used to gain experimental access to the disordered ensemble of a protein with a composition of L-amino acids and pattern of side chains identical to those of a conventional folded protein [42]. Using staphylococcal nuclease for these studies, the unaltered wild type adopts a stable native structure consisting of three α-helices and a five-stranded, barrel-shaped β-sheet [97]. The protein variant with reversed sequence directionality, Retro-nuclease, was found to be an elongated monomer, and exhibits the structural characteristics of intrinsic disorder [42]. At 25 °C, the mean R_h of Retro-nuclease was found to be 34.0 ± 0.5 Å by DLS techniques. Sedimentation analysis by analytical ultracentrifugation (AUC) and SEC methods gave similar results under similar conditions (33.0 Å at 20 °C by AUC, and 33.7 Å at ~23 °C by SEC). Equation (2), for comparison, predicts 33.1 Å using the Retro-nuclease primary sequence, which is close to the observed experimental values.

The hydrodynamic size of Retro-nuclease is highly sensitive to temperature changes (Figure 8A), which is consistent with observations from other IDPs [39–41]. The enthalpy and entropy of the PPII to nonPPII transition have been measured in short alanine peptides by monitoring heat effects on structure over a broad temperature range [46]. The results from CD spectroscopy, which monitored the change in the CD signal at 215 nm, gave

ΔH_{PPII} and ΔS_{PPII} of ~10 kcal mol^{-1} and 32.7 cal mol^{-1} K^{-1}, respectively, while NMR measurements, using heat effects on $^3J_{HN\alpha}$, gave ~13 kcal mol^{-1} and 40.9 cal mol^{-1} K^{-1}.

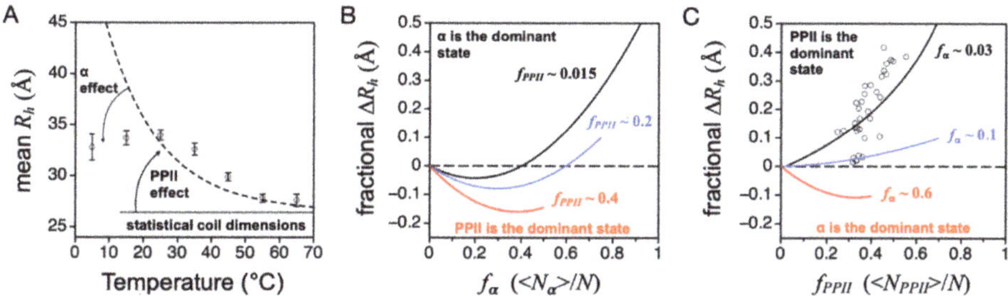

Figure 8. Temperature, α, and PPII effects on DSE hydrodynamic size. (**A**) Open circles show Retro-nuclease mean R_h measured using DLS methods from 5 to 65 °C. The dashed line was calculated with Equation (2) and modeling temperature effects on the intrinsic PPII propensities by Equation (3) and with ΔH_{PPII} = 13 kcal mol^{-1}. Temperature-dependent changes to the amino acid PPII propensities, from Equation (3), cause the Equation (2)-predicted R_h to change accordingly. (**B**,**C**) Simulated effects on population-weighted size from α and PPII bias. Fractional change in mean R_h (i.e., <L>/2) was used to normalize simulation results for chain length. In panel C, open circles represent experimental R_h measured for IDPs and normalized relative to the simulated size of an unbiased ensemble [37], as explained in the main text. f_{PPII} for each IDP was calculated from sequence using the IDP experimental PPII scale (blue bars, Figure 6C).

Because the PPII bias is noncooperative [46] and locally determined [72], the effect from temperature changes can be modeled at the level of individual residue positions using the integrated van't Hoff equation,

$$ln(K_{PPII}(T)) = \left(\frac{\Delta H_{PPII}}{R}\right)\left(\frac{1}{(298\ K)} - \frac{1}{T}\right) + ln(K_{PPII}(298\ K)) \tag{3}$$

where K_{PPII} is the equilibrium between PPII and nonPPII states, T is temperature, and R is the gas constant. ΔH_{PPII} is assumed to be constant. If PPII is the lone dominant conformation, then K_{PPII} for each amino acid type can be estimated from experimental PPII propensities at 25 °C as $K_{PPII,i} = (1 - PPII_i)/PPII_i$. The importance of Equation (3) is that it provides another check on the ability of the DSE to be described from the results of peptide studies. Moreover, these two values, ΔH_{PPII} and $PPII_i$, give access to the entropy from the relationship $(\partial G/\partial T)_P = -S$. Using IDP-measured intrinsic PPII propensities (blue bars, Figure 6C), we found that ΔH_{PPII}~13 kcal mol^{-1} captures the decrease in Retro-nuclease mean R_h from 25 to 65 °C (Figure 8A). For alanine, using its IDP-measured PPII propensity at 25 °C (0.32) and ΔH_{PPII} = 13 kcal mol^{-1} yields ΔS_{PPII} = 45.1 cal mol^{-1} K^{-1}.

Although the predicted and experimental mean R_h agree at 25 and 65 °C, experimental and Equation (2)-predicted values at 5, 15, 35, and 45 °C show obvious differences (Figure 8A). At 35 and 45 °C, the experimental mean R_h values were larger than predicted, whereas at 5 and 15 °C, they were smaller. The analysis of heat effects on R_h using Equation (3) assumed PPII to be the lone dominant DSE conformation, which is not necessarily correct. Indeed, the Retro-nuclease CD spectrum reported a cold-induced local minimum at 222 nm for T < 25 °C [42], revealing temperature-dependent population of the α backbone conformation. By including the effects of an α bias in simulations of DSE hydrodynamic size, both the over- and underpredictions of mean R_h at 5, 15, 35, and 45 °C can be explained.

Briefly, preferential sampling of main chain dihedral angles for Φ and Ψ associated with α-helix can cause changes in the structural dimensions of the DSE [38]. Monitored from the population-weighted mean size, R_h ~ <L>/2, computer-generated ensembles that sample (Φ, Ψ) in the α region (−64° ± 10°, −41° ± 10°) show compaction under modest preferences, and elongated sizes at higher α sampling rates (Figure 8B). Specifically, when (Φ, Ψ) sampling for α is weakly preferred, the probability of contiguous stretches of residues in the α state is low, and turn structures are more likely than helical segments that form when the α bias is higher. Because the effect of the α bias on the mean R_h of the DSE can be accentuated by the PPII bias, whereby ensembles with high PPII propensities show increased sensitivity to changes in the α bias, the consequences of both the α and PPII biases for mean R_h must be considered. For example, the average chain propensity for PPII in our IDP database is ~0.4 when estimated from sequence. Thus, the IDP trend of mean R_h with α bias should follow the red line in Figure 8B, and not the black line. Likewise, the effect of PPII bias on mean R_h is codependent on the α bias (Figure 8C). When PPII is the dominant conformation, the structural dimensions of the denatured state follow the relationship given by Equation (1) (black line in Figure 8C). If, instead, PPII is not the dominant conformation, and moderate α preferences are present, then the R_h dependence on PPII bias changes. More precisely, the result of increasing the chain preference for α is to suppress the effect of PPII on mean R_h (blue line in Figure 8C). When the α bias is stronger than the PPII bias (i.e., α is the dominant conformation), then the effect of the PPII bias is compaction (red line in Figure 8C).

The comparison of experimental IDP R_h to the curves in Figure 8C (open circles in the figure) confirms that PPII is the dominant backbone conformation in IDP ensembles [37]. Here, fractional ΔR_h was calculated as (experimental R_h—simulated R_h)/simulated R_h, where simulated R_h refers to the size of an unbiased ensemble that has been corrected for net charge effects. In the figure, a majority of the IDPs are found to have experimental mean R_h values slightly larger than expected based upon the sequence-calculated value of f_{PPII}. This suggests that the amino acid preferences for PPII may be underestimated by the IDP-based scale, and the values for f_{PPII} in this figure should be shifted to the right. The possibility of a larger intrinsic PPII bias cannot be eliminated because PPII effects on mean R_h are suppressed by the presence of an α bias. The magnitude and sequence dependence to the α bias in the protein DSE is currently unknown, although it has been estimated in short alanine-rich peptides [22].

The idea that PPII propensities are underestimated possibly explains some of the Retro-nuclease data shown in Figure 8A. An underestimated PPII bias gives an underestimated predicted mean R_h at 35 and 45 °C. At 5 and 15 °C, the disagreement between theory and experiment is likely caused by the α bias detected in the Retro-nuclease CD spectrum [37,38]. To obtain the sequence dependence of both the α and PPII biases in the DSE and test these assumptions, the analysis of sequence effects on IDP mean R_h reviewed above could simply be repeated at both colder and warmer temperatures. Higher temperatures reduce α effects on mean R_h and isolate the effects of the PPII bias. Colder temperatures give access to the α bias. Just as the sequence dependence of mean R_h at $T \geq 25$ °C yields the amino acid-specific biases for PPII from the comparison of experimental R_h to simulated coil values that omit PPII effects, the sequence dependence of mean R_h at T < 25 °C can yield the amino acid bias for the α conformation via comparison to the theoretical treatment that omits the α effects.

5. Temperature Dependence of Intrinsic α-Helix and PPII Propensities

If we assume Tiffany and Krimm are correct, and the DSE is composed of three main structural states (PPII, α-helix, and unordered), then the PPII and α-helix propensities given in Tables 1 and 2 can be used to model how PPII, α-helix, and unordered populations change with temperature for a generic polypeptide. This is shown in Figure 9A, where populations at different temperatures were modeled by using the integrated van't Hoff equation (Equation (3)), a transition enthalpy for PPII to nonPPII (ΔH_{PPII}), and a transition enthalpy for α to non-α (ΔH_α). As discussed above, peptide [46] and IDP studies [37,42] both indicate ΔH_{PPII} is ~10 kcal mol^{-1}. Calorimetric studies using alanine-rich peptides that adopt α-helix by Bolen and coworkers indicate ΔH_α is ~1 kcal mol^{-1} [98]. In this model, because ΔH_{PPII} >> ΔH_α, PPII populations are highly sensitive to temperature changes, while α-helix populations show reduced temperature sensitivity. Moreover, also because ΔH_{PPII} >> ΔH_α, PPII populations dominate at very cold temperatures. Unfortunately, the model predicts α-helix populations that decrease with decreasing temperatures, in stark contrast to the known stabilities of peptide and protein structures.

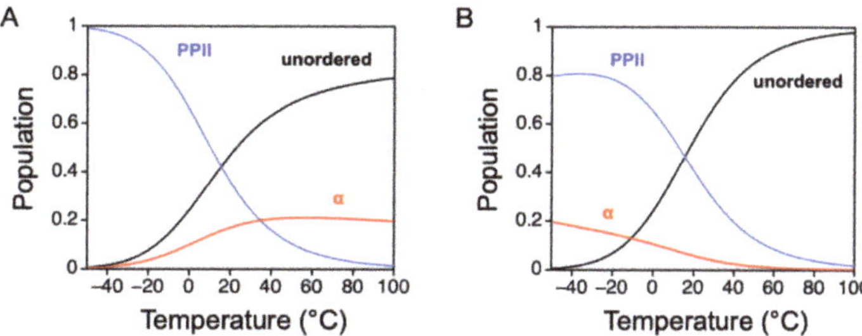

Figure 9. Temperature effects on PPII, α-helix, and unordered populations in an unfolded polypeptide. (**A**) ΔH_{PPII} = 10 kcal mol^{-1} and ΔH_{α} = 1 kcal mol^{-1}. (**B**) ΔH_{PPII} = 10 kcal mol^{-1} and ΔH_{α} = 11 kcal mol^{-1}. To model a generic polypeptide, the PPII and α-helix propensities used the average value from the propensities measured by Hilser and coworkers (column 4, Table 1) and Baldwin and coworkers (column 3, Table 2). Specifically, the PPII propensity was 0.35 at 25 °C, while the α-helix propensity was 0.29 at 0 °C. To calculate populations, the partition function was determined from $Q = 1 + e^{-\Delta GPPII/RT} + e^{-\Delta G\alpha/RT}$, with the unordered state as the reference. ΔG_{PPII} and ΔG_{α} were calculated from the propensities by $-RTln(PPII/1 - PPII)$ and $-RTln(\alpha/1 - \alpha)$, and the temperature dependence of the propensities was calculated with Equation (3). The unordered, α-helix, and PPII populations thus were $1/Q$, $e^{-\Delta G\alpha/RT}/Q$, and $e^{-\Delta GPPII/RT}/Q$.

If, instead, ΔH_{α} is given a value comparable to ΔH_{PPII}, the model yields temperature-dependent populations that reasonably agree with experimental results (Figure 9B). Specifically, both PPII and α-helix populations decrease to low levels at high temperatures. Moreover, under cold conditions, PPII dominates, but α-helix is also populated at non-negligible levels that gradually increase as heat is removed from the system. This result from the model can be explained by assuming that the calorimetry measured ΔH_{α} is the net heat associated with forming α-helix at the cost of disrupting PPII (i.e., $\Delta H_{\alpha} {\sim} \Delta H_{cal,\alpha} + \Delta H_{PPII} {\sim} 1$ kcal mol^{-1} + 10 kcal mol^{-1} = 11 kcal mol^{-1}). In Figure 9B, the transition enthalpies are modeled as 10 kcal mol^{-1} and 11 kcal mol^{-1} for ΔH_{PPII} and ΔH_{α}, respectively. This model is supported by the experimental data obtained for Retro-nuclease (Figure 8). The observed temperature dependence of the Retro-nuclease hydrodynamic size revealed PPII and α-helix intrinsic propensities that changed with temperature in a manner similar to the Figure 9B model.

6. Discussion

Structural and energetic characterization of the DSE is required for a molecular-level understanding of both protein stability and fold specificity. Historically, short peptides [11–13] and the protein coil library [27–30] have been used as the principal models from which to investigate the DSE. For these two models, there is good quantitative agreement in the sense that the protein coil library, when compared to peptide results, has been found to reproduce the intrinsic conformational preferences of the amino acids for helix, sheet, and PPII [29], as well as the effects on the conformational preferences from neighboring residues [31]. This agreement between two independent models indicates that the magnitudes and types of intrinsic biases in unstructured polypeptides are reasonably well-known. The role of the temperature in describing DSE structure, however, is less well understood. Heat indeed modulates the populations of unstructured states, which is evidenced by the large temperature-dependent changes in hydrodynamic size exhibited by IDPs [39–41]. Moreover, the ability of a protein to fold [2], phase separate [99], or recognize its binding partner [69] is also temperature-dependent.

Recently, we demonstrated that the enthalpy, entropy, and magnitude of DSE conformational bias can be elucidated by analyzing heat effects on the mean R_h of IDPs [37].

The sequence dependence of IDP hydrodynamic size yields an independent measure of the intrinsic bias for PPII, because PPII-rich structures are extended [43]. Additionally, as the PPII bias is driven by a favorable enthalpy [46], the effect of increased temperature is to populate nonPPII states at the expense of PPII. Thus, the enthalpy and entropy of the PPII–nonPPII transition can be determined from the heat-induced changes to the mean R_h. Our analysis of the sequence dependence on IDP hydrodynamic size revealed amino acid-specific preferences for PPII that are in good quantitative agreement with both calorimetry-measured values from short peptides and those inferred by a survey of the protein coil library (Figure 7). Modeling the effects of heat on IDP hydrodynamic size yields an enthalpy and entropy of PPII formation that were quantitatively similar to the peptide-measured values [37,46]. It is important to note that these three DSE models (i.e., peptides, the coil library, and IDPs) universally report that the allowed regions of Ramachandran space are unevenly sampled, and that PPII is the predominant denatured state conformation under normal conditions.

When interpreting the effects of the PPII bias on the mean R_h of unstructured proteins, the population of the α backbone conformation has consequences that must be considered. The α basin of the Ramachandran map of Φ and Ψ dihedral angles is among the most populated regions in the coil library distribution [27,30], and is shared with turn structures [29]. Because of the backbone geometry of the α configuration, whereby sparse sampling at dispersed positions can produce turns, and heavy sampling among contiguous positions yields helices, the effect of a PPII bias on the mean R_h can be either compaction or expansion. This is demonstrated in Figure 8C. The codependence of DSE mean R_h on both the α and PPII biases predicts that intrinsic α preferences, and its corresponding thermodynamic parameters, can be estimated from low-temperature studies that compare experimental R_h to computer-simulated DSE trends (Figure 8A). Specifically, for some unstructured proteins, the intrinsic α bias at low temperatures may be sufficiently strong that its magnitude, sequence dependence, and enthalpy and entropy of formation can be measured from the effect on the mean R_h. It remains to be seen if this strategy can be successful, and if the resultant intrinsic α propensities as measured in IDPs compare favorably to those obtained from short peptides (Table 2) and surveys of the protein coil library [27–30].

Author Contributions: S.T.W. conceived and coordinated this work; E.A.P., K.A.L. and S.T.W. co-wrote the manuscript. All authors have read and agreed to the published version of the manuscript.

Funding: This work was funded by the National Institutes of Health, grant numbers R15GM115603, R25GM102783, and R15GM119096.

Institutional Review Board Statement: Not applicable.

Informed Consent Statement: Not applicable.

Data Availability Statement: Data presented in this study are openly available and cited in the references.

Conflicts of Interest: The authors declare no conflict of interests.

References

1. Pace, C.N.; Hermans, J. The Stability of Globular Protein. *CRC Crit. Rev. Biochem.* **1975**, *3*, 1–43. [CrossRef] [PubMed]
2. Guo, M.; Xu, Y.; Gruebele, M. Temperature Dependence of Protein Folding Kinetics in Living Cells. *Proc. Natl. Acad. Sci. USA* **2012**, *109*, 17863–17867. [CrossRef] [PubMed]
3. McCarney, E.R.; Kohn, J.E.; Plaxco, K.W. Is There or Isn't There? The Case for (and Against) Residual Structure in Chemically Denatured Proteins. *Crit. Rev. Biochem. Mol. Biol.* **2005**, *40*, 181–189. [CrossRef] [PubMed]
4. Kim, Y.H.; Stites, W.E. Effects of Excluded Volume upon Protein Stability in Covalently Cross-Linked Proteins with Variable Linker Lengths. *Biochemistry* **2008**, *47*, 8804–8814. [CrossRef] [PubMed]
5. Serrano, L. Comparison between the Phi Distribution of the Amino Acids in the Protein Database and NMR Data Indicates That Amino Acids Have Various Phi Propensities in the Random Coil Conformation. *J. Mol. Biol.* **1995**, *254*, 322–333. [CrossRef]
6. Manson, A.; Whitten, S.T.; Ferreon, J.C.; Fox, R.O.; Hilser, V.J. Characterizing the Role of Ensemble Modulation in Mutation-Induced Changes in Binding Affinity. *J. Am. Chem. Soc.* **2009**, *131*, 6785–6793. [CrossRef]

7. Pan, H.; Lee, J.C.; Hilser, V.J. Binding Sites in Escherichia Coli Dihydrofolate Reductase Communicate by Modulating the Conformational Ensemble. *Proc. Natl. Acad. Sci. USA* **2000**, *97*, 12020–12025. [CrossRef]
8. Wong, K.-B.; Clarke, J.; Bond, C.J.; Neira, J.L.; Freund, S.M.V.; Fersht, A.R.; Daggett, V. Towards a Complete Description of the Structural and Dynamic Properties of the Denatured State of Barnase and the Role of Residual Structure in Folding. *J. Mol. Biol.* **2000**, *296*, 1257–1282. [CrossRef]
9. Kazmirski, S.L.; Wong, K.-B.; Freund, S.M.V.; Tan, Y.-J.; Fersht, A.R.; Daggett, V. Protein Folding from a Highly Disordered Denatured State: The Folding Pathway of Chymotrypsin Inhibitor 2 at Atomic Resolution. *Proc. Natl. Acad. Sci. USA* **2001**, *98*, 4349–4354. [CrossRef]
10. Shortle, D. Staphylococcal Nuclease: A Showcase of m-Value Effects. In *Advances in Protein Chemistry*; Anfinsen, C.B., Richards, F.M., Edsall, J.T., Eisenberg, D.S., Eds.; Protein Stability; Academic Press: Cambridge, MA, USA, 1995; Volume 46, pp. 217–247.
11. Shi, Z.; Olson, C.A.; Rose, G.D.; Baldwin, R.L.; Kallenbach, N.R. Polyproline II Structure in a Sequence of Seven Alanine Residues. *Proc. Natl. Acad. Sci. USA* **2002**, *99*, 9190–9195. [CrossRef]
12. Rucker, A.L.; Pager, C.T.; Campbell, M.N.; Qualls, J.E.; Creamer, T.P. Host-Guest Scale of Left-Handed Polyproline II Helix Formation. *Proteins* **2003**, *53*, 68–75. [CrossRef]
13. Ferreon, J.C.; Hilser, V.J. The Effect of the Polyproline II (PPII) Conformation on the Denatured State Entropy. *Protein Sci.* **2003**, *12*, 447–457. [CrossRef] [PubMed]
14. Mezei, M.; Fleming, P.J.; Srinivasan, R.; Rose, G.D. Polyproline II Helix Is the Preferred Conformation for Unfolded Polyalanine in Water. *Proteins Struct. Funct. Bioinform.* **2004**, *55*, 502–507. [CrossRef] [PubMed]
15. Pappu, R.V.; Rose, G.D. A simple model for polyproline II structure in unfolded states of alanine-based peptides. *Protein Sci.* **2002**, *11*, 2437–2455. [CrossRef] [PubMed]
16. Shi, Z.; Chen, K.; Liu, Z.; Ng, A.; Bracken, W.C.; Kallenbach, N.R. Polyproline II Propensities from GGXGG Peptides Reveal an Anticorrelation with Beta-Sheet Scales. *Proc. Natl. Acad. Sci. USA* **2005**, *102*, 17964–17968. [CrossRef] [PubMed]
17. Elam, W.A.; Schrank, T.P.; Campagnolo, A.J.; Hilser, V.J. Evolutionary Conservation of the Polyproline II Conformation Surrounding Intrinsically Disordered Phosphorylation Sites. *Protein Sci.* **2013**, *22*, 405–417. [CrossRef]
18. Eker, F.; Cao, X.; Nafie, L.; Schweitzer-Stenner, R. Tripeptides Adopt Stable Structures in Water. A Combined Polarized Visible Raman, FTIR, and VCD Spectroscopy Study. *J. Am. Chem. Soc.* **2002**, *124*, 14330–14341. [CrossRef]
19. Eker, F.; Griebenow, K.; Schweitzer-Stenner, R. Stable Conformations of Tripeptides in Aqueous Solution Studied by UV Circular Dichroism Spectroscopy. *J. Am. Chem. Soc.* **2003**, *125*, 8178–8185. [CrossRef]
20. Weise, C.F.; Weisshaar, J.C. Conformational Analysis of Alanine Dipeptide from Dipolar Couplings in a Water-Based Liquid Crystal. *J. Phys. Chem. B* **2003**, *107*, 3265–3277. [CrossRef]
21. Hinderaker, M.P.; Raines, R.T. An Electronic Effect on Protein Structure. *Protein Sci.* **2003**, *12*, 1188–1194. [CrossRef]
22. Chakrabartty, A.; Kortemme, T.; Baldwin, R.L. Helix Propensities of the Amino Acids Measured in Alanine-Based Peptides without Helix-Stabilizing Side-Chain Interactions. *Protein Sci.* **1994**, *3*, 843–852. [CrossRef] [PubMed]
23. Reiersen, H.; Clarke, A.R.; Rees, A.R. Short Elastin-like Peptides Exhibit the Same Temperature-Induced Structural Transitions as Elastin Polymers: Implications for Protein Engineering. *J. Mol. Biol.* **1998**, *283*, 255–264. [CrossRef] [PubMed]
24. Minor, D.L.; Kim, P.S. Context Is a Major Determinant of β-Sheet Propensity. *Nature* **1994**, *371*, 264–267. [CrossRef] [PubMed]
25. Dyson, H.J.; Bolinger, L.; Feher, V.A.; Osterhout, J.J., Jr.; Yao, J.; Wright, P.E. Sequence Requirements for Stabilization of a Peptide Reverse Turn in Water Solution. *Eur. J. Biochem.* **1998**, *255*, 462–471. [CrossRef] [PubMed]
26. Hutchinson, E.G.; Thornton, J.M. A Revised Set of Potentials for Beta-Turn Formation in Proteins. *Protein Sci.* **1994**, *3*, 2207–2216. [CrossRef]
27. Swindells, M.B.; MacArthur, M.W.; Thornton, J.M. Intrinsic φ,ψ Propensities of Amino Acids, Derived from the Coil Regions of Known Structures. *Nat. Struct. Mol. Biol.* **1995**, *2*. [CrossRef]
28. Fitzkee, N.C.; Fleming, P.J.; Rose, G.D. The Protein Coil Library: A Structural Database of Nonhelix, Nonstrand Fragments Derived from the PDB. *Proteins Struct. Funct. Bioinform.* **2005**, *58*, 852–854. [CrossRef]
29. Jha, A.K.; Colubri, A.; Zaman, M.H.; Koide, S.; Sosnick, T.R.; Freed, K.F. Helix, Sheet, and Polyproline II Frequencies and Strong Nearest Neighbor Effects in a Restricted Coil Library. *Biochemistry* **2005**, *44*, 9691–9702. [CrossRef]
30. Perskie, L.L.; Street, T.O.; Rose, G.D. Structures, Basins, and Energies: A Deconstruction of the Protein Coil Library. *Protein Sci.* **2008**, *17*, 1151–1161. [CrossRef]
31. Griffiths-Jones, S.R.; Sharman, G.J.; Maynard, A.J.; Searle, M.S. Modulation of Intrinsic Phi, Psi Propensities of Amino Acids by Neighbouring Residues in the Coil Regions of Protein Structures: NMR Analysis and Dissection of a Beta-Hairpin Peptide. *J. Mol. Biol.* **1998**, *284*, 1597–1609. [CrossRef]
32. Smith, L.J.; Bolin, K.A.; Schwalbe, H.; MacArthur, M.W.; Thornton, J.M.; Dobson, C.M. Analysis of Main Chain Torsion Angles in Proteins: Prediction of NMR Coupling Constants for Native and Random Coil Conformations. *J. Mol. Biol.* **1996**, *255*, 494–506. [CrossRef] [PubMed]
33. Mantsyzov, A.B.; Shen, Y.; Lee, J.H.; Hummer, G.; Bax, A. MERA: A Webserver for Evaluating Backbone Torsion Angle Distributions in Dynamic and Disordered Proteins from NMR Data. *J. Biomol. NMR* **2015**, *63*, 85–95. [CrossRef] [PubMed]
34. Shen, Y.; Roche, J.; Grishaev, A.; Bax, A. Prediction of Nearest Neighbor Effects on Backbone Torsion Angles and NMR Scalar Coupling Constants in Disordered Proteins. *Protein Sci.* **2018**, *27*, 146–158. [CrossRef] [PubMed]

35. Eliezer, D. Biophysical Characterization of Intrinsically Disordered Proteins. *Curr. Opin. Struct. Biol.* **2009**, *19*, 23–30. [CrossRef] [PubMed]
36. Kohn, J.E.; Millett, I.S.; Jacob, J.; Zagrovic, B.; Dillon, T.M.; Cingel, N.; Dothager, R.S.; Seifert, S.; Thiyagarajan, P.; Sosnick, T.R.; et al. Random-Coil Behavior and the Dimensions of Chemically Unfolded Proteins. *Proc. Natl. Acad. Sci. USA* **2004**, *101*, 12491–12496. [CrossRef] [PubMed]
37. English, L.R.; Voss, S.M.; Tilton, E.C.; Paiz, E.A.; So, S.; Parra, G.L.; Whitten, S.T. Impact of Heat on Coil Hydrodynamic Size Yields the Energetics of Denatured State Conformational Bias. *J. Phys. Chem. B* **2019**, *123*, 10014–10024. [CrossRef]
38. English, L.R.; Tilton, E.C.; Ricard, B.J.; Whitten, S.T. Intrinsic α Helix Propensities Compact Hydrodynamic Radii in Intrinsically Disordered Proteins. *Proteins* **2017**, *85*, 296–311. [CrossRef]
39. Kjaergaard, M.; Nørholm, A.-B.; Hendus-Altenburger, R.; Pedersen, S.F.; Poulsen, F.M.; Kragelund, B.B. Temperature-Dependent Structural Changes in Intrinsically Disordered Proteins: Formation of α-Helices or Loss of Polyproline II? *Protein Sci.* **2010**, *19*, 1555–1564. [CrossRef]
40. Wuttke, R.; Hofmann, H.; Nettels, D.; Borgia, M.B.; Mittal, J.; Best, R.B.; Schuler, B. Temperature-Dependent Solvation Modulates the Dimensions of Disordered Proteins. *Proc. Natl. Acad. Sci. USA* **2014**, *111*, 5213–5218. [CrossRef]
41. Langridge, T.D.; Tarver, M.J.; Whitten, S.T. Temperature Effects on the Hydrodynamic Radius of the Intrinsically Disordered N-Terminal Region of the P53 Protein. *Proteins* **2014**, *82*, 668–678. [CrossRef]
42. English, L.R.; Tischer, A.; Demeler, A.K.; Demeler, B.; Whitten, S.T. Sequence Reversal Prevents Chain Collapse and Yields Heat-Sensitive Intrinsic Disorder. *Biophys. J.* **2018**, *115*, 328–340. [CrossRef] [PubMed]
43. Cowan, P.M.; McGAVIN, S. Structure of Poly-L-Proline. *Nature* **1955**, *176*, 501–503. [CrossRef]
44. Perez, R.B.; Tischer, A.; Auton, M.; Whitten, S.T. Alanine and Proline Content Modulate Global Sensitivity to Discrete Perturbations in Disordered Proteins. *Proteins* **2014**, *82*, 3373–3384. [CrossRef] [PubMed]
45. Tomasso, M.E.; Tarver, M.J.; Devarajan, D.; Whitten, S.T. Hydrodynamic Radii of Intrinsically Disordered Proteins Determined from Experimental Polyproline II Propensities. *PLoS Comput. Biol.* **2016**, *12*, e1004686. [CrossRef] [PubMed]
46. Chen, K.; Liu, Z.; Kallenbach, N.R. The Polyproline II Conformation in Short Alanine Peptides Is Noncooperative. *Proc. Natl. Acad. Sci. USA* **2004**, *101*, 15352–15357. [CrossRef] [PubMed]
47. Ramachandran, G.N.; Ramakrishnan, C.; Sasisekharan, V. Stereochemistry of Polypeptide Chain Configurations. *J. Mol. Biol.* **1963**, *7*, 95–99. [CrossRef]
48. Richards, F.M. Areas, Volumes, Packing and Protein Structure. *Annu. Rev. Biophys. Bioeng.* **1977**, *6*, 151–176. [CrossRef]
49. Whitten, S.T.; Yang, H.-W.; Fox, R.O.; Hilser, V.J. Exploring the Impact of Polyproline II (PII) Conformational Bias on the Binding of Peptides to the SEM-5 SH3 Domain. *Protein Sci.* **2008**, *17*, 1200–1211. [CrossRef]
50. Levinthal, C. How to Fold Graciously. *Mössbaun Spectrosc. Biol. Syst. Proc.* **1969**, *67*, 22–24.
51. Brooks, C.L.; Gruebele, M.; Onuchic, J.N.; Wolynes, P.G. Chemical Physics of Protein Folding. *Proc. Natl. Acad. Sci. USA* **1998**, *95*, 11037–11038. [CrossRef]
52. Craig, P.O.; Lätzer, J.; Weinkam, P.; Hoffman, R.M.B.; Ferreiro, D.U.; Komives, E.A.; Wolynes, P.G. Prediction of Native-State Hydrogen Exchange from Perfectly Funneled Energy Landscapes. *J. Am. Chem. Soc.* **2011**, *133*, 17463–17472. [CrossRef] [PubMed]
53. Bai, Y.; Englander, S.W. Future Directions in Folding: The Multi-State Nature of Protein Structure. *Proteins* **1996**, *24*, 145–151. [CrossRef]
54. Englander, S.W.; Mayne, L. The Case for Defined Protein Folding Pathways. *Proc. Natl. Acad. Sci. USA* **2017**, *114*, 8253–8258. [CrossRef] [PubMed]
55. Tiffany, M.L.; Krimm, S. Circular Dichroism of Poly-L-Proline in an Unordered Conformation. *Biopolymers* **1968**, *6*, 1767–1770. [CrossRef]
56. Tiffany, M.L.; Krimm, S. New Chain Conformations of Poly(Glutamic Acid) and Polylysine. *Biopolymers* **1968**, *6*, 1379–1382. [CrossRef]
57. Tiffany, M.L.; Krimm, S. Effect of Temperature on the Circular Dichroism Spectra of Polypeptides in the Extended State. *Biopolymers* **1972**, *11*, 2309–2316. [CrossRef]
58. Mattice, W.L. The Effect of Temperature and Salt Concentration on the Circular Dichroism Exhibited by Unionized Derivatives of L-Alanine in Aqueous Solution. *Biopolymers* **1974**, *13*, 169–183. [CrossRef]
59. Woody, R. Circular Dichroism and Conformation of Unordered Polypeptides. *Adv. Biophys. Chem.* **1992**, *2*, 37–79.
60. Woody, R. Optical Rotatory Properties of Biopolymers. *J. Polym. Sci. Macromol. Rev.* **1977**, *12*, 181–321. [CrossRef]
61. Karplus, M. Contact Electron-Spin Coupling of Nuclear Magnetic Moments. *J. Chem. Phys.* **1959**, *30*, 11–15. [CrossRef]
62. Asher, S.A.; Mikhonin, A.V.; Bykov, S. UV Raman Demonstrates That Alpha-Helical Polyalanine Peptides Melt to Polyproline II Conformations. *J. Am. Chem. Soc.* **2004**, *126*, 8433–8440. [CrossRef] [PubMed]
63. Ding, L.; Chen, K.; Santini, P.A.; Shi, Z.; Kallenbach, N.R. The Pentapeptide GGAGG Has PII Conformation. *J. Am. Chem. Soc.* **2003**, *125*, 8092–8093. [CrossRef] [PubMed]
64. McColl, I.H.; Blanch, E.W.; Hecht, L.; Kallenbach, N.R.; Barron, L.D. Vibrational Raman Optical Activity Characterization of Poly(l-Proline) II Helix in Alanine Oligopeptides. *J. Am. Chem. Soc.* **2004**, *126*, 5076–5077. [CrossRef]
65. Schweitzer-Stenner, R.; Eker, F.; Griebenow, K.; Cao, X.; Nafie, L.A. The Conformation of Tetraalanine in Water Determined by Polarized Raman, FT-IR, and VCD Spectroscopy. *J. Am. Chem. Soc.* **2004**, *126*, 2768–2776. [CrossRef] [PubMed]

66. Chakrabartty, A.; Kortemme, T.; Padmanabhan, S.; Baldwin, R.L. Aromatic Side-Chain Contribution to Far-Ultraviolet Circular Dichroism of Helical Peptides and Its Effect on Measurement of Helix Propensities. *Biochemistry* **1993**, *32*, 5560–5565. [CrossRef]
67. Krittanai, C.; Johnson, W.C. Correcting the Circular Dichroism Spectra of Peptides for Contributions of Absorbing Side Chains. *Anal. Biochem.* **1997**, *253*, 57–64. [CrossRef]
68. Chen, K.; Liu, Z.; Zhou, C.; Shi, Z.; Kallenbach, N.R. Neighbor Effect on PPII Conformation in Alanine Peptides. *J. Am. Chem. Soc.* **2005**, *127*, 10146–10147. [CrossRef]
69. Hamburger, J.B.; Ferreon, J.C.; Whitten, S.T.; Hilser, V.J. Thermodynamic Mechanism and Consequences of the Polyproline II (PII) Structural Bias in the Denatured States of Proteins. *Biochemistry* **2004**, *43*, 9790–9799. [CrossRef]
70. Lim, W.A.; Richards, F.M.; Fox, R.O. Structural Determinants of Peptide-Binding Orientation and of Sequence Specificity in SH3 Domains. *Nature* **1994**, *372*, 375–379. [CrossRef]
71. Shi, Z.; Chen, K.; Liu, Z.; Kallenbach, N.R. Conformation of the Backbone in Unfolded Proteins. *Chem. Rev.* **2006**, *106*, 1877–1897. [CrossRef]
72. Creamer, T.P. Left-Handed Polyproline II Helix Formation Is (Very) Locally Driven. *Proteins* **1998**, *33*, 218–226. [CrossRef]
73. Berman, H.M.; Westbrook, J.; Feng, Z.; Gilliland, G.; Bhat, T.N.; Weissig, H.; Shindyalov, I.N.; Bourne, P.E. The Protein Data Bank. *Nucleic Acids Res.* **2000**, *28*, 235–242. [CrossRef] [PubMed]
74. Stapley, B.J.; Creamer, T.P. A Survey of Left-Handed Polyproline II Helices. *Protein Sci.* **1999**, *8*, 587–595. [CrossRef] [PubMed]
75. Wang, S.; Gu, J.; Larson, S.A.; Whitten, S.T.; Hilser, V.J. Denatured-State Energy Landscapes of a Protein Structural Database Reveal the Energetic Determinants of a Framework Model for Folding. *J. Mol. Biol.* **2008**, *381*, 1184–1201. [CrossRef] [PubMed]
76. Marsh, J.A.; Forman-Kay, J.D. Sequence Determinants of Compaction in Intrinsically Disordered Proteins. *Biophys. J.* **2010**, *98*, 2383–2390. [CrossRef]
77. Momany, F.A.; McGuire, R.F.; Burgess, A.W.; Scheraga, H.A. Energy Parameters in Polypeptides. VII. Geometric Parameters, Partial Atomic Charges, Nonbonded Interactions, Hydrogen Bond Interactions, and Intrinsic Torsional Potentials for the Naturally Occurring Amino Acids. *J. Phys. Chem.* **1975**, *79*, 2361–2381. [CrossRef]
78. Mandel, N.; Mandel, G.; Trus, B.L.; Rosenberg, J.; Carlson, G.; Dickerson, R.E. Tuna Cytochrome c at 2.0 A Resolution. III. Coordinate Optimization and Comparison of Structures. *J. Biol. Chem.* **1977**, *252*, 4619–4636. [CrossRef]
79. MacArthur, M.W.; Thornton, J.M. Influence of Proline Residues on Protein Conformation. *J. Mol. Biol.* **1991**, *218*, 397–412. [CrossRef]
80. Lovell, S.C.; Word, J.M.; Richardson, J.S.; Richardson, D.C. The Penultimate Rotamer Library. *Proteins* **2000**, *40*, 389–408. [CrossRef]
81. Iijima, H.; Dunbar, J.B.; Marshall, G.R. Calibration of effective van der Waals atomic contact radii for proteins and peptides. *Proteins Struct. Funct. Bioinform.* **1987**, *2*, 330–339. [CrossRef]
82. Baldwin, R.L. Temperature Dependence of the Hydrophobic Interaction in Protein Folding. *Proc. Natl. Acad. Sci. USA* **1986**, *83*, 8069–8072. [CrossRef] [PubMed]
83. Murphy, K.P.; Freire, E. Thermodynamics of Structural Stability and Cooperative Folding Behavior in Proteins. *Adv. Protein Chem.* **1992**, *43*, 313–361. [CrossRef] [PubMed]
84. Murphy, K.P.; Bhakuni, V.; Xie, D.; Freire, E. Molecular Basis of Co-Operativity in Protein Folding: III. Structural Identification of Cooperative Folding Units and Folding Intermediates. *J. Mol. Biol.* **1992**, *227*, 293–306. [CrossRef]
85. Lee, K.H.; Xie, D.; Freire, E.; Amzel, L.M. Estimation of Changes in Side Chain Configurational Entropy in Binding and Folding: General Methods and Application to Helix Formation. *Proteins* **1994**, *20*, 68–84. [CrossRef] [PubMed]
86. Xie, D.; Freire, E. Structure Based Prediction of Protein Folding Intermediates. *J. Mol. Biol.* **1994**, *242*, 62–80. [CrossRef] [PubMed]
87. Gómez, J.; Hilser, V.J.; Xie, D.; Freire, E. The Heat Capacity of Proteins. *Proteins Struct. Funct. Bioinform.* **1995**, *22*, 404–412. [CrossRef]
88. D'Aquino, J.A.; Gómez, J.; Hilser, V.J.; Lee, K.H.; Amzel, L.M.; Freire, E. The Magnitude of the Backbone Conformational Entropy Change in Protein Folding. *Proteins* **1996**, *25*, 143–156. [CrossRef]
89. Habermann, S.M.; Murphy, K.P. Energetics of Hydrogen Bonding in Proteins: A Model Compound Study. *Protein Sci.* **1996**, *5*, 1229–1239. [CrossRef]
90. Luque, I.; Mayorga, O.L.; Freire, E. Structure-Based Thermodynamic Scale of Alpha-Helix Propensities in Amino Acids. *Biochemistry* **1996**, *35*, 13681–13688. [CrossRef]
91. Das, R.K.; Pappu, R.V. Conformations of Intrinsically Disordered Proteins Are Influenced by Linear Sequence Distributions of Oppositely Charged Residues. *Proc. Natl. Acad. Sci. USA* **2013**, *110*, 13392–13397. [CrossRef]
92. Dunker, A.K.; Obradovic, Z.; Romero, P.; Garner, E.C.; Brown, C.J. Intrinsic Protein Disorder in Complete Genomes. *Genome Inform. Ser. Workshop Genome Inform.* **2000**, *11*, 161–171. [PubMed]
93. Wootton, J.C.; Federhen, S. Statistics of Local Complexity in Amino Acid Sequences and Sequence Databases. *Comput. Chem.* **1993**, *17*, 149–163. [CrossRef]
94. Wootton, J.C. Non-Globular Domains in Protein Sequences: Automated Segmentation Using Complexity Measures. *Comput. Chem.* **1994**, *18*, 269–285. [CrossRef]
95. Romero, P.; Obradovic, Z.; Li, X.; Garner, E.C.; Brown, C.J.; Dunker, A.K. Sequence Complexity of Disordered Protein. *Proteins Struct. Funct. Bioinform.* **2001**, *42*, 38–48. [CrossRef]
96. Uversky, V.N. Natively Unfolded Proteins: A Point Where Biology Waits for Physics. *Protein Sci.* **2002**, *11*, 739–756. [CrossRef] [PubMed]

97. Hynes, T.R.; Fox, R.O. The Crystal Structure of Staphylococcal Nuclease Refined at 1.7 A Resolution. *Proteins* **1991**, *10*, 92–105. [CrossRef]
98. Scholtz, J.M.; Marqusee, S.; Baldwin, R.L.; York, E.J.; Stewart, J.M.; Santoro, M.; Bolen, D.W. Calorimetric Determination of the Enthalpy Change for the Alpha-Helix to Coil Transition of an Alanine Peptide in Water. *Proc. Natl. Acad. Sci. USA* **1991**, *88*, 2854–2858. [CrossRef]
99. Dignon, G.L.; Zheng, W.; Kim, Y.C.; Mittal, J. Temperature-Controlled Liquid—Liquid Phase Separation of Disordered Proteins. *ACS Cent. Sci.* **2019**, *5*, 821–830. [CrossRef]

Article

Dynamic Preference for NADP/H Cofactor Binding/Release in *E. coli* YqhD Oxidoreductase

Rajni Verma [1,*], Jonathan M. Ellis [2] and Katie R. Mitchell-Koch [1,*]

[1] Department of Chemistry, McKinley Hall, Wichita State University, 1845 Fairmount, Wichita, KS 67260, USA
[2] Department of Chemistry, University of Wisconsin-Madison, 1101 University Avenue, Madison, WI 53706, USA; jmellis4@wisc.edu
* Correspondence: rajnixverma@gmail.com (R.V.); katie.mitchell-koch@wichita.edu (K.R.M.-K.); Tel.: +1-316-978-7372 (R.V.); +1-316-978-7371 (K.R.M.-K.)

Abstract: YqhD, an *E. coli* alcohol/aldehyde oxidoreductase, is an enzyme able to produce valuable bio-renewable fuels and fine chemicals from a broad range of starting materials. Herein, we report the first computational solution-phase structure-dynamics analysis of YqhD, shedding light on the effect of oxidized and reduced NADP/H cofactor binding on the conformational dynamics of the biocatalyst using molecular dynamics (MD) simulations. The cofactor oxidation states mainly influence the interdomain cleft region conformations of the YqhD monomers, involved in intricate cofactor binding and release. The ensemble of NADPH-bound monomers has a narrower average interdomain space resulting in more hydrogen bonds and rigid cofactor binding. NADP-bound YqhD fluctuates between open and closed conformations, while it was observed that NADPH-bound YqhD had slower opening/closing dynamics of the cofactor-binding cleft. In the light of enzyme kinetics and structural data, simulation findings have led us to postulate that the frequently sampled open conformation of the cofactor binding cleft with NADP leads to the more facile release of NADP while increased closed conformation sampling during NADPH binding enhances cofactor binding affinity and the aldehyde reductase activity of the enzyme.

Keywords: NAD(P)H-dependent oxidoreductase; zinc-containing alcohol dehydrogenase; cofactor binding and release; interdomain cleft dynamics; molecular dynamics simulations

Citation: Verma, R.; Ellis, J.M.; Mitchell-Koch, K.R. Dynamic Preference for NADP/H Cofactor Binding/Release in *E. coli* YqhD Oxidoreductase. *Molecules* **2021**, *26*, 270. https://doi.org/10.3390/molecules26020270

Academic Editors: Marilisa Leone
Received: 29 November 2020
Accepted: 4 January 2021
Published: 7 January 2021

Publisher's Note: MDPI stays neutral with regard to jurisdictional claims in published maps and institutional affiliations.

Copyright: © 2021 by the authors. Licensee MDPI, Basel, Switzerland. This article is an open access article distributed under the terms and conditions of the Creative Commons Attribution (CC BY) license (https://creativecommons.org/licenses/by/4.0/).

1. Introduction

YqhD is a homo-dimeric, NADP(H)-dependent *E. coli* oxidoreductase identified in 2003 [1]. Since its discovery, it has received considerable attention for bioengineering efforts [2], being targeted for its utility in biomass conversion to biofuels [3] and chemical feedstocks [4]. The enzyme was initially identified during the *E. coli* structural genomics program and was shown to be widely distributed throughout the bacteria kingdom [5]. YqhD enzyme works as a homodimer with Zn^{2+} in the active site, catalyzing the interconversion of alcohols and aldehydes with NADP and NADPH, respectively, as a cofactor. Although YqhD binds Zn^{2+}, it has structural similarities to the iron-dependent (group III) alcohol dehydrogenase enzymes [6]. Initially, YqhD was proposed as NADP-dependent alcohol dehydrogenase with a weak affinity toward short- and medium- chain alcohols [7]. Later studies characterized this enzyme as a NADPH-dependent aldehyde reductase with a broad range of substrates, such as 3-hydroxypropionaldehyde [8], propanaldehyde, isobutylaldehyde [3], acrolein, and malondialdehyde [2,9]. The enzyme's biological role was also evident in the reduction of various reactive aldehydes derived from membrane lipid peroxidation [9] and mitigation of furfural toxicity [10–12]. Other biosynthetic efforts with YqhD have involved its utilization in the production of aromatic alcohols [13], ethylene glycol, 1-butanol [14], 1,4-butanediol, acetol [15,16], 1,2-propanediol [17] and, in a pathway using CO_2, 1-butanol [18].

Just a few years after YqhD's characterization, it became a candidate for protein engineering studies toward the biological production of useful chemicals and fuels, including valorization of biomass-derived precursors. Protein engineering approaches have produced variants of YqhD with improved affinity and higher catalytic efficiency toward 3-hydroxypicolinic acid for the production of 1,3-propanediol [19]. Engineered strains for the production of 1,3-propanediol from D-glucose [20] and glycerol [21,22] have also utilized YqhD. The broad enzyme substrate scope, combined with this report on the dependence of enzyme conformational dynamics on cofactor oxidation state, makes YqhD useful as a model system to better understand NADP-dependent zinc bound oxidoreductases [23–26]. Despite the abundance of structural and kinetics data for the YqhD oxidoreductase, there are no reports (experimental or simulations) until now on the structure and dynamics of this enzyme in solution.

Herein, we report results from a set of molecular dynamics (MD) simulations of YqhD enzyme with both oxidized and reduced cofactors (NADP and NADPH) in aqueous solution. The study aims to rationalize the enzyme preference toward NADPH over NADP as a cofactor. The crystal structure [7] (PDB ID: 1OJ7) is the starting point for our simulations. In the crystal structure of the YqhD holoenzyme, the bound NADP cofactor is modified to NADPH(OH)$_2$ due to oxidative stress during the crystallization process [7]. Figure 1 shows the dimeric structure of YqhD, which crystallizes as a tetramer in the asymmetric unit, as well as geometries of the modified and native cofactors (NADPH(OH)$_2$, NADP, and NADPH). YqhD shows a lower Michaelis constant for NADPH [11] (K_M = 0.008 mM) than NADP [27] (K_M = 0.15 mM) and, concomitant with cofactor reactivity, a lower Michaelis constant for 1-butyaldehyde [9] (K_M = 0.67 mM) compared to 1-butanol [7] (K_M = 36 mM). Our simulation work of the YqhD homodimer has found that the cofactor oxidation state has a profound effect on enzyme structure and dynamics, which is consistent with differences in enzyme efficacy toward alcohols versus aldehydes.

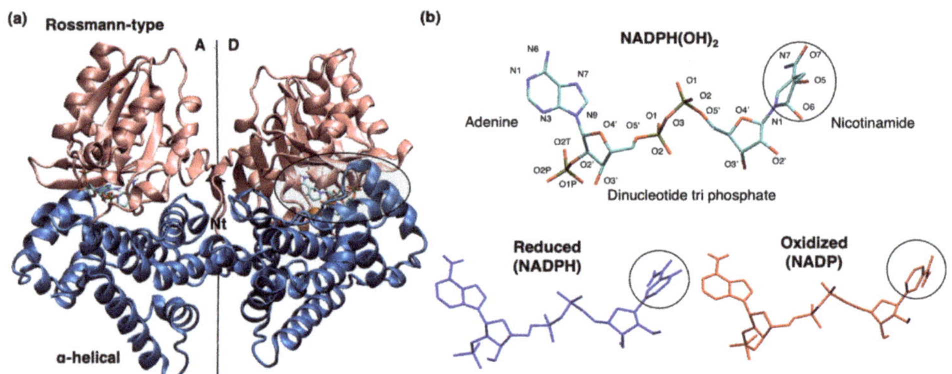

Figure 1. (**a**) YqhD dimer (comprised of monomers A and D) is in cartoon representation with the α-helical and Rossmann-type domains colored in blue and pink, respectively. The shaded region shows the cofactor-binding site of monomer D. Zn^{2+} is located at the interdomain face within the cofactor-binding site and is shown in orange colored VDW representation. The NADP cofactor is in licorice representation colored by elements (nitrogen, oxygen, hydrogen, carbon and, phosphorus in blue, red, white, carbon, and tan color, respectively). (**b**) NADP cofactor is present in the crystal structure as NADPH(OH)$_2$ (with modification at C5 and C6 position in the nicotinamide moiety colored by elements and N and O atomic labeling) and in the MD simulations as NADPH and NADP in blue and red color, respectively. The black ovals highlight the nicotinamide moiety.

It has been observed previously that differences in cofactor oxidation states induce protein conformational changes [25,28] and also influence the cofactor binding affinities [29]. Along these lines, structural differences in cofactor binding for FucO oxidoreductase, a NADH-dependent enzyme with similarities to YqhD, were recently reported [30], while conformational changes in cofactor were shown to be critical in the pathway toward the transition state for horse liver alcohol dehydrogenase [31]. Recent work on product release, a rate-limiting step in dihydrofolate reductase (DHFR), indicates that NADPH induces product release through steric repulsion during conformational sampling of the closed excited state [32]. There is some variation in conformational dynamics (rates between exchange of states) when the DHFR-product complex is bound to oxidized (NADP) vs. reduced (NADPH) cofactor, but conformational exchange rates measured by nmR are on the same order of magnitude (1890 $\pm$ 80 s^{-1} bound to NADPH vs. 1420 $\pm$ 70 s^{-1} when bound to $NADP^+$).

The present work has gone beyond structural effects to computationally study how dynamics are affected by cofactor oxidation/reduction. To gain insight into dynamical preference of the YqhD protein towards NADPH over NADP cofactor, the article is organized as follows: the details of MD simulations are reported in the Methods section; the results section has the analysis of MD trajectories, with the main focus on structural and dynamical properties of YqhD enzyme and NADP/H cofactor binding; next, we discuss our results, focusing on the effect of cofactor oxidation/reduction on structure-dynamics and activity of the enzyme.

2. Results

2.1. *Structural and Dynamical Properties of YqhD Enzyme*

As mentioned in the Introduction section, the crystal structure has a modified $NADPH(OH)_2$ cofactor, but in this work, simulations were performed for the YqhD homodimer with oxidized NADP and reduced NADPH cofactors (separately) for comparison (see Figure 1 for details). First, we performed an equilibration step involving 20 ns MD simulation in isothermal-isobaric (NPT) ensemble with each state. During the equilibration step, the YqhD protein adopts a conformation suitable for the binding of NADP and NADPH cofactors. Starting with equilibrated structures, we performed five 200 ns long independent MD simulations for each state, starting with different initial velocities (generated through a random number seed) to check the structural convergence. First, analysis of structure and dynamics was done using the crystal structure as reference for three different types of structural data subsets within each state of the NADP/H-cofactor: the entire dimer, the monomers comprising dimer, and the domains (α-helical and Rossmann-type domains) within each monomer, in addition to dynamics during the simulations (see Figure 1 for cofactor-bound crystal structure of YqhD).

2.1.1. YqhD Dimer

During the simulations, the populated dimer conformations have an average root mean square deviation (RMSD) of 0.28 $\pm$ 0.05 nm and 0.24 $\pm$ 0.04 nm when bound with NADP and NADPH cofactors, respectively. Radius of gyration (Rg) of dimer for the crystal structure is 2.71 nm. During the simulations, the YqhD dimer has conformations with an average 2.75 $\pm$ 0.01 nm Rg, which is slightly less compact than the crystal structure and is consistent with solution phase dynamical sampling. The cofactor binding and protonation states affect the dimer compactness during the simulations resulting in a less compact structure with NADPH and NADP cofactors than with $NADPH(OH)_2$.

2.1.2. YqhD Domain

A Rossmann-type domain is involved in NADP cofactor binding. With NADP-bound, the domain shows an average RMSD of 0.17 $\pm$ 0.03 nm with slightly higher variations than NADPH-bound (average RMSD of 0.15 $\pm$ 0.02 nm). The α-helical domain is a metal-binding domain and has an average RMSD of 0.11 $\pm$ 0.02 nm for NADP-bound and 0.13 $\pm$ 0.03 nm for NADPH-bound YqhD. Monomers bound to reduced cofactor show slightly higher deviations than monomers bound to oxidized cofactor. Both domains within each monomer become slightly more compact in the simulations with Rg values of 1.57 $\pm$ 0.01 for Rossmann-type and 1.63 $\pm$ 0.01 for α-helical domains during the simulations, compared to 1.60 nm for Rossmann-type and 1.65 nm for α-helical domains in the crystal structure.

2.1.3. YqhD Monomer

When comparing monomers in the 2 μs aggregated trajectory, it can be seen that the oxidation state of the NADP/H cofactor affects the overall monomer conformation significantly. NADP-bound monomer shows an average backbone RMSD of 0.27 ± 0.06 nm with slightly higher variations than NADPH-bound monomer, which has an average RMSD of 0.23 ± 0.05 nm. The distribution of backbone RMSD and Rg values for YqhD monomers is reported in Figure 2. YqhD monomers bound to NADP show higher deviation from the reference crystal structure. The RMSD distribution ranges from 0.13–0.45 nm, with a main peak at 0.31 nm and two additional peaks at 0.23 and 0.17 nm. NADPH-bound YqhD has a higher probability of remaining closer to the reference structure, showing a main peak at 0.19 nm and two additional sharp peaks at 0.25 and 0.15 nm; however, the RMSD distribution is broader, ranging from 0.09–0.47 nm. The Rg value of monomer is 2.07 nm in the crystal structure. NADP-bound monomers show an average Rg of 2.14 ± 0.02 nm, while NADPH-bound monomers have an average Rg of 2.11 ± 0.03 nm. For NADPH-bound monomers, Rg values show a broad distribution ranges from 2.04–2.27 nm with three peaks at 2.07, 2.10 and 2.13 nm. NADP-bound monomer exhibits a narrower Rg distribution, ranging from 2.04–2.25 nm, with a main peak at 2.15 nm and a wider shoulder at 2.09 nm. Overall, YqhD monomer bound with reduced NADPH cofactor populates conformations (two peaks) with lower RMSD (<0.2 nm) and Rg (<2.1 nm) values, i.e., close to the crystal structure that has the modified NADPH(OH)$_2$ cofactor. Thus, some of the structural differences in YqhD monomer relative to the crystal structure can be attributed to the changes in the oxidation states of the cofactor (reflecting relevant, functional states of NADP/H bound to the enzyme); the molecular details are discussed later.

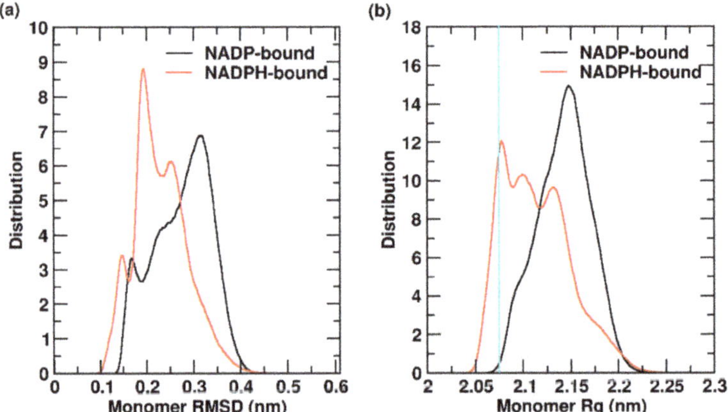

Figure 2. The distribution of backbone (**a**) root mean square deviation (RMSD) and (**b**) radius of gyration (Rg) values are shown for the YqhD monomers. The vertical, cyan-colored bar shows the Rg value observed in the crystal structure.

It is worthwhile to keep in mind that the RMSD and Rg values are calculated using the crystal structure as a reference, which may not reflect the functional enzyme structure due to its modified cofactor (NADPH(OH)$_2$). Still, the crystal structure gives a good starting point to witness how the enzyme samples different conformations during the binding of oxidized and reduced cofactors. In the simulations, YqhD shows slightly higher structural flexibility, populating diverse conformations with a more compact yet broader, distinctive distribution of Rg values when bound to NADPH (required for reductase function) compared to NADP (released for enzyme turnover). The population of conformations towards lower RMSD and Rg values in the simulations with reduced cofactor indicates the slight change in protein conformation in NADPH binding relative to NADP binding.

Figure 3a shows the per-residue backbone RMSD for the YqhD domains illustrated in Figure 3b. High deviations are restricted mainly to loop regions and the N- and C-terminus. Both the monomers show significant deviations in the loops that are present at the dimer interface (between monomers) and the interface between domains within each monomer (marked by the cyan colored horizontal bar in Figure 3a). Monomers bound to NADPH show a higher deviation in loop regions α6/α7, α7/α8, α9/α10 and α12/α13 of the metal-binding helical domain then the NADP-bound monomer.

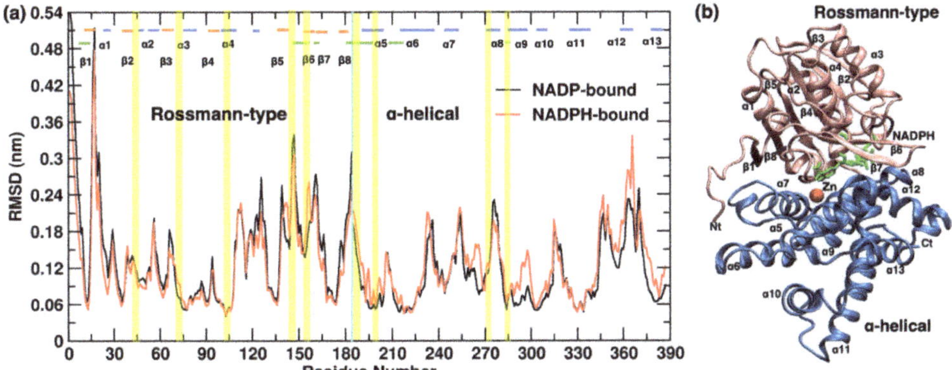

Figure 3. (**a**) Backbone RMSD per residue for the domains of monomers bound to NADP (black) and NADPH (red) with respect to the crystal structure. Horizontal bars show α-helices (blue), β-sheets (orange), and the residues involved in dimer and monomer interactions (cyan). The cofactor-binding region in Rossmann-type domain and metal-binding regions within the α-helical domain are shown in yellow vertical bars. (**b**) YqhD monomer is in cartoon representation with bound NADPH (green licorice representation) and Zn^{2+} (red sphere) in the active site. Helices, beta sheets, and C and N terminals are labeled.

2.2. Hydrogen Bonding in YqhD Enzyme

Inter-monomer and interdomain hydrogen bonds were calculated in the crystal structure and during the simulations of YqhD bound to NADP/H. Hydrogen bonds were calculated using a distance between acceptor and hydrogen donor of 3.5 Å and an angle of <30° among acceptor, donor, and hydrogen as criteria. The dimer has ~8 inter-monomer hydrogen bonds, and monomers A and D have ~7 interdomain hydrogen bonds during the simulations (averaging over data collected every 10 ps). The hydrogen bonds discussed below are present for a cumulative total of >30% of the time points analyzed in the trajectories.

2.2.1. Inter-Monomer Hydrogen Bonding

During the simulations, eight unique inter-monomer hydrogen bonds were observed in the dimer bound to NADP/H involving residues of Nt, β1, α5/α6, α6, and α7. The crystal structure also has seven unique hydrogen bonds between monomers, involving the residue pairs Leu1-Tyr238/Asp239 (Nt- α7), Asn3-Lys16 (Nt- β1), Phe4-Phe14 (β1- β1), Asn5-Arg11 (β1- β1), Leu6-Ile12 (β1- β1), Asp209-Asn243 (α5/α6- α7), and Lys211-Glu226/Asp227/Asn243 (α6- α6, α7). During the simulations, these hydrogen bonds were also observed in YqhD bound to NADP/H involving the residue pairs: Leu1-Tyr238 (Nt- α7), Asn3-Lys16 (β1- β1), Phe4-Phe14 (β1- β1), Asn5-Arg11 (β1- β1), Leu6-Ile12 (β1- β1), Asp209-Asn243 (α5/α6- α7), Lys211-Asp227/Asn243 (α6- α6/α7), and Arg215-Glu226 (α6- α6). NADP/H binding results in a dimer conformation that is less compact than the crystal structure with a slightly higher Rg value, but it maintains the hydrogen bonds observed in the crystal structure.

2.2.2. Interdomain Hydrogen Bonding

Seven interdomain hydrogen bonds were observed in the crystal structure of monomers bound to NADPH(OH)$_2$ involving the residue pairs Asn2-Gly259 (Nt- α7/α8), Lys16-238Tyr (β1/α1- α7), Asn145-Thr249 (β5/β6- α7/α8), Thr157-Lys359 (β6/β7- α12), Asp159-Lys359/His363 (β7- α12, α12/α13), Thr180-Arg241 (β8/α5- α7) and Pro184-Val188 (β8/α5- α5). During the simulations, five interdomain hydrogen bonds (with both NADP, NADPH bound) involve the residues of Nt [Leu1-Asn208/Asp209 (α5/α6) and Asn2-Gly259 (α7/α8)], β1 [Thr8-Asn253 (α7)], and β8/α5 [Thr180-Arg241 (α7), Pro184-Gln187/Val188 (α5)]. In addition, other hydrogen bonds involving residue pairs Thr142-Asp194 and Asp176-Tyr238 were observed in NADP-bound monomer. On average, three interdomain hydrogen bonds observed in the crystal structure between Asn2-Gly259, Thr180-Arg241 and Pro184-Val188 occurred frequently during the simulations. Observed changes in the hydrogen bonding interactions evidence conformational changes in the monomer to accommodate NADP/H cofactors.

2.3. Cluster Analysis of YqhD Enzyme

Conformational differences observed due to the cofactor oxidation/reduction state were further quantified using cluster analysis on monomers (see Section 4.4 for details). In the combined trajectory, a total of 54 (NADP) and 43 (NADPH) clusters were obtained for the monomer using an RMSD cutoff of 0.13 nm. Figure 4 represents the first three clusters of NADP-bound and NADPH-bound monomers in red, blue and green colored ribbons, respectively. These first three clusters comprise 62.7% (30.8%, 19.3%, and 12.6%) and 72.9% (46.0%, 13.8%, and 13.1%) of the total monomer conformations occupied by YqhD for NADP- and NADPH-bound enzyme, respectively.

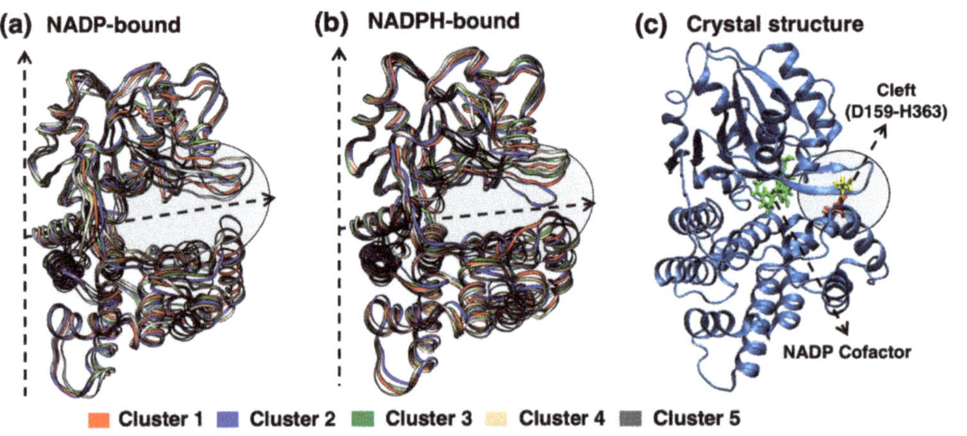

Figure 4. The representative conformations of the first five clusters (colored by red, blue, green, tan, and gray color, respectively) of homodimer (**a**) NADP-bound and (**b**) NADPH-bound are superimposed and represented as ribbons. The dotted arrows show the monomer- and domain- interface. The region involved in the opening/closing of the interdomain cleft is highlighted. (**c**) Interdomain cleft residues D159 and H363 are represented in the NADPH-bound monomer of crystallographic structure. Cluster 1 (in red color) shows partially closed conformations in both NADP/H bound- Yqhd monomer, while Cluster 2 (in blue color) is open confirmation in NADP-bound and closed in NADPH-bound monomer.

By looking at the superimposed mean cluster structures and crystal structure illustrated in Figure 4, it can be seen that the primary differences in monomer conformations arise from the position of the β6/β7 loop region (circled) of the Rossmann-type domain and the α8/α12 helix of α-helical domain. Figure 5a shows the superimposed crystallographic structures of apo- and holo- proteins. The β6/β7 loop region is in open conformation in the apo form of the YqhD crystal structure [7] (and closed in the holoenzyme), indicating that they comprise an interdomain cleft which opens and closes for cofactor binding and release. In this side-by-side comparison of the monomer, it can be seen that, with both cofactors, these loops are significantly dynamic. The highly populated conformations of NADPH-bound involve both monomers within the dimer remaining with a partially closed cleft (cluster 1), closed cleft (cluster 2), and open cleft (cluster 3). On the other hand, when oxidized NADP cofactor is bound, each monomer within the dimer samples more conformations with open cleft. All three of the highly populated conformations of NADP-bound enzyme are open cleft, indicating that cofactor oxidation state has an effect on the structures and populations of the open and closed conformation. Figure 5b represents the distribution of distances using the center of mass of two domains in the simulations as a global measure of the opening and closing of the domains. Domain distances observed in the crystal structures are shown in Figure 5b by vertical cyan and blue colored lines for the holo- and apo- enzymes, respectively. The NADPH-bound monomer remains in more closed and partially closed conformation, indicated by two main peaks for the distances between domains at 2.68 nm and 2.73 nm, respectively. A third peak is also observed at 2.83 nm, which is close to the one observed in the apoprotein, at 2.86 nm. In contrast, the NADP-bound monomer has the main peak at 2.90 nm for distances between domains, indicating a population of more open conformations, and a second small peak at 2.72 nm for sampling partially-closed structures; the molecular details are discussed later.

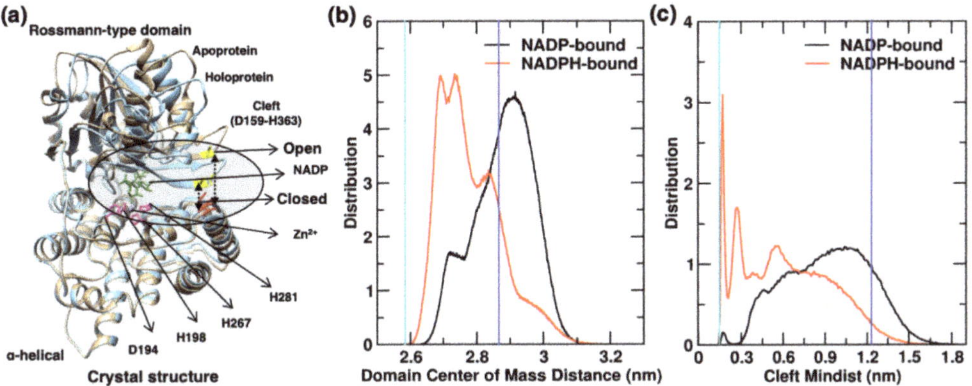

Figure 5. (**a**) Crystallographic structure of the YqhD monomer as apoprotein and holoprotein is in tan and sky-blue colored cartoon representation, respectively, with bound NADP (green licorice representation) and Zn^{2+} (violent sphere) in the active site [7]. Cleft residues and Zn^{2+} binding residues are labeled. (**b**) The distribution of distances between Rossmann-type and α-helical domains is shown using the center of mass. (**c**) The distribution for cleft minimum distances using residue pair D159-H363 is shown for NADP- and NADPH-bound monomers during the simulations. The cleft distances observed in the apo- and holo- YqhD protein crystal structures are marked with blue (apo) and cyan (holo) bars.

2.4. Interdomain Cleft

The cofactor binding site in each YqhD monomer spans the interdomain face. The residues of β6/β7 loop region act as the clamp at the mouth of the cofactor binding region, which remains in open conformation in the apo form [7] and closed in the holoenzyme crystal structures. The closed conformation of holoenzyme has hydrogen bond Asp159-Lys359/His363 involving residues D159 of β6/β7 loop region in the Rossmann-type domain and residues K359/H363 of α12 helix in the α-helical domain. For a better understanding of cleft opening/closing and its relationship with cofactor binding/release, we defined interdomain cleft using residue pair D159-H363 at the mouth of the cofactor binding site, on the interface of the Rossmann-type and α-helical domains (see Figure 5a). In the crystal structure, the holoenzyme has a distance of 0.16 nm for cleft, while the apo-enzyme has distances of 1.25 nm for cleft [7], which indicates the involvement of cleft opening/closing in cofactor binding and release. Figure 5a shows the residue pair D159-H363 which is involved in interdomain cleft formation and clamps the NADP/H cofactor in the binding site. The interdomain cleft shows hydrogen bonding between residue pairs D159-K359/H363 and K160-H271/E272 in the crystal structure and simulations. YqhD has ~5% D159-H363 hydrogen bond existence in the NADPH-bound monomer trajectory; however, its occurrence is negligible in NADP-bound monomers.

Figure 5c shows the distribution of cleft distances using the residue pair D159-H363. Cleft distances observed in the crystal structures are shown in Figure 5c by vertical cyan and blue colored lines for the holo- and apo-enzymes, respectively. NADP-bound monomer shows a wider distribution range of cleft distances, from 0.14–2.0 nm, compared to those in NADPH: 0.14–1.7 nm. It can be seen in Figure 5c that the distribution of cleft distances shifts to the right with oxidized cofactor, reflecting wider openings of the interdomain cleft in NADP-bound monomers. Monomers sample conformations with interdomain cleft distances using residue pairs D159-H363: (i) less than 0.55 nm, representing closed cleft as observed in holoenzyme, (ii) ranges from 0.55–1.0 nm, for partially-closed cleft representing transition states and (iii) more than 1.0 nm, having open cleft as observed in protein. NADPH-bound monomers have two peaks in the distance distribution (Figure 5c) at 0.16 and 0.28 nm for cleft distances representing closed confirmations, and the third peak at 0.58 nm showing partially-closed conformations. However, NADP-bound monomers show a peak at 0.17 and a broader distribution of cleft distances, encompassing partially-closed and, mainly, open cleft confirmations. Changes in interdomain cleft distances occurred mainly due to a major shift in the positions of helix α12 and beta sheets β6 and β7 (see the circled region in Figure 5a), but it is a cooperative motion (involving the cleft forming regions of both domains). Cleft distance data indicate that residue D159 of cleft works as the clamp to monitor cleft opening and closing for cofactor binding and release, and its movement is affected by cofactor oxidation state.

2.4.1. Interdomain Opening-Closing Cleft Dynamics

To characterize the cleft dynamics, we evaluated the minimum distance data of cleft opening/closing residues D159 and H363. For Yqhd monomer, we set a distance of <0.55 nm for closed states and >1.0 nm for open states. By assigning the open, partially-closed, and closed states a value of 0, 0.5 and 1, respectively, we were able to generate a graph of state vs time, shown in Figure 6a,b. The state-time analysis is performed on the 2 μs long aggregated trajectory involving 10 sets of NADP/H-bound monomer data. Within the aggregated data, the NADP-bound monomer is in the open state for 41% of the time, partially closed 49%, and closed for 10%. In contrast, the NADPH-bound monomer highly populates the closed state (44% of time points) and exists in the transitional conformation 39% and open state for 17% of the time points in the aggregated trajectories. The state population data indicate an efficient sampling of protein conformations during the 10 independent sets of simulations. From the simulation data, the effect of cofactor oxidation state is clearly evident on the occurrence of one state over another (see Figure 6a,b). For example, the NADP-bound monomer is prone to access open states, while NADPH-bound monomer remains in closed states more frequently. We also evaluated cleft minimum distance data to calculate waiting times for the enzyme to remain in a conformation before switching cleft state (i.e., switching conformation from closed to open, and vice versa). Figure 6c,d shows the cleft dynamics in NADP/H-bound monomers as the number of events and wait time for each instance required for switching from closed to open state and vice versa. The equilibrated starting structure of YqhD homodimer is in the closed state in both NADP/H-bound monomer. During the simulations, the monomer rapidly fluctuates between open and closed conformations, resulting in waiting times ranging from picosecond to nanoseconds for switching of states. The interdomain cleft remains more dynamic in NADP-bound monomer with a total of 750 events of cleft opening/closing, compared to NADPH-bound domain with 318 events of state switching. The waiting time for state switching remains under 1 ns in 75% of cleft opening/closing events of NADP-bound monomers and 80% of NADPH bound monomers. For switching from closed to open state, the average wait time was 2.4 ns in NADPH-bound monomer and 409 ps in NADP-bound monomer. The longest wait time of switching from closed to open state was 8.3 ns in NADP-bound monomer and 68 ns in NADPH-bound monomer. However, switching from open to closed state occasionally required even longer average waiting times of 2.9 ns (maximum 56 ns) for NADPH-bound and of 1.4 ns (maximum 43 ns) for NADP-bound monomer. The cleft dynamics data evidences the effect of cofactor oxidation state on enzyme dynamics. Simulation results indicate that (a) the cleft dynamics are influenced by the cofactor oxidation state and (b) opening of the interdomain cleft facilitates the release of NADP cofactor (below), which is consistent with the experimentally observed properties of the biocatalysts such as the higher affinity of YqhD enzyme for NADPH cofactor indicated by its lower K_M value of 0.008 mM [11] over NADP cofactor, K_M = 0.150 mM [21].

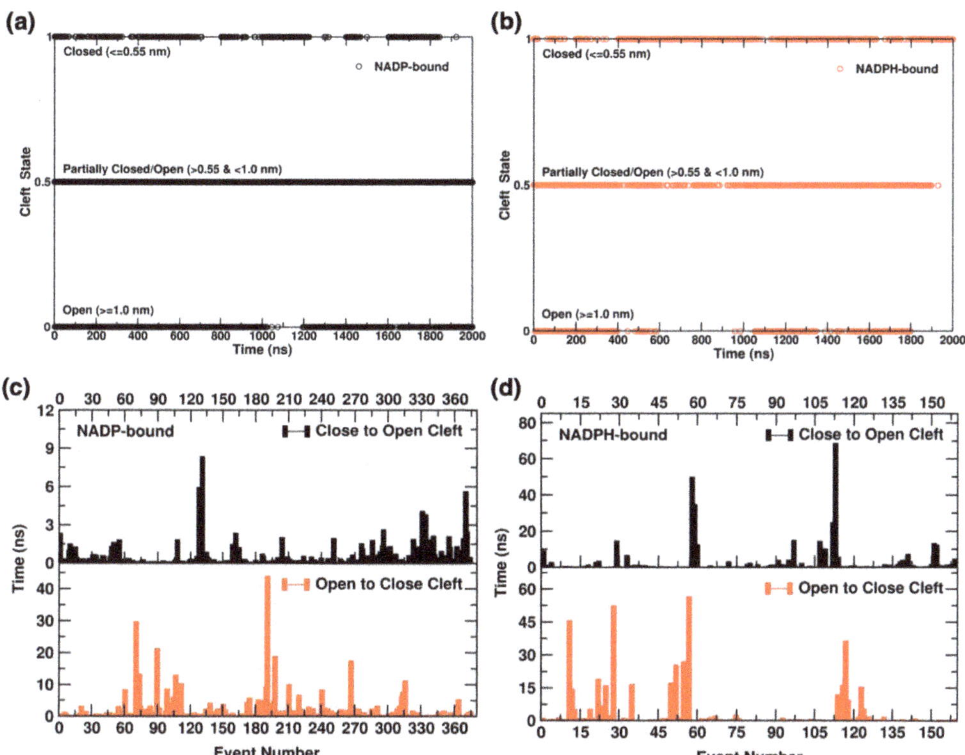

Figure 6. Open, partially closed/open, and closed states of the interdomain cleft observed using residue pair D159-H363 in YqhD monomer bound to (**a**) NADP and (**b**) NADPH cofactors during the simulations. Waiting time is for each instance of cleft opening and closing in YqhD monomer bound to (**c**) NADP and (**d**) NADPH.

2.4.2. Cofactor Binding and Release in YqhD Enzyme

The cofactor was found to be quite flexible during simulations, which can be quantified with variations in cofactor RMSD and Rg values. Figure 7 shows the distribution of RMSD and Rg values in the 2 μs of aggregated trajectory data, analyzing NADP/H-bound within each monomer of the YqhD dimer. The NADP cofactor shows a broader distribution of

RMSD values, with five peaks at 0.40, 0.30, 0.21, 0.13 and 0.50 nm, compared to NADPH with four peaks at 0.15, 0.30, 0.22 and 0.42 nm. The RMSD distribution for NADP cofactor is shifted to the right, sampling conformations having higher RMSD values relative to the holoprotein crystal structure. Rg values of the NADP/H cofactors indicate its compactness during the simulations. The NADPH cofactor shows a main peak at 0.73 nm that is close to the Rg of cofactor observed in the crystal structure, representing an extended cofactor confirmation bound in the intradomain region of each monomer. Additional peaks are observed at 0.60 nm and small peaks at 0.44 and 0.49 nm. However, the Rg distribution of the NADP cofactor is shifted to the left (more compact), with three main peaks at 0.58, 0.60 and 0.49 nm. During the simulations, extended confirmations of cofactor were sampled more frequently by NADPH than NADP. This detail is in agreement with the previous observation that the NADP-bound monomers sample more open and partially closed conformations, having a looser cofactor binding, with the open cleft conformation facilitating cofactor release.

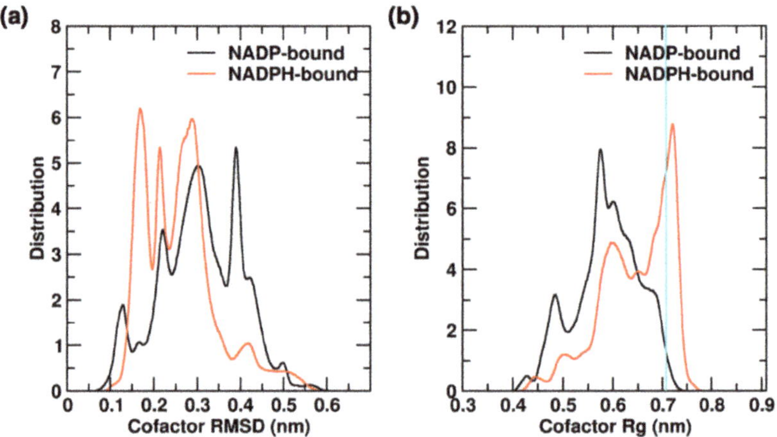

Figure 7. Distributions of (**a**) RMSD and (**b**) Rg values are shown for NADP and NADPH cofactors in the aggregated trajectories of monomers. Cyan colored horizontal bar represents cofactor Rg in the crystal structures.

Subsequently, we observed an average minimum distance in the monomer trajectories between protein and cofactor, and cofactor and Zn^{2+} of 0.16 and 0.45 nm, respectively, to the crystal structure (see Figure 8a,b). The distances between cofactor and protein had significant differences during the simulations and are related to the binding of cofactor in each monomer. Zn^{2+} has an average distance of 0.25 nm from residues Asp194, His198, His267 and His281 during the simulations as observed in the crystal structure (see Figure 5a). A minimum of one water and Ala141 remain within 0.19 nm of the Zn^{2+} over the course of both NADP and NADPH simulations. Within 0.35 nm distances from Zn^{2+}, an average of 12 ± 3 and 10 ± 3 water molecules were present in NADP and NADPH-bound YqhD monomer, respectively. A higher number of water molecules were present close to Zn^{2+} in open conformations than the closed one: 15 ± 2 water molecules for NADPH and 15 ± 7 for NADP-bound YqhD during the time-interval 800–1000 ns, which shows the population of open conformations indicated by higher distances between cofactor and protein in Figure 8a.

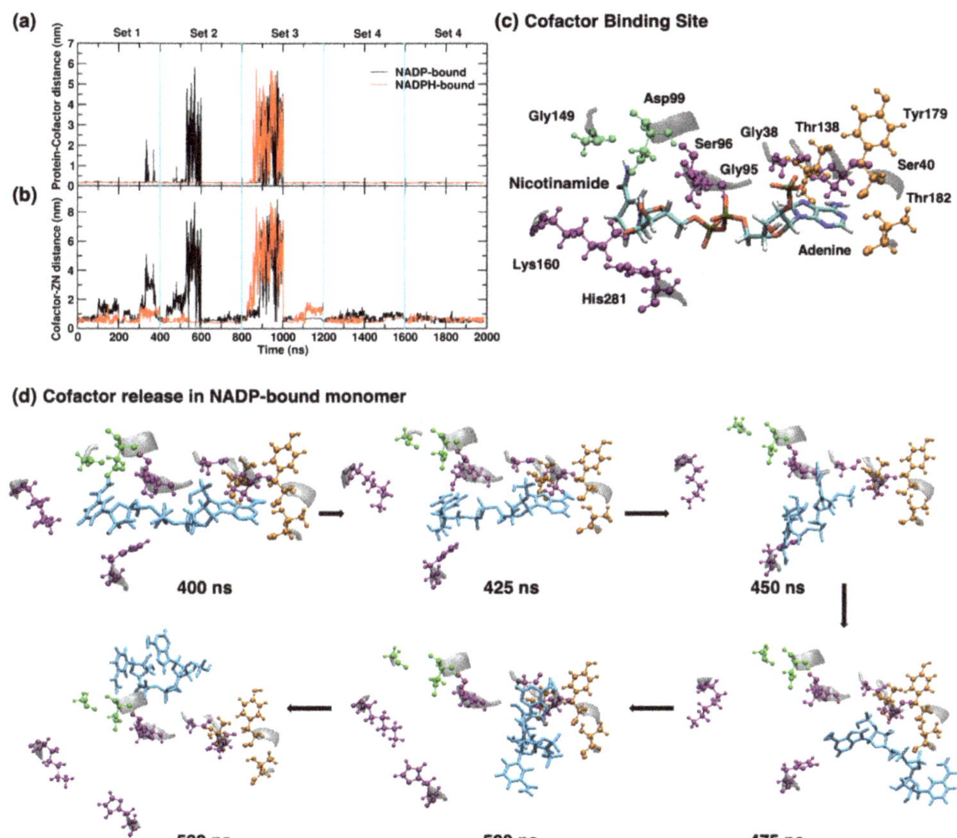

Figure 8. The minimum distance is in nm between (**a**) protein and cofactor, and (**b**) cofactor and zinc ion as a function of time in the aggregated trajectory of monomer bound to NADP and NADPH. Large distances correspond to cofactor release events. (**c**) Cofactor binding site shows residues forming hydrogen bonds in the starting structure of YqhD monomer. Cofactor binding residues are colored based on their hydrogen bonding partner, being adenine moiety in orange, nicotinamide in green color, and nucleotide or phosphate group in purple colors. (**d**) Snapshots of NADP-bound YqhD monomer showing cofactor binding site starting from the equilibrated structure in Set 2 trajectory followed by every 25 ns interval up to cofactor release event in the second set of YqhD simulation.

NADP-bound YqhD showed cofactor release in three out of five sets of simulation trajectories, while NADPH-bound monomer exhibited cofactor release only once, in Set 3. The NADPH-bound monomer has a more rigid cofactor binding, which coincides with a conformational change in the cofactor binding pocket. The changes in the cofactor distances from the protein are related to the dynamic nature of cofactor in the binding site, as observed earlier with deviations in RMSD and Rg values of monomer and cofactor during the simulations. A total of 15 hydrogen bonds were observed in the starting structure of the cofactor binding site to the cofactor. Figure 8c shows the residues involved in hydrogen bonding with NADP/H cofactor. Hydrogen bonds involve the adenine moiety with residues Thr138, Tyr179, and Thr182; the dinucleotide moiety and phosphate groups with residues Gly38, Ser40, Gly95, Ser96, His281, and Lys160; and the nicotinamide moiety with residues Asp99 and Gly149. During the simulations, Lys160 has no hydrogen bond with NADP/H cofactor, while His281 shows hydrogen bond existence in 5% of NADP-bound monomer trajectory frames. Excluding these latter two hydrogen bonds present in the starting structure, NADP/H-bound structures have conserved hydrogen bonds involving the adenine moiety, nucleotide and phosphate group in 50% of the aggregated trajectory frames. The NADP-bound monomer has only one hydrogen bond involving Leu279 and nicotinamide moiety for 6% existence in the aggregated trajectory. However, NADPH bound monomer forms three potential hydrogen bonds involving Asp99, Ser144, and Gly149 in ~30% of the aggregated trajectory. In both NADP/H-bound structures, hydrogen bonds involving the adenine moiety were observed more frequently (50% time points) than with nicotinamide moiety (6% in NADP and 30% in NADPH-bound monomers), which also supports our earlier observations of lower Rg and indicates the population of a bent conformation of NADP/H cofactor in open cleft states before release shown in Figure 8d. Overall, these results indicate that cofactor release is associated with an increase in the interdomain cleft (>0.55 nm), followed by loosened cofactor binding at cleft distances of 0.55–1.0 nm and loss of hydrogen bonds involving nicotinamide moiety, a change in cofactor conformation to a more compact structure (lower Rg), and, finally, release.

3. Discussion and Conclusions

Molecular dynamics simulations were performed on the YqhD dimer, with oxidized and reduced NADP/H cofactors bound in aqueous solution. The starting point of the simulations came from the only solved crystal structure of holo YqhD [7], which has a modified $NADPH(OH)_2$ cofactor, and Zn^{2+} in the active site of only one of the monomers within the functional dimer unit. Simulations were run on the functional structure of YqhD, with Zn^{2+} present in each monomer, and with $NADP^+$ and NADPH cofactors, using careful methodology to prepare these structures (i.e., repairing the crystal structure artifacts). Good conformational sampling of YqhD bound to NADP/H was obtained by performing five sets of simulations that were assigned different initial velocities from a Maxwell-Boltzmann velocity distribution at 300 K. The structures remain conserved throughout the simulations, regardless of the oxidation state of NADP/H cofactor. So, we compiled a combined trajectory of the 200 ns from the set of five simulations with both NADP/H cofactors and obtained sampling of various conformations representing opening and closing of the cofactor-binding interdomain cleft. Each monomer of the YqhD dimer showed coordinated cleft opening/closing, with both oxidized and reduced NADP/H, via the movement of $\beta6/\beta7$ and $\alpha12$ regions in the interdomain cleft. The cleft remains open in the apo form of the YqhD crystal structure [7], indicating that it opens and closes for cofactor binding. Observations of cofactor release during MD simulations confirm the movement to an open-cleft conformation prior to the release of both NADP and NADPH. The sampling of open-cleft conformations depends on the cofactor oxidation state. NADP-bound monomers tend to sample more open conformations, with cleft distances ranging from ~0.7–1.8 nm (cf. ~0.7–1.5 for NADPH, 1.25 nm for apoprotein crystal structure [7]).

The dynamics of cleft opening/closing were also found to depend on the oxidation/reduction of the cofactor. With both cofactors, YqhD undergoes periods of rapid cleft opening and closing, sampling open conformations similar to the apoenzyme [7]. NADP-bound monomer scarcely populates conformations with a closed cleft in 10% of time points and remains in a partially closed state for 49% of time points or adopts open cleft conformation in 41% of time points. NADPH-bound monomers remain in closed conformation for 44% and spend only 17% in open state conformations. The maximum waiting time of transition from closed to open state was observed to be ~8 ns in NADP-bound monomer and 68 ns in NADPH-bound monomer. However, NADP-bound monomer underwent a total of 750 transitions between open and closed states, compared to 318 total cleft state transitions observed for NADPH-bound monomer, indicating a higher propensity for cleft opening dynamics when the enzyme is bound to the oxidized cofactor. Out of five independent simulations, cofactor release was observed in three sets of NADP-bound and only one set of NADPH-bound structures. This suggests an entropic, dynamics-based preference for cofactor release when NADP is bound.

Significant differences were observed even in the conformation of NADP/H cofactor, concomitant with conformational differences in the monomers. The differences in cofactor-enzyme interactions and YqhD conformational dynamics may rationalize differences in the Michaelis constant, K_M, which depends on cofactor binding (k_f) and release (k_r) substrate-binding rates as $K_M = (k_r + k_{cat})/k_f$. The higher number of cofactor-monomer hydrogen bonds between NADPH-bound vs. NADP-bound YqhD may lead to higher binding rates and/or slower release rates for NADPH, resulting in the higher affinity for NADPH indicated by its lower K_M value (0.008 mM [11] vs. 0.150 mM [21] for NADP). Furthermore, the less frequent sampling of open conformations with NADPH-bound YqhD may hinder NADPH release, dropping k_r and subsequently K_M for NADPH. Meanwhile, conformational dynamics appear to promote the more facile release of NADP, following hydride transfer from NADPH to aldehydes. Thus, the conformational differences induced by cofactor oxidation state, dynamical effects of more-frequent cofactor cleft opening with NADP, and differences in hydrogen bond motifs may lead to preferential kinetics for the aldehyde reductase activity of YqhD. These findings raise questions about whether enzymes with higher alcohol dehydrogenase activity show a dynamic preference for the release of reduced cofactor. Results of this study enhance our basic understanding toward this class of enzyme, with the possible application of guiding the rational design of YqhD to enhance substrate affinity and biocatalyst efficiency.

4. Materials and Methods

4.1. Starting Coordinates

The crystal structure of YqhD (PDB ID: 1OJ7, 2.0 Å resolution) [7] was used to extract the starting coordinates for setting up MD simulations. The starting coordinates include a dimer (crystallographic monomers A and D, shown in Figure 1a) with bound Zn^{2+}, modified NADP cofactor as $NADPH(OH)_2$ due to oxidative stress, and crystallographic waters. In the crystal structure, only one monomer (D) of the YqhD homodimer has Zn^{2+} in the active site. Herein, monomer D is used to model the starting structure of YqhD homodimer with bound Zn^{2+} and $NADPH(OH)_2$ cofactor using PyMOL molecular graphics system [33]. Monomer D was duplicated and superimposed onto monomer A using the root mean square deviation (RMSD) minimization criteria of PyMOL to generate a homodimer structure with two bound Zn^{2+} ions for MD simulations.

4.2. Modeling of NADP/H Cofactor

In the crystal structure of YqhD homodimer, [7] NADP cofactor is present as $NADPH(OH)_2$ (see Figure 1b), with modification at the C5 and C6 positions of the nicotinamide moiety. The native oxidized cofactor (NADP) was modeled by removing the hydroxyl moieties at the fifth and sixth positions in the nicotinamide ring. The reduced cofactor (NADPH) was prepared by modifying the oxidation state of the C4 atom of the nicotinamide ring. Forcefield parameters come from CHARMM 36 force field [34] for Zn^{2+}, NADP, and NADPH cofactors. Throughout this paper, NADP/H is used to signify the NADPH cofactor generally, whether in its oxidized ($NADP^+$) or reduced (NADPH) state. When a specific oxidation state is indicated, NADP and NADPH are specified.

4.3. Molecular Dynamics Simulations

The CHARMM 36 force field [34] was used for the simulations summarized in Table 1 using the GROMACS software package version 5.1.3 [35]. The YqhD dimer bound to Zn^{2+} and NADP/H cofactors was centered in a cubic periodic box (~11 nm^3) and set to have a distance larger than 1 nm from any side of the box. Solvent molecules having any atom within 0.15 nm from the protein were removed. TIP3P model [36,37] was used for explicit water solvent. Sodium counter ions were added by replacing the solvent molecules at the sites of most negative electrostatic potential to provide the box with a total charge of zero. The protonation state of residues was assumed to be the same as that of the isolated amino acids in solution at pH 7. The LINCS [38] algorithm was used to constrain all bond lengths and the SETTLE [39] algorithm was used for the water molecules. Electrostatic interactions were calculated using the Particle Mesh Ewald method [40]. For the calculation of long-range interactions, a grid spacing of 0.12 nm combined with a fourth-order B-spline interpolation was used to compute the potential and forces between grid points. A non-bonded pair-list cutoff of 1.4 nm was used and updated at every five time-steps. V-rescale thermostat [41] was used to keep the temperature at 300 K through a weak coupling of the system to an external thermal bath with a relaxation time constant $\tau = 0.1$ ps. The pressure of the system was kept at 1 bar using Berendsen's barostat [42] with a time constant of 1 ps. A time step of 2 fs was used to integrate the equations of motion.

Table 1. Simulation summary for the YqhD homodimer in aqueous solution.

Enzyme	Cofactor	Atoms	Water Molecules	Counter Ions	No. of Simulations	Time (ns)
YqhD Dimer	NADP	128,728	40,363	14 Na^+	5	200
YqhD Dimer	NADPH	128,723	40,360	16 Na^+	5	200

First, the simulated systems were energy minimized, using the steepest descent algorithm, for at least 5000 steps to remove clashes between atoms that were too close. After energy minimization, all atoms were given an initial velocity obtained from a Maxwell-Boltzmann velocity distribution at 300 K to start the MD simulations. The system was initially equilibrated by 30 ps with position restraints on the heavy atoms of the dimer to allow relaxation of the solvent molecules. After the equilibration procedure, position restraints were removed, and the system was gradually heated from 50 K to 300 K during 200 ps of simulation. The equilibrated structure was used to perform MD simulations in the NPT ensemble for 20 ns at 300 K. The final conformation of the 20 ns isothermal-isobaric NPT ensemble was used to set up five sets of 200 ns independent simulations in the canonical NVT ensemble initiating with five different velocities.

The starting coordinates were adopted from the crystal structure which has modified cofactor ($NADPH(OH)_2$); hence, the equilibrated conformations of YqhD bound to NADP and NADPH cofactors were obtained after a 20 ns NPT simulation, which was used as the starting point for the final production run of 200 ns long NVT simulations. The starting crystallographic coordinates of the YqhD homodimer were used as the reference structure for the analysis of the trajectories to characterize the conformational changes induced by a change in the oxidation state of NADP/H cofactor. The analysis was focused on three sets/sub-structures within the simulation data, considering dimer, monomer, and domain in a 1 μs aggregated trajectory for each cofactor oxidation state (NADP and NADPH). Structural analysis includes the root mean square deviation (RMSD), root mean square fluctuation (RMSF), radius of gyration (Rg), and cofactor binding interactions (Zn^{+2} and NADP/H), comparing these with respect to the crystal structure.

4.4. Cluster Analysis

The conformational diversity of the structures generated during the MD simulations was characterized using the Gromos [43] clustering algorithm. In this method, an RMSD cutoff criterion is used to assign a structure in a cluster based on the root-mean-square differences of selected atoms among the conformations obtained from the simulations. An RMSD cutoff of 0.13 nm was used to determine neighboring backbone atom conformations of YqhD monomers to discriminate among the less varying conformations. For the cluster analysis, a total of 52,815 structures were evaluated for each monomer within the dimer over the 2 μs aggregated trajectory, using time intervals of 100 ps. [33]

Author Contributions: R.V. and K.R.M.-K. designed the project and wrote the manuscript. R.V. and J.M.E. performed the simulations. R.V. analyzed the data. All authors have read and agreed to the published version of the manuscript.

Funding: This material is based upon work supported by the National Science Foundation under Award No. EPS-0903806 and matching support from the State of Kansas through the Kansas Board of Regents. This project was supported by grants from the National Institute of General Medical Sciences

(P20 GM103418) from the National Institutes of Health. The content is solely the responsibility of the authors and does not necessarily represent the official views of the National Institute of General Medical Sciences or the National Institutes of Health.

Data Availability Statement: The data presented in this study are available on request from the corresponding author.

Acknowledgments: This research is supported by Wichita State University (WSU), Department of Chemistry and Fairmount College of Liberal Arts and Sciences. This work was performed using the High Performance Computing Center at WSU, which was supported by the National Science Foundation under Grant No. EIA-0216178, and Grant No. EPS-0236913, with matching support from the State of Kansas and WSU.

Conflicts of Interest: The authors declare that they have no conflicts of interest with the contents of this article.

Sample Availability: Not applicable.

References

1. Emptage, M.; Haynie, S.L.; Laffend, L.A.; Pucci, J.P.; Whited, G. Process for the Biological Production of 1,3-Propanediol with High Titer. U.S. Patent 6,514,733B1, 4 February 2003.
2. Jarboe, L.R. YqhD: A broad-substrate range aldehyde reductase with various applications in production of biorenewable fuels and chemicals. *Appl. Environ. Microbiol.* **2011**, *89*, 249–257. [CrossRef] [PubMed]
3. Atsumi, S.; Wu, T.Y.; Eckl, E.M.; Hawkins, S.D.; Buelter, T.; Liao, J.C. Engineering the isobutanol biosynthetic pathway in *Escherichia coli* by comparison of three aldehyde reductase/alcohol dehydrogenase genes. *Appl. Microbiol. Biotechnol.* **2010**, *85*, 651–657. [CrossRef] [PubMed]
4. Vaidyanathan, H.; Kandasamy, V.; Gopal Ramakrishnan, G.; Ramachandran, K.; Jayaraman, G.; Ramalingam, S. Glycerol conversion to 1, 3-Propanediol is enhanced by the expression of a heterologous alcohol dehydrogenase gene in *Lactobacillus reuteri*. *AMB Express* **2011**, *1*, 37. [CrossRef] [PubMed]
5. Abergel, C.; Coutard, B.; Byrne, D.; Chenivesse, S.; Claude, J.B.; Deregnaucourt, C.; Fricaux, T.; Gianesini-Boutreux, C.; Jeudy, S.; Lebrun, R.; et al. Structural genomics of highly conserved microbial genes of unknown function in search of new antibacterial targets. *J. Struct. Funct. Genomics* **2003**, *4*, 141–157. [CrossRef] [PubMed]
6. Montella, C.; Bellsolell, L.; Perez-Luque, R.; Badia, J.; Baldoma, L.; Coll, M.; Aguilar, J. Crystal structure of an iron-dependent group III dehydrogenase that interconverts L-lactaldehyde and L-1,2-propanediol in *Escherichia coli*. *J. Bacteriol.* **2005**, *187*, 4957–4966. [CrossRef]
7. Sulzenbacher, G.; Alvarez, K.; Van Den Heuvel, R.H.; Versluis, C.; Spinelli, S.; Campanacci, V.; Valencia, C.; Cambillau, C.; Eklund, H.; Tegoni, M. Crystal structure of *E.coli* alcohol dehydrogenase YqhD: Evidence of a covalently modified NADP coenzyme. *J. Mol. Biol.* **2004**, *342*, 489–502. [CrossRef]
8. Chen, Z.; Liu, H.J.; Liu, D.H. Decrease the accumulation of 3-hydroxypropionaldehyde for 1,3-propanediol production by expressing the Yqhd gene in *Klebsiella pneumonia*. *J. Biotechnol.* **2008**, *136*, S354. [CrossRef]
9. Perez, J.M.; Arenas, F.A.; Pradenas, G.A.; Sandoval, J.M.; Vasquez, C.C. Escherichia coli YqhD exhibits aldehyde reductase activity and protects from the harmful effect of lipid peroxidation-derived aldehydes. *J. Biol. Chem.* **2008**, *283*, 7346–7353. [CrossRef]
10. Turner, P.C.; Miller, E.N.; Jarboe, L.R.; Baggett, C.L.; Shanmugam, K.T.; Ingram, L.O. YqhC regulates transcription of the adjacent *Escherichia coli* genes yqhD and dkgA that are involved in furfural tolerance. *J. Ind. Microbiol. Biotechnol.* **2011**, *38*, 431–439. [CrossRef]
11. Miller, E.N.; Jarboe, L.R.; Yomano, L.P.; York, S.W.; Shanmugam, K.T.; Ingram, L.O. Silencing of NADPH-dependent oxidoreductase genes (yqhD and dkgA) in furfural-resistant ethanologenic *Escherichia coli*. *Appl. Environ. Microbiol.* **2009**, *75*, 4315–4323. [CrossRef]
12. Wang, X.; Yomano, L.P.; Lee, J.Y.; York, S.W.; Zheng, H.B.; Mullinnix, M.T.; Shanmugam, K.T.; Ingram, L.O. Engineering furfural tolerance in *Escherichia coli* improves the fermentation of lignocellulosic sugars into renewable chemicals. *Proc. Natl. Acad. Sci. USA* **2013**, *110*, 4021–4026. [CrossRef] [PubMed]
13. Koma, D.; Yamanaka, H.; Moriyoshi, K.; Ohmoto, T.; Sakai, K. Production of Aromatic Compounds by Metabolically Engineered *Escherichia coli* with an Expanded Shikimate Pathway. *Appl. Environ. Microbiol.* **2012**, *78*, 6203–6216. [CrossRef] [PubMed]
14. Iwayanagi, T.; Miyamoto, S.; Konno, T.; Mizutani, H.; Hirai, T.; Shigemoto, Y.; Gojobori, T.; Sugawara, H. TP Atlas: Integration and dissemination of advances in Targeted Proteins Research Program (TPRP)—Structural biology project phase II in Japan. *J. Struct. Funct. Genomics* **2012**, *13*, 145–154. [CrossRef] [PubMed]
15. Voelker, F.; Dumon-Seignovert, L.; Soucaille, P. Mutant Yqhd Enzyme for the Production of a Biochemical by Fermentation. U.S. Patent 8,969,053B2, 3 March 2015.
16. Zhu, H.L.; Yi, X.Y.; Liu, Y.; Hu, H.B.; Wood, T.K.; Zhang, X.H. Production of acetol from glycerol using engineered *Escherichia coli*. *Bioresour. Technol.* **2013**, *149*, 238–243. [CrossRef] [PubMed]

17. Clomburg, J.M.; Gonzalez, R. Metabolic Engineering of *Escherichia coli* for the Production of 1,2-Propanediol From Glycerol. *Biotechnol. Bioeng.* **2011**, *108*, 867–879. [CrossRef] [PubMed]
18. Lan, E.I.; Liao, J.C. ATP drives direct photosynthetic production of 1-butanol in cyanobacteria. *Proc. Natl. Acad. Sci. USA* **2012**, *109*, 6018–6023. [CrossRef]
19. Li, H.; Chen, J.; Li, Y. Enhanced activity of yqhD oxidoreductase in synthesis of 1,3-propanediol by error-prone PCR. *Pro. Nat. Sci.* **2008**, *18*, 1519–1524. [CrossRef]
20. Rao, Z.; Ma, Z.; Shen, W.; Fang, H.; Zhuge, J.; Wang, X. Engineered *Saccharomyces cerevisiae* that produces 1,3-propanediol from d-glucose. *J. Appl. Microbiol.* **2008**, *105*, 1768–1776. [CrossRef]
21. Tang, X.M.; Tan, Y.S.; Zhu, H.; Zhao, K.; Shen, W. Microbial Conversion of Glycerol to 1,3-Propanediol by an Engineered Strain of *Escherichia coli*. *Appl. Environ. Microbiol.* **2009**, *75*, 1628–1634. [CrossRef]
22. Wang, F.H.; Qu, H.J.; Zhang, D.W.; Tian, P.F.; Tan, T.W. Production of 1,3-propanediol from glycerol by recombinant *E. coli* using incompatible plasmids system. *Mol. Biotechnol.* **2007**, *37*, 112–119. [CrossRef]
23. Elleuche, S.; Fodor, K.; von der Heyde, A.; Klippel, B.; Wilmanns, M.; Antranikian, G. Group III alcohol dehydrogenase from *Pectobacterium atrosepticum*: Insights into enzymatic activity and organization of the metal ion-containing region. *Appl. Environ. Microbiol.* **2014**, *98*, 4041–4051. [CrossRef] [PubMed]
24. Elleuche, S.; Klippel, B.; von der Heyde, A.; Antranikian, G. Comparative analysis of two members of the metal ion-containing group III-alcohol dehydrogenases from *Dickeya zeae*. *Biotechnol. Lett.* **2013**, *35*, 725–733. [CrossRef] [PubMed]
25. Verma, R.; Schwaneberg, U.; Roccatano, D. Conformational Dynamics of the FMN-Binding Reductase Domain of Monooxygenase P450BM-3. *J. Chem. Theory Comput.* **2013**, *9*, 96–105. [CrossRef] [PubMed]
26. Sellés Vidal, L.; Kelly, C.L.; Mordaka, P.M.; Heap, J.T. Review of NAD(P)H-dependent oxidoreductases: Properties, engineering and application. *Biochimica Biophys. Acta (BBA)* **2018**, *1866*, 327–347. [CrossRef] [PubMed]
27. Zhu, J.-G.; Li, S.; Ji, X.-J.; Huang, H.; Hu, N. Enhanced 1,3-propanediol production in recombinant *Klebsiella pneumoniae* carrying the gene yqhD encoding 1,3-propanediol oxidoreductase isoenzyme. *World J. Microbiol. Biotechnol.* **2009**, *25*, 1217. [CrossRef]
28. Verma, R.; Mitchell-Koch, K. In Silico Studies of Small Molecule Interactions with Enzymes Reveal Aspects of Catalytic Function. *Catalysts* **2017**, *7*, 212. [CrossRef]
29. Cummins, P.L.; Ramnarayan, K.; Singh, U.C.; Gready, J.E. Molecular dynamics/free energy perturbation study on the relative affinities of the binding of reduced and oxidized NADP to dihydrofolate reductase. *J. Am. Chem. Soc.* **1991**, *113*, 8247–8256. [CrossRef]
30. Blikstad, C.; Dahlstrom, K.M.; Salminen, T.A.; Widersten, M. Substrate scope and selectivity in offspring to an enzyme subjected to directed evolution. *FEBS J.* **2014**, *281*, 2387–2398. [CrossRef]
31. Luo, J.; Bruice, T.C. Dynamic Structures of Horse Liver Alcohol Dehydrogenase (HLADH): Results of Molecular Dynamics Simulations of HLADH-NAD^+-$PhCH_2OH$, HLADH-NAD^+-$PhCH_2O^-$, and HLADH-NADH-PhCHO. *J. Am. Chem. Soc.* **2001**, *123*, 11952–11959. [CrossRef]
32. Oyen, D.; Fenwick, R.B.; Stanfield, R.L.; Dyson, H.J.; Wright, P.E. Cofactor-Mediated Conformational Dynamics Promote Product Release From *Escherichia coli* Dihydrofolate Reductase via an Allosteric Pathway. *J. Am. Chem. Soc.* **2015**, *137*, 9459–9468. [CrossRef]
33. *The PyMOL Molecular Graphics System, Version 1.8*; Schrodinger LLC.: New York, NY, USA, 2015.
34. Best, R.B.; Zhu, X.; Shim, J.; Lopes, P.E.; Mittal, J.; Feig, M.; Mackerell, A.D., Jr. Optimization of the additive CHARMM all-atom protein force field targeting improved sampling of the backbone phi, psi and side-chain chi(1) and chi(2) dihedral angles. *J. Chem. Theory Comput.* **2012**, *8*, 3257–3273. [CrossRef] [PubMed]
35. Hess, B.; Kutzner, C.; van der Spoel, D.; Lindahl, E. GROMACS 4: Algorithms for highly efficient, load-balanced, and scalable molecular simulation. *J. Chem. Theory. Comput.* **2008**, *4*, 435–447. [CrossRef] [PubMed]
36. Jorgensen, W.L.; Chandrasekhar, J.; Madura, J.D.; Impey, R.W.; Klein, M.L. Comparison of simple potential functions for simulating liquid water. *J. Chem. Phys.* **1983**, *79*, 926–935. [CrossRef]
37. Neria, E.; Fischer, S.; Karplus, M. Simulation of activation free energies in molecular systems. *J. Chem. Phys.* **1996**, *105*, 1902–1921. [CrossRef]
38. Hess, B.; Bekker, H.; Berendsen, H.J.C.; Fraaije, J.G.E.M. LINCS: A linear constraint solver for molecular simulations. *J. Comput. Chem.* **1997**, *18*, 1463–1472. [CrossRef]
39. Miyamoto, S.; Kollman, P.A. Settle - an Analytical Version of the Shake and Rattle Algorithm for Rigid Water Models. *J. Comput. Chem.* **1992**, *13*, 952–962. [CrossRef]
40. Darden, T.; York, D.; Pedersen, L. Particle Mesh Ewald - an N.Log(N) Method for Ewald Sums in Large Systems. *J. Chem. Phys.* **1993**, *98*, 10089–10092. [CrossRef]
41. Bussi, G.; Donadio, D.; Parrinello, M. Canonical sampling through velocity rescaling. *J. Chem. Phys.* **2007**, *126*, 014101. [CrossRef]
42. Berendsen, H.J.C.; Postma, J.P.M.; Vangunsteren, W.F.; Dinola, A.; Haak, J.R. Molecular-Dynamics with Coupling to an External Bath. *J. Chem. Phys.* **1984**, *81*, 3684–3690. [CrossRef]
43. Daura, X.; Gademann, K.; Jaun, B.; Seebach, D.; van Gunsteren, W.F.; Mark, A.E. Peptide folding: When simulation meets experiment. *Angew. Chem. Int. Ed.* **1999**, *38*, 236–240. [CrossRef]

Article

Binding Ensembles of *p53*-MDM2 Peptide Inhibitors by Combining Bayesian Inference and Atomistic Simulations

Lijun Lang and Alberto Perez *

Chemistry Department, University of Florida, Gainesville, FL 32611, USA; lijunlang@chem.ufl.edu
* Correspondence: perez@chem.ufl.edu; Tel.: +1-352 3927009

Abstract: Designing peptide inhibitors of the *p53*-MDM2 interaction against cancer is of wide interest. Computational modeling and virtual screening are a well established step in the rational design of small molecules. But they face challenges for binding flexible peptide molecules that fold upon binding. We look at the ability of five different peptides, three of which are intrinsically disordered, to bind to MDM2 with a new Bayesian inference approach (MELD×MD). The method is able to capture the folding upon binding mechanism and differentiate binding preferences between the five peptides. Processing the ensembles with statistical mechanics tools depicts the most likely bound conformations and hints at differences in the binding mechanism. Finally, the study shows the importance of capturing two driving forces to binding in this system: the ability of peptides to adopt bound conformations ($\Delta G_{conformation}$) and the interaction between interface residues ($\Delta G_{interaction}$).

Keywords: IDP 1; binding 2; molecular dynamics 3; MELD×MD 4; advanced sampling 5; *p53* 6; MDM2 7

Citation: Lang, L.; Perez, A. Binding Ensembles of *p53*-MDM2 Peptide Inhibitors by Combining Bayesian Inference and Atomistic Simulations. *Molecules* **2021**, *26*, 198. https://doi.org/10.3390/molecules26010198

Academic Editor: Marilisa Leone
Received: 13 November 2020
Accepted: 28 December 2020
Published: 2 January 2021

Publisher's Note: MDPI stays neutral with regard to jurisdictional claims in published maps and institutional affiliations.

Copyright: © 2020 by the authors. Licensee MDPI, Basel, Switzerland. This article is an open access article distributed under the terms and conditions of the Creative Commons Attribution (CC BY) license (https://creativecommons.org/licenses/by/4.0/).

1. Introduction

Peptide molecule inhibitors have the potential to bind to proteins classified as "undruggable" by small molecules thanks to their flexibility and complementary nature to proteins [1,2]. Rational drug design of small molecules via computational tools (e.g., docking of virtual libraries) is a common practice in the drug discovery process. However, these tools are not well suited to handle the flexible nature of peptide molecules, many of which are intrinsically disordered and only adopt stable structures in the presence of their binding partners [3].

Modeling the binding of flexible molecules continues to be a grand challenge in computational structure prediction. In recent years, with the increase of peptide therapeutics in the market there has been a continuous development and adaptation of docking tools to capture protein-peptide interactions [3,4]. Docking programs address the flexibility of peptides by two main routes: (1) using homology models, PDB (Protein Data Bank)structural motifs, or other sources of structures for docking [5–8]; and (2) provide peptide flexibility for folding upon binding [9–14]. Initial peptide conformations for docking could come from computationally expensive molecular dynamics (MD) simulations of the free peptide. However, many such peptides are intrinsically disordered (IDP), limiting their use [7]. Full exploration of folding upon binding through standard molecular dynamics becomes too computationally demanding [15], requiring advanced sampling strategies to efficiently sample the energy landscape.

In this work, we take a look at binding and free-peptide ensembles (simulating the peptide in isolation) for different peptides to better understand the nature of the *p53*-MDM2 interaction. *p53* is called the guardian of the genome, triggering programmed death (apoptosis) when cells misbehave. MDM2 down-regulates *p53* limiting its tumor suppressor activity. Thus, inhibitors of the *p53*-MDM2 and the closely related MDMX interaction have long been a cancer drug target [16–18]. Multiple studies of the native interaction [19–22] and the ability to design inhibitors that simultaneously block MDM2 and MDMX [23–25] provide a wealth of data to assess new computational tools. Since binding simulations are more computationally demanding than free peptide simulations, our goal is to identify peptide properties that might make the peptide a better binder–leading to faster computational screening of peptide therapeutics.

The *p53*-MDM2 interaction is characterized by three hydrophobic residues (Phe19, Trp23 and Leu26) from the peptide which anchor into a deep cavity in MDM2. In order for the three hydrophobic residues to align with the pocket, the *p53* epitope adopts a helical conformation. This is in contrast with the IDP nature of the peptide in isolation. We use noisy information to guide binding using our previously developed Bayesian inference approach (MELD×MD [26]) to identify the subset of data that is most compatible with the force field and the resulting bound conformations (see Figure 1). To further test the methodology, we simulated five different peptides, including the peptide epitope from *p53*, two inhibitors, and two alanine-based peptides that we do not expect to be good binders, as control. The work highlights the ability of molecular dynamics tools to capture the two driving forces behind binding: preferences of the peptides to adopt bound-like conformations and the use of binding simulations to differentiate binding preferences.

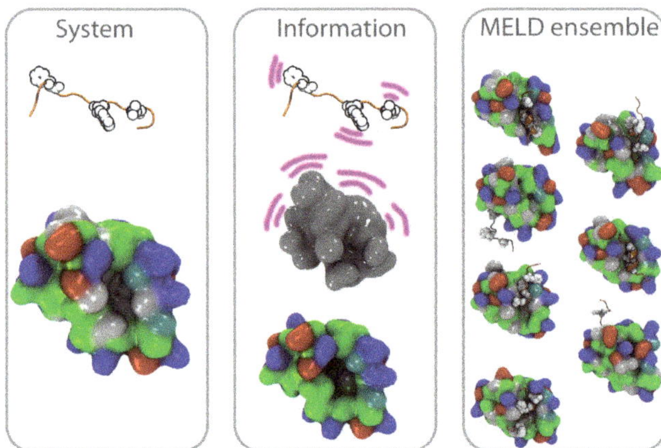

Figure 1. Outline of the MELD×MD setup. We start with the peptide far from MDM2 (system). We use noisy information to favor sampling of binding/unbinding events (middle panel). A statistical mechanics of the posterior distribution coming from the MELD ensemble identifies conformations that are most consistent with the force field and a subset of the data, and we compare these to the experimental structure.

2. Results

2.1. Free Peptide Simulation Ensembles Show the IDP Nature of p53

We simulated five peptides in their free form (see Methods and Table 1), capturing their intrinsic degree of disorder. All peptides are able to visit multiple states with short life times. A 2D-RMSD clustering of the ensemble reveals many clusters with low populations for the *p53* and two control peptides, consistent with their intrinsically disordered nature (see unrestrained Molecular Dynamics (MD) column in Table 2). The peptide *pdiq* adopts stable helix conformations for a significant amount of time, while *ATSP*-7041 is an outlier in this analysis, adopting very stable helical conformations due to the presence of a chemical staple. We used these simulations to define a common reference frame to compare simulations for all peptides in their free and binding simulations (see Methods). Each peptide ensemble was projected onto the corresponding eigenvectors that showed a good separation between helical and non-helical states—as those are the states required for binding (see left panels in Figures 2 and A1–A4). Clustering on the space defined by the top 14 eigenvectors shows that only *ATSP*-7041 and, to a lesser extent, *pdiq* adopt stable helical structures—consistent with the IDP nature of the other three peptides.

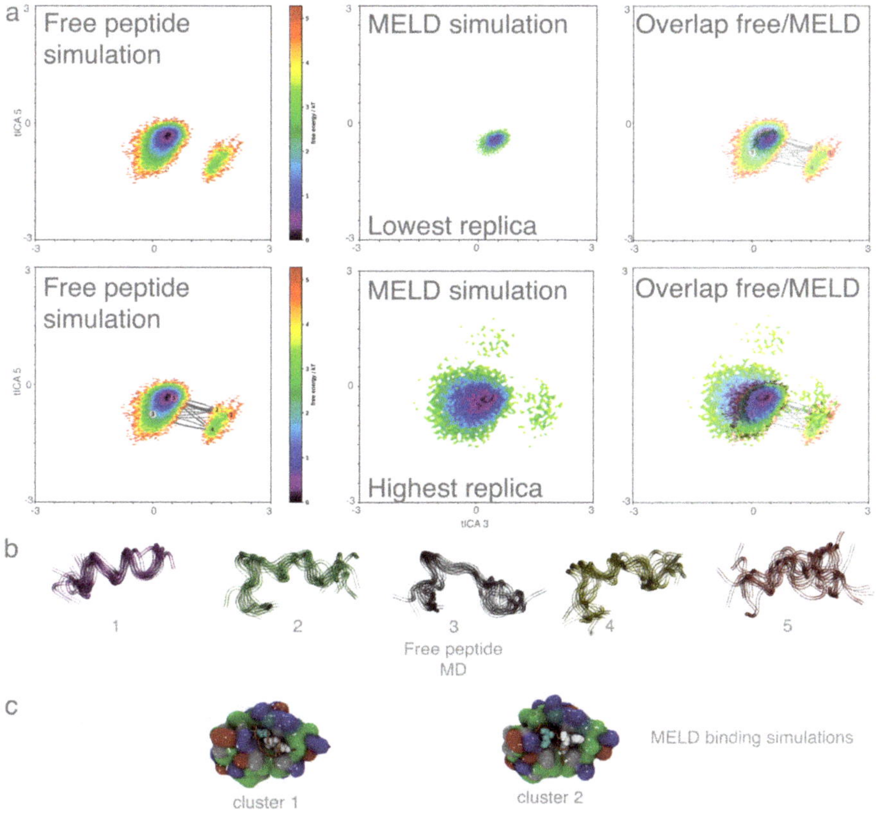

Figure 2. Comparison of the conformational space for free peptide versus binding simulations for the *ATSP*-7041 peptide. (**a**) The peptide ensembles are projected onto the third and fifth tICA eigenvectors common to all five peptides. (**b**) The metastable states sampled for the free peptide. (**c**) Top clusters by population from MELD×MD binding simulations.

Table 1. Peptides used in the current work. Bold letters represent the anchoring residues.

Name	Sequence																
p53	S	Q	E	T	**F**	S	D	L	**W**	K	L	**L**	P	E	N		
pdiq			E	T	**F**	E	H	W	**W**	S	Q	**L**	L	S			
Ala1			A	A	**F**	A	A	A	**W**	A	A	**L**	A	A			
Ala2			A	A	A	A	A	A	A	A	A	A	A	A			
ATSP-7041		ACE	L	T	**F**	R8	E	Y	**W**	A	Q	**Cba**	S5	S	A	A	NHE

Table 2. Populations for peptides in free and MELD binding simulations. Clustering is done on the lowest temperature replica using hierarchical clustering with $\epsilon = 1.5$.

Name	Peptide Population (% Top Cluster)		
	Unrestrained MD	MELD×MD (Peptide Align)	MELD×MD (Protein Align)
p53	0.6	70.6	46.1
pdiq	24.0	97.6	95.3
Ala1	1.4	54.7	16.0
Ala2	0.2	31.3	17.5
ATSP-7041	69.5	97.8	91.6

2.2. MELD×MD Simulations Balance Exploration and Exploitation of the Binding Energy Landscape

Figure 3 provides a visual outlook on the binding process explored by the MELD×MD replica exchange procedure in terms of the relative position of the peptide with respect the protein and the peptide's intrinsic conformational preferences. At high replica indexes, the force constants for the restraints are set to zero and the temperature is high (see methods). In these conditions, the peptide samples conformations far away from the active site, distributed uniformly around the protein. During the binding process, the MDM2 flexibility allows for the opening and closing of the binding cavity (see right panel in Figure A5). As the replica index decreases, the temperature decreases and the biasing restraints towards the protein become active, producing a frustrated energy landscape. Under these conditions, the peptide samples conformations on the surface of the protein, identifying early on the MDM2 hydrophobic pocket as the most likely region for binding. Sampling is concentrated in the binding pocket at the lowest replica. Thus, at the highest replica, the protocol favors full exploration of the energy landscape, while, at the lowest replica, it favors full exploitation by sampling around a particular binding region near the protein. The nature of MELD×MD enhances binding/unbinding events by allowing replicas to explore different Hamiltonian and temperature conditions, leading to a different balance of exploration and exploitation [27,28].

2.3. Peptides Become More Structured in Proximity to MDM2

MELD×MD binding simulations show a higher fraction of helical conformations for all peptides with respect to their free simulations (see middle column in Table 2). However, the increase in helical content for the peptide is not always associated with binding at the correct binding site (right column in Table 2). A 2D-RMSD clustering calculation on all replicas (see Figures 4 and A6) reveals the funneled nature of binding for three peptides. We can identify three broadly defined regions in the funneling plots based on the RMSD distribution: between 0–5 Å (high accuracy binding), 5–15 Å (pre-bound), ~15–30 Å (misbound), and a fourth region for unbound conformations sampled by higher replicas (see Figure A5). All five peptides identify the binding pocket as the binding site, but the two control sequences bind through multiple backbone conformations with little structural preference. *ATSP*-7041 exhibits the most funneled behavior, rapidly converging onto a large high accuracy native-like cluster. Both *pdiq* and *p53* exhibit a similar behavior, in which all three regions are explored even at the lower replicas, with funneling to one major state. For *p53*, the native configuration is sampled, but is not identified as the most populated cluster. The observed binding mode introduces a kink in the backbone between the helical and non-helical region that is not observed in the experimental structure. The control *Ala1* sequence contains the three anchoring residues present in *p53*, but exhibits a binding profile more similar to control *Ala2*, which lacks the anchoring residues. Thus, the control sequences show that the MELD×MD setup is not over-constraining the peptide to bind in the binding pocket or in the binding conformation, and large cluster populations are reflective of significant binding.

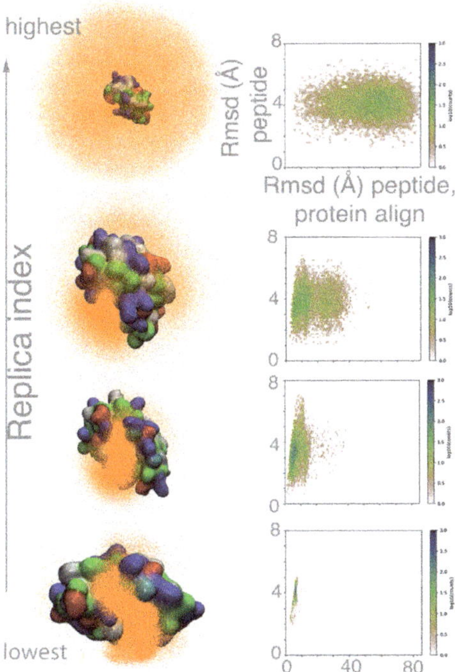

Figure 3. MELD×MD simulations explore unbound states, and different possible binding regions of *p53* on the surface of the protein. The left panel shows a superposition of all peptide conformations (heavy atoms of each conformation are drawn as orange dots) sampled at different replicas. The right panel shows the internal backbone RMSD of the peptide with respect the experimental conformation versus the RMSD of the peptide when aligning to the protein.

We compare all peptide binding ensembles on equal footing by projecting them on the same eigenvector space as the free peptides. Figure 2 compares the free peptide ensemble with those produced from MELD×MD at the lowest/highest replica index (bound/unbound) for *ATSP*-7041. The figure also shows the clusters arising from the free peptide ensemble, as well as the highest population clusters, from the binding simulations. The preferred conformation for *ATSP*-7041 in its free peptide is the same conformation needed for binding, resulting in significant binding observed throughout the simulations. A similar behavior is observed for *pdiq*, where the ensemble of the free peptide is larger due to the absence of the chemical staple (see Figure A1). For the three IDP peptides (see Figures A2–A4), the ensembles are even broader than for *pdiq* resulting in a larger number of clusters. The free peptide clusters for these IDP peptides are low in population and lack agreement with the preferred binding mode. However, in binding simulations at low temperature, *p53* explores a narrow conformational ensemble similar to *pdiq* and very different from the broader ensembles sampled by the control sequences. For the two control sequences, the minima of the free ensemble distribution is displaced with respect to the three other peptides, disfavoring bound-like conformations, which result in broader ensembles for the two control peptides during MELD×MD binding simulations.

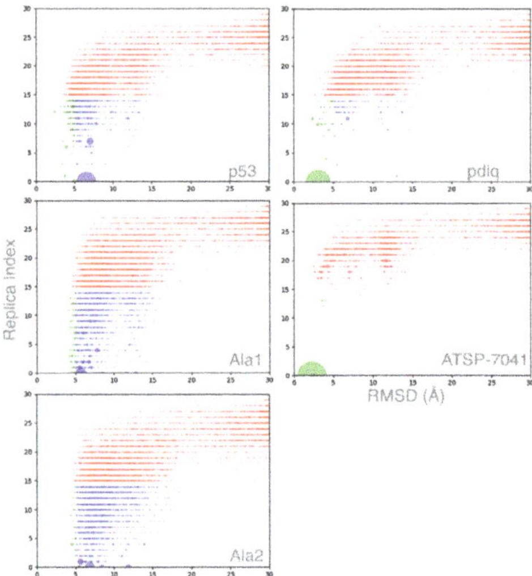

Figure 4. Funneling binding plot for the five peptides. Each dot corresponds to a cluster center from a 2D-RMSD based on all replicas. The larger the circle the larger the population of the cluster. Each circle is plotted at the average RMSD inside that cluster with respect to the native conformation and the mode of the index replica in that cluster. The color code is green ($RMSD < 5$) or blue ($RMSD > 5$) when the mode of the replica index is lower than 15, and red otherwise.

Complementary knowledge for the binding process emerges from looking at the internal structure of the peptide (radius of gyration) with respect to the position of the peptide to MDM2 (RMSD, see right column in Figures 3 and A7). At high replica index, all peptides sample conformations far from the protein, with large fluctuations in the radius of gyration (between 5 and 12 Å except for the *ATSP*-7041 peptide, where the chemical staple prevents conformations with a radius of gyration above 9 Å). When binding in the MDM2 hydrophobic pocket, the peptide adopts compact conformations with a radius of gyration around 7 Å. This happens early in the binding process (higher replica index) for the *pdiq* and *ATSP*-7041 peptides and is not observed for the *Ala2* control due to the lack of anchoring residues.

2.4. Helical Propensities Show Different Binding Patterns

The binding ensembles produce a higher helical content with respect to the free-peptide simulations. Figure A8 shows the three anchoring residues to be predominantly in helical conformations for *pdiq* and *ATSP*-7041. *p53* is well known to make a helix in the N-terminal region of the peptide which our simulations reproduce. The *Ala1* control sequence, which has three anchoring residues, is also able to adopt helical conformations, to a lesser extent. The spacing of the anchoring residues in the sequence (residues i, $i+4$ and $i+7$) and the size of the binding site favor helical conformations for the simultaneous interaction inside the active site. However, the *Ala2* control, adopts very small amounts of helical conformations for two of the three anchoring regions, consistent with the lack of anchoring residues to stabilize those conformations. Not surprisingly, the peptides with larger helical content also have narrower conformational ensembles at low replica index (*ATSP*-7041 and *pdiq*), whereas *p53*, *Ala1*, and *Ala2* have progressively larger ensembles at the lowest temperature replica.

2.5. MDM2 Exhibits a sMall Conformational Change upon Binding

There is a small conformational rearrangement of the backbone (1.9 Å RMSD) between the apo (unbound) and holo (bound) crystal structures (1z1m [29] and 1ycr [30], respectively), which opens up the cavity for binding. Sidechain rearrangement of the surface residues happens on a faster timescale, changing the surface accessibility to the binding cavity. In MELD×MD binding, the conformational freedom of MDM2 by using flat-bottom harmomic restraints on the C_α around the holo structure to prevent unfolding in the replica exchange ladder (see methods). At high temperatures, we find that the protein is sampling conformations between 2–3.5 Åbackbone RMSD from the holo structure and a similar range (2.5–4.0 Å) with respect to the apo MDM2 structure (see Figure A9). At low temperatures, the thermal ensemble is narrower, with an RMSD in between 1 to 2 Åfrom the holo structure and 2–4 Å from the apo structure. The presence of the peptide binding for a significant amount of time to the active site further shifts the RMSD to lower values (see *pdiq* and *ATSP*-7041 in Figure A9). There are no restraints on the sidechains, in which fluctuations determine the open/closed state of the cavity. Reorientation of these sidechains is fast and adapts to the presence of the peptide near the active site.

3. Discussion

The debate between conformational selection and induced fit mechanisms of binding is being reconciled into a mixture of the two [7], with different balance of each depending of the system. In the MDM2/p53 system, the protein undergoes a small conformational change from its apo (unbound) to holo (bound) structure, whereas the IDP peptide folds upon binding to the active site. MD simulations of the free *p53* sequence shows its intrinsically disordered nature, with little propensity for helical conformations. Querying the *p53* binding motif in the PDB returns twelve structures, covering four different protein targets: MDM2 [30], MDMX [20], p300 [31], and the CREB-binding protein [32]. In all cases, the *p53* N-terminal domain adopts a helical conformation, but with different sidechain rotameric states [32]. Our binding simulations reproduce these trend, with the size of the conformational ensemble considerably shrinking upon binding (see Figures A2 and A7). Our MELD×MD simulations lose the kinetic information of binding, but, following a particular replica as it goes up and down the replica ladder, we can observe the series of events that lead to binding. In general, we observe a higher helical content for *p53* near the protein surface, consistent with excluded volume effects [33]. The spacing of the anchoring residues (i, $i+4$ and $i+7$), combined with the size of the hydrophobic pocket favors binding in helical conformations. The funneling towards the binding site driven by hydrophobic interactions in MELD responds to our knowledge that the hydrophobic anchoring residues were vital for binding. Hydrophobic residues on the surface of the protein are more likely in the active site, hence accelerating binding towards it. A different choice of information (e.g., using polar residues) would have resulted in less directed sampling, as polar and charged residues are frequent in the protein surface. Visual inspection shows binding through different modes, with one of the hydrophobic residues anchoring in the cavity and driving the rest; although the peptide does not bind as a helix, it quickly adopts a partial helical structure (see Figure A8) consistent with experiments. The tryptophan seems to be critical for adopting the correct experimental binding conformations: we observe many instances of the peptide bound in the cavity as a helix with the bulky tryptophan interacting with MDM2 hydrophobic sidechains not in the binding pocket, leading to kinks in the backbone structure (see cluster 1 in Figure A2). These incorrect bindings have a long life time in our simulations and require at least partial unbinding before correctly binding according to the experimental motif, which, in our simulations, is sampled but is not the predominant state.

The *pdiq* inhibitor binds experimentally with longer helical motifs covering all amino acids. Analyzing PDB codes 1ycr and 3jzs reveals differences in the secondary structure (using STRIDE [34] as incorporated in VMD (Visual Molecular Dynamics) [35]). The difference in helicity affects the last anchoring residue (leucine), which is in a coil rotameric state for *p53* and in a helical state for *pdiq*. Our simulations show that *pdiq* forms a significant amount of helix in its free form, which favors binding significantly in our binding simulations (see Figures A1 and A8). For *pdiq*, we observe pre-binding in helical conformations, shifted with respect to the experimental binding site, and fast rearrangement of the peptide, sometimes involving rapid helix unfolding and refolding in the binding site leading to the experimental structure. The helical behavior is further accentuated for the *ATSP*-7041 inhibitor, where all binding takes place through helical conformations thanks to the chemical staple. Rearrangements in the active site involve displacements of the alpha helix to allow better interaction of the alpha helices; this is done through partial unbinding of the helix without loosing the helical character. Both control sequences access the binding site unfolded and explore many possible conformations. *Ala1* can sample the helical conformations which favor strong binding as seen from the top clusters (see Figure A3), but they have significantly lower population than the other three binding peptides (see Figure 4). On the contrary, *Ala2* has no anchoring residues and is rarely observed adopting helical conformations in the binding site (see Figure A4). These observations are supported by looking at the narrow conformational binding ensembles sampled at the lowest temperature replica for the peptides that bind (see Figures 2, A1, and A2), in contrast with the broader ensembles of the control peptides (see Figures A3 and A4). Taken together, the results show that the anchoring residues are necessary to adopt the helical conformations associated with good binding to MDM2 but not enough on their own to promote this helical state.

The ensembles at different replica index depict the nature of the binding/unbinding dynamics. The control peptides rapidly lose any memory of the bound conformation as the replica index increases (see Figures A7 and A10). For the other three peptides, we observe a bimodal distribution of states: for *p53* centered at 2/4 Å and at 1/2 Å for *pdiq* and *ATSP*-7041. As we increase the temperature in the replica ladder, the higher RMSD state becomes more prevalent. By replica 20, all peptides are sampling broad distributions with very low populations of the 1 RMSD state for the peptide, except for *ATSP*-7041, which, due to the chemical staple, even at high temperatures can sample conformations close to the bound conformation. However, at the highest replica, all peptides have lost memory of the bound conformation. Overall, we can distinguish three regions: an unbound conformation in which the peptide explores conformations far from its bound conformation, a pre-bound conformation, and a bound conformation. Both pre-bound and bound conformation lock the protein close to its holo conformation. For the poly-ala peptide, the bound state is rarely seen, while, for the poly-ala with binding side-chains, it is easier but not the predominant state. The pre-bound state for *pdiq* and *ATSP*-7041 is very close to the bound conformations, while, for *p53*, it is further away.

The binding free energy can be separated into a contribution coming from the conformational preferences of the peptide and protein systems, as well as an interaction contribution ($\Delta G_{bind} = \Delta G_{conformation} + \Delta G_{interaction}$), where $\Delta G_{conformation}$ can further be separated into the protein and peptide contributions ($\Delta G_{conformation} = \Delta G^{peptide}_{conformation} + \Delta G^{protein}_{conformation}$). $\Delta G_{interaction}$ is dictated by the specific interactions between the protein and peptide, which, in this case, arises from the three anchoring residues highlighted in Table 1, as shown by alanine scan mutagenesis studies [25]. Given a force field, MELD×MD samples multiple binding/unbinding events, effectively capturing both free energy contributions, even if it cannot decompose the contribution of each. Using the same protocol for all peptides allows us to identify differences in their binding preferences and peptide conformations. The main advantage is that, in this process, the peptide is completely flexible, free to adopt conformations in response to the environment. We observe the active site changing in response to the presence—and conformation—of the peptide.

Our studies hint that the binding mechanisms for *pdiq* and *ATSP*-7041 both favor initial binding as helices, with different mechanisms for rearrangement. Since kinetics are lost in our replica exchange ladder, testing this hypothesis will require future work in which the states discovered from our ensembles can be used for seeding unbiased simulations to construct markov models that show the binding pathways [36–38]. The chemical staple successfully increases the helical content, but it also plays a role in reducing side chain rotamer freedom through the steric volume it occupies (see Figure A11). Thus, *ATSP*-7041 is predisposed to make helical conformations, and to establish the right interactions. Figure 2 shows only two clusters: a major cluster binding as a helix with the three anchoring residues in the active site and a minor one with the staple in the active site. For *pdiq*, we see a higher number of minor clusters (see Figure A1) exhibiting helical conformations, in which at least one anchoring residue is not in the active site.

Thus, for accurate modeling of the *p53*-MDM2 interaction, we need to capture: (1) the intrinsic peptide propensity to helical conformations and (2) type and alignment of the anchoring residues inside the binding cavity. Peptides that, in their free form, favor helices seem to favor binding (given the same interface residues) by reducing the $\Delta G^{peptide}_{conformation}$. However, even when shifting the helical propensities, binding simulations are needed as the binding mode can change (as we see for *ATSP*-7041 and *pdiq*).

4. Materials and Methods

4.1. Choice of Peptide Systems

We chose a set of five peptides for this study: the sequence from the *p53* binding epitope, two high affinity inhibitors (*pdiq* [39] and *ATSP*-7041) [25] and two control peptides, based on the poly-Ala sequence (*Ala1* and *Ala2*; see Table 1). Of the two control sequences, *Ala1* sequence conserves the set of hydrophobic residues that allow binding, and *Ala2* does not. *ATSP*-7041 is a stapled peptide using three non-standard amino acids, where one of the three anchoring residues (Leucine) is substituted by a non-canonical amino acid.

For *p53* and *pdiq*, we used crystal structures of the peptides binding to MDM2 (PDB codes 1ycr [30] and 3jzs [39]. For *ATSP*-7041, we used the structure bound to MDMX (PDB code 4n5t [25]) and superposed the active site onto MDM2 to have the reference structure of the peptide on the active site of MDM2. For the two control peptides bases on poly-ALA, there is no native structure. We compare it to the *p53*-MDM2 conformation for those two peptides. Parameters for the *ATSP*-7041 peptide are derived from the general amber force field (GAFF) [40], deriving charges based on the AM1 model [41].

4.2. Free Peptide Simulations

We used the ff14SB force field for amino acid sidechains [42] and the ff99SB force field for backbone parameters [43], using the GBneck2 implicit solvent model (igb = 8) [44] to improve sampling efficiency. We ran the simulations for 2 μs using hydrogen mass repartitioning [45] with a 4fs timestep using the Amber molecular dynamics package [46]. A concern with implicit solvents is the bias towards some secondary structure [47]. However, this combination of force field with implicit solvent has shown to be reliable in reproducing the folding of peptide and protein systems [44,48,49].

4.3. MELD × MD Binding Simulations

We ran 1 μs-long H,T-REMD simulations using OpenMM [50] with the MELD plugin [26]. MELD allows us to incorporate noisy information to increase the sampling in regions of interest [51,52]. In this case, our interest was in observing the peptide-protein association. We required that there were at least five heavy-atom contacts between the three anchoring residues in the peptide (F, W, and L in *p53*) and any other hydrophobic residue in MDM2, andthe pool of possible contacts was selected from the combinatorics of both sets. The restraints were imposed using flat-bottom harmonic restraints. The flat region was defined as a pair of residues closer than 5 from each other, the restraints increased quadratically up to 7 and linearly beyond, with a force constant of 250 J/K/mol. At every timestep, all possible restraints are evaluated, sorted by energy, and only the lowest 5 in restraint energy are used until the next timestep. In this way, no information is lost as the simulation progresses.

The H,T-REMD protocol includes 30 replicas, where the change in Hamiltonian affects the force constant of the restraints. The 30 replicas are mapped to a value of alpha (α) between 0 (lowest replica) and 1 (highest replica). The Hamiltonian and temperature have defined values of the restraint force constant and temperature as a function of alpha.

The temperature increases geometrically from 300K ($\alpha = 0$) to 500K ($\alpha = 0.5$) and is kept at this temperature for higher values of alpha. The force constants for the restraints is set to $0\,\mathrm{J/K/nm^2}$ at $\alpha = 1$ and is gradually increased to the value of $250\,\mathrm{J/K/nm^2}$ for $\alpha \leq 0.6$. Exchanges between active restraints are more likely at higher index replicas.

4.4. Clustering Analysis

We use 2D-RMSD hierarchical clustering using a single linkage scheme within cpptraj [53] and report the centroid structure of each cluster and its population as representative of the clusters. We used the last 500ns of each replica, aligning on the protein (C_α and C_β atoms) and clustering on the overlapping peptide residues (C_α and C_β atoms). For Figure 2, the lowest temperature replica and all replicas for the clustering of the funnel plots, both with ϵ =1.5. For those in Figures 2 and A1–A4, we increase ϵ to 2.0, to depict more diverse clusters.

4.5. Projections Onto a Common Feature Space

We used pyEMMA [54] to featurize our system according to phi and psi dihedrals by choosing a common set of residues on all peptide systems resulting in 22 dihedrals, and we used dihedral shifting to reduce discontinuities in the distibution rather than using sine and cosines on the dihedrals [55]. The ensembles from free *p53*, *pdiq*, and *Ala1* were chosen as a common ensemble before dimensionality reduction of the system by using time-independent coordinate analysis [56] with a lag time of 10 ns, from which we extracted the top 14 eigenvectors that account for 95% of the variance. We then projected each peptide ensemble (from free and bound simulations) into the top eigenvectors. Finally, we performed clustering of the free peptide ensembles in the space defined by the top 14 eigenvectors to produce Figures 2 and A1–A4. Since the vectors were calculated for intrinsically disordered ensembles of the free form of the peptides, they are not representative of the slowest transitions during the binding process, which we cannot extract from the MELD-biased ensembles. Nonetheless, they provide a common set of vectors to represent all free and bound peptide systems studied. For these plots, we decided to project onto the third and fifth eigenvectors since these offered the best separation between clusters for the relevant states during binding.

5. Conclusions

Predicting bound structures for IDP peptides that fold upon binding is a computational grand challenge. We have shown that possible peptide inhibitors do not necessarily bind with the same binding mode, requiring modeling approaches that allow identification of the correct binding pose. The method successfully reproduces the binding of the two inhibitors and the *p53* epitope, while showing that the two control peptides are unsuccessful binders. We further show that, by changing the intrinsic properties (e.g., helical propensity, in this case), we can identify better binders; this simplifies the design of peptide inhibitors into two distinct tasks: optimizing interface residues and optimize structural propensities. The first task requires knowing the binding mode, and the second one can be assessed by MD simulations on the free peptide, at a lower computational cost than the binding simulations. Finally, we have shown that MELD$\times$MD is a useful tool to handle flexible binding and helps to ensure that the designed binders indeed bind and what their preferred binding mode is.

Author Contributions: Conceptualization, L.L. and A.P.; methodology, A.P.; software, L.L., A.P.; validation, L.L. and A.P.; formal analysis, L.L. and A.P.; investigation, L.L. and A.P.; resources, A.P.; data curation, L.L. and A.P.; writing—original draft preparation, A.P.; writing—review and editing, L.L., A.P.; visualization, L.L., A.P.; supervision, A.P.; project administration, A.P.; funding acquisition, A.P. All authors have read and agreed to the published version of the manuscript.

Funding: This research received no external funding.

Data Availability Statement: The MELD code used to run binding simulations is available to download from github: https://github.com/maccallumlab/meld .

Acknowledgments: This research was supported by startup funds from the Chemistry department at the University of Florida.

Conflicts of Interest: The authors declare no conflict of interest.

Appendix A

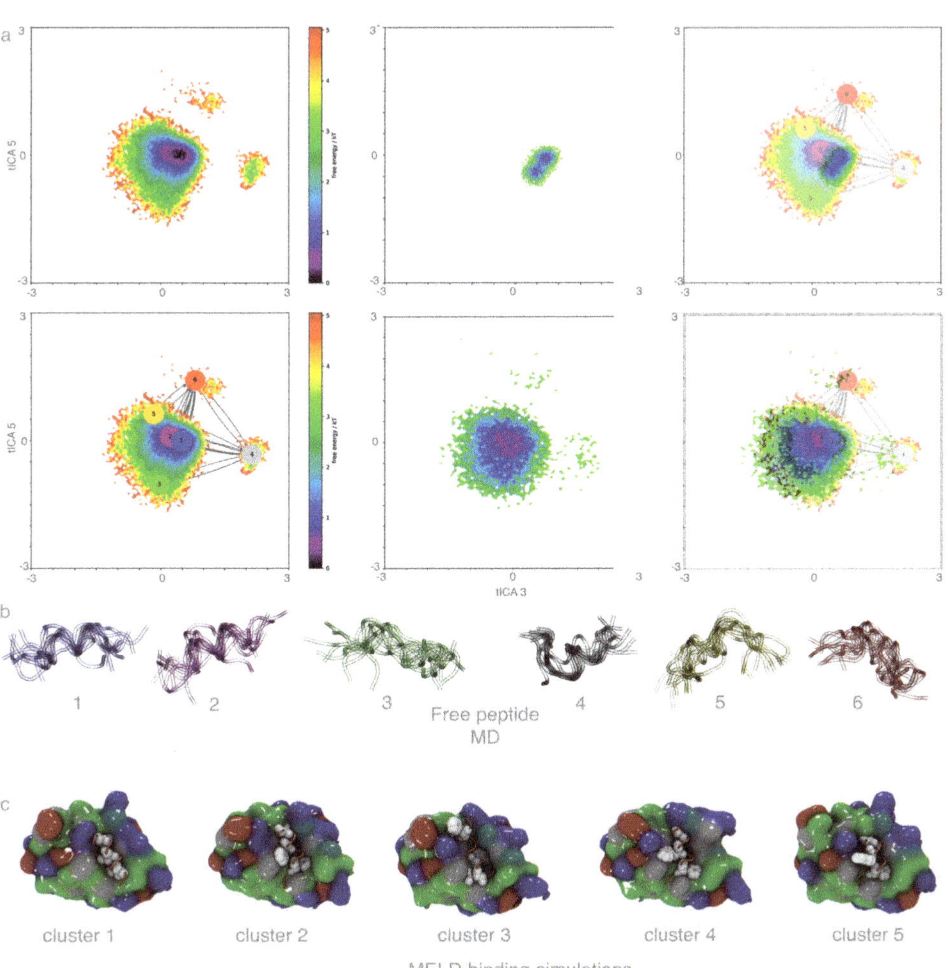

Figure A1. Comparison of the conformational space for free peptide versus binding simulations for the *pdiq* peptide. (**a**) The peptide ensembles are projected onto two tICA eigenvectors common to all five peptides. (**b**) The metastable states sampled for the free peptide. (**c**) Top clusters by population from MELD×MD binding simulations.

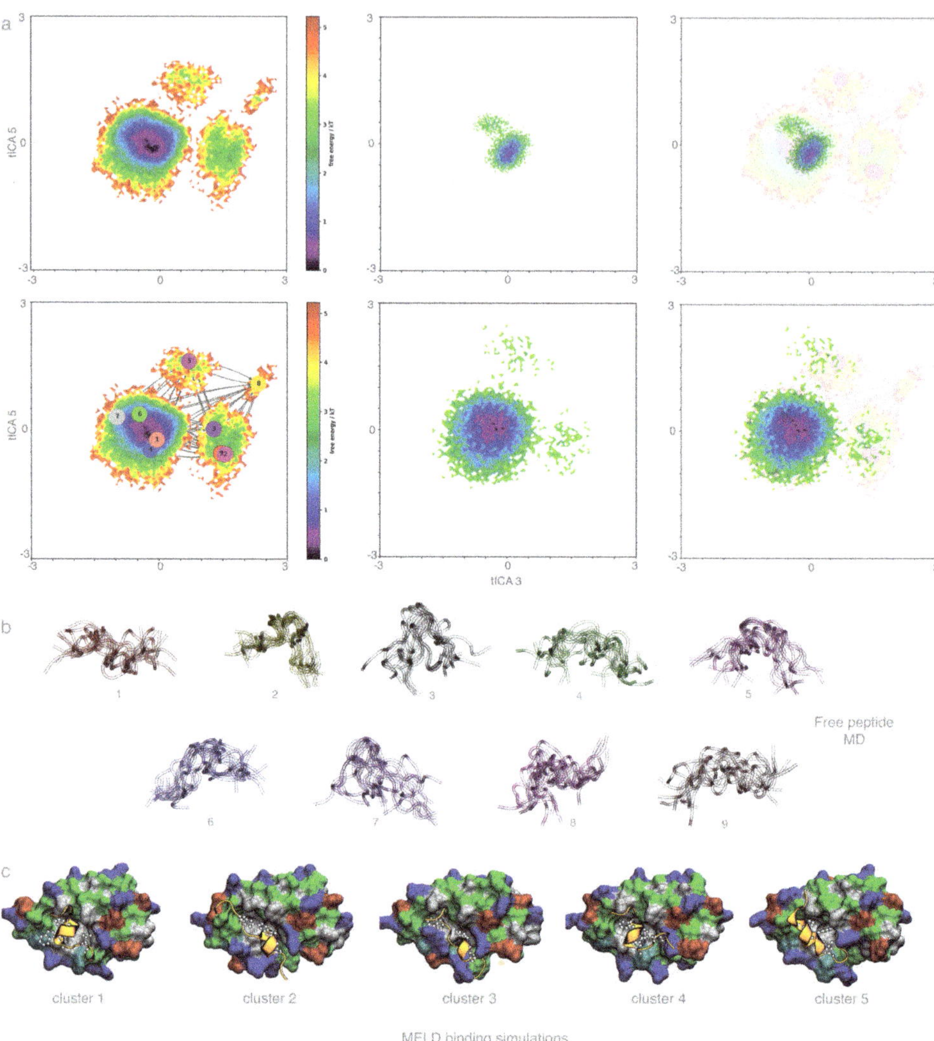

Figure A2. Comparison of the conformational space for free peptide versus binding simulations for the *p53* epitope. (**a**) The peptide ensembles are projected onto two tICA eigenvectors common to all five peptides. (**b**) The metastable states sampled for the free peptide. (**c**) Top clusters by population from MELD×MD binding simulations.

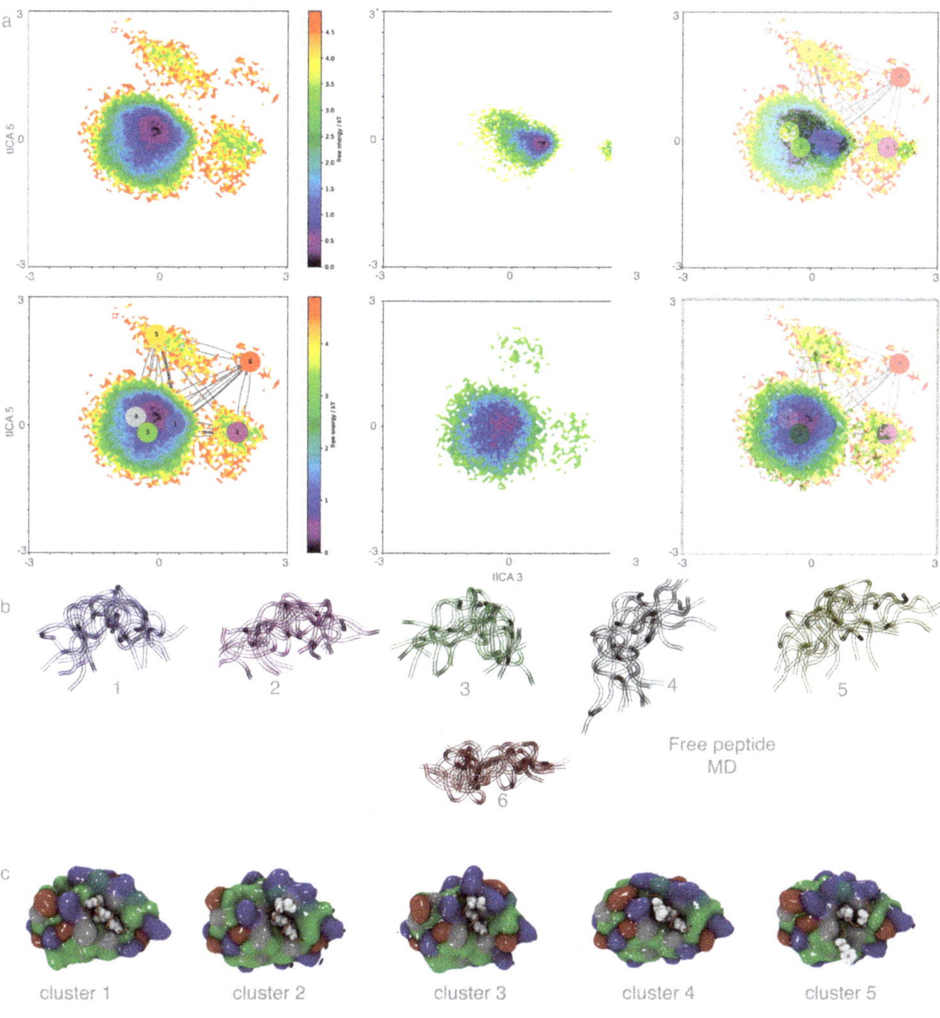

Figure A3. Comparison of the conformational space for free peptide versus binding simulations for *Ala1*. (**a**) The peptide ensembles are projected onto two tICA eigenvectors common to all five peptides. (**b**) The metastable states sampled for the free peptide. (**c**) Top clusters by population from MELD×MD binding simulations.

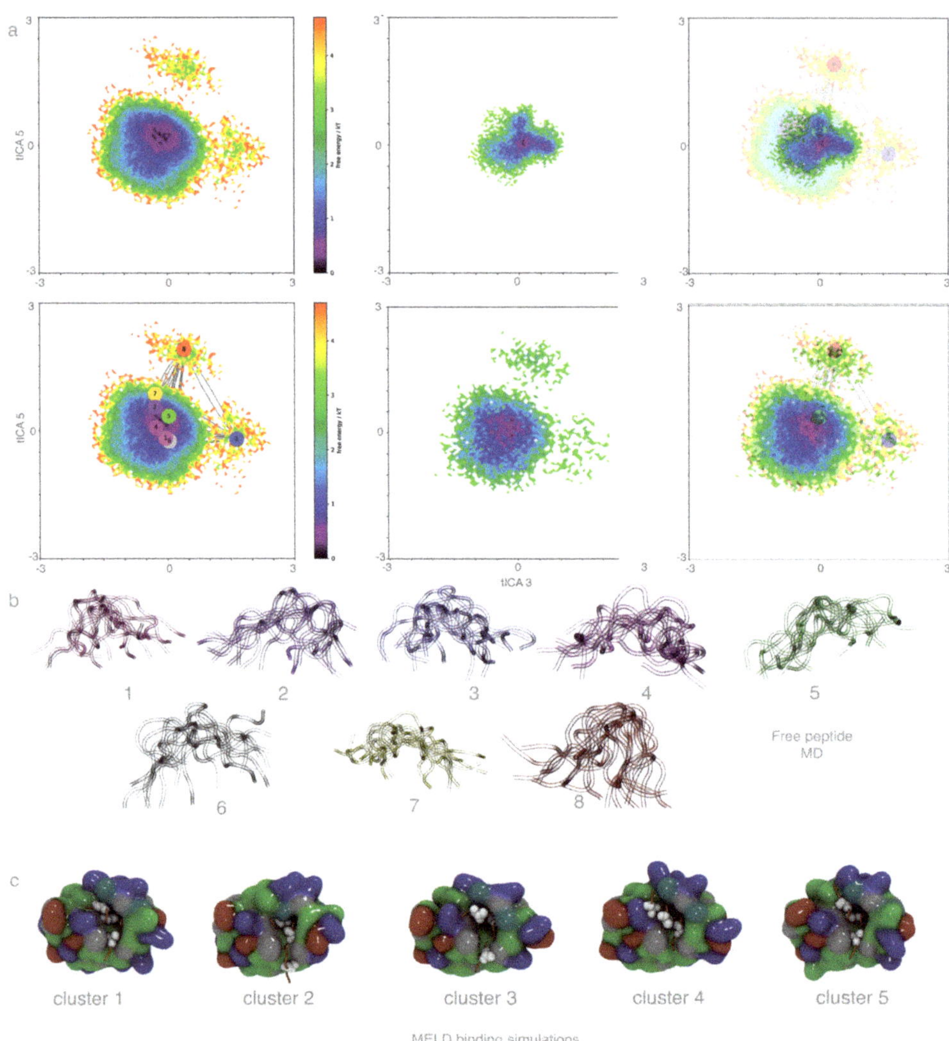

Figure A4. Comparison of the conformational space for free peptide versus binding simulations for *Ala2*. (**a**) The peptide ensembles are projected onto two tICA eigenvectors common to all five peptides. (**b**) The metastable states sampled for the free peptide. (**c**) Top clusters by population from MELD×MD binding simulations.

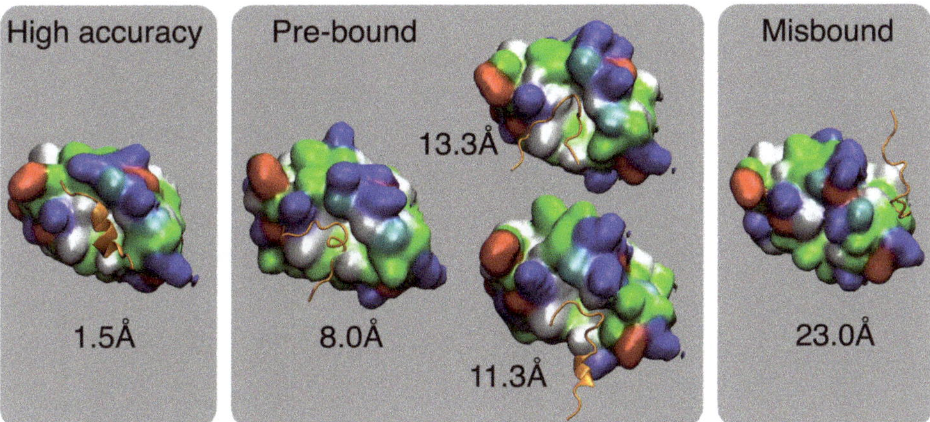

Figure A5. Clustering MELD×MD ensembles correctly identifies the MDM2 pocket as the binding site. Clustering is done by aligning on the MDM2 protein and using that alignment and the RMSD of the peptide between conformations as a distance metric. We define high accuracy binding (left) as the peptide binding in the right pocket with the right conformation (RMSD < 5 Å). For many clusters, the peptide is at least partially occupying the experimental binding site, with incorrect peptide conformations. Our approach samples these conformations at higher replica index, with a few progressing to the experimentally bound conformation. Finally, we find some conformations interacting with MDM2 at different sites in the protein (right panel).

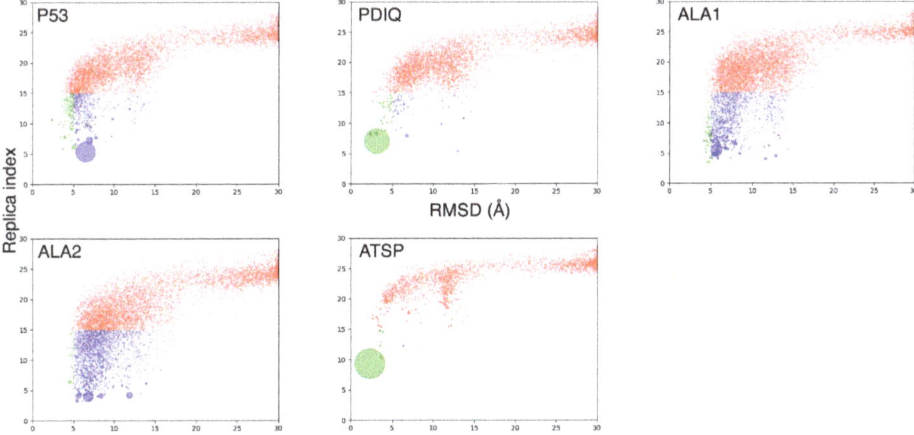

Figure A6. Funneling binding plot for the five peptides. Each dot corresponds to a cluster center from a 2D-RMSD based on all replicas. The larger the circle the larger the population of the cluster. Each circle is plotted at the average RMSD inside that cluster with respect to the native conformation and the average of the index replica in that cluster. The color code is green ($RMSD < 5$) or blue ($RMSD > 5$) when the average of the replica index is lower than 15, and red otherwise.

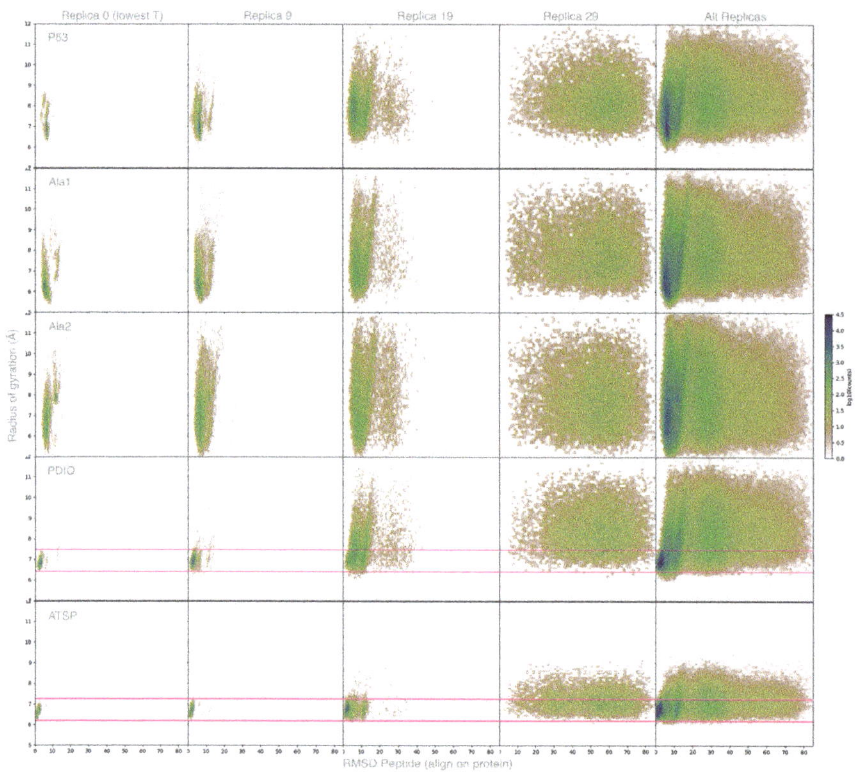

Figure A7. Radius of gyration versus RMSD of the peptide (protein align) for different replicas in MELD×MD for all peptides studied. The red lines for peptides *pdiq* and *ATSP*-7041 depict the region of the radius of gyration explored at the lowest temperature replica.

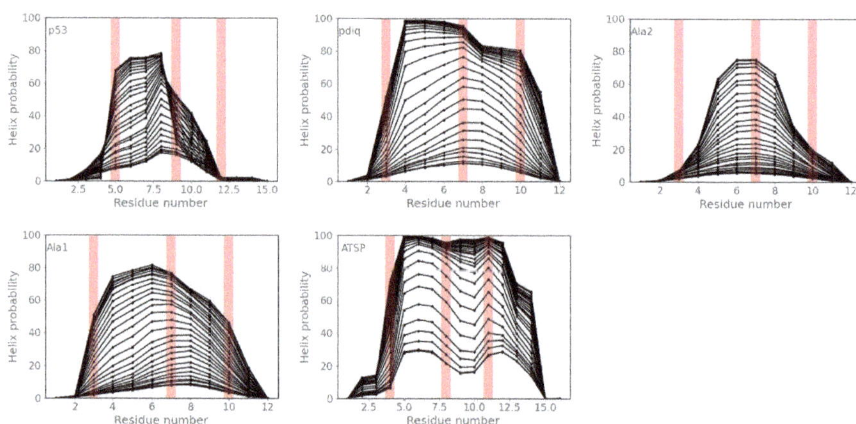

Figure A8. Secondary structure profiles for each peptide in MELD×MD runs. Each line represents a different replica (different H and T). Replicas with higher percentage of secondary structure are sampling at low temperature and ambiguous restraints guiding the peptide to the protein at full strength. Red bars represent the location of the three anchoring residues for each peptide.

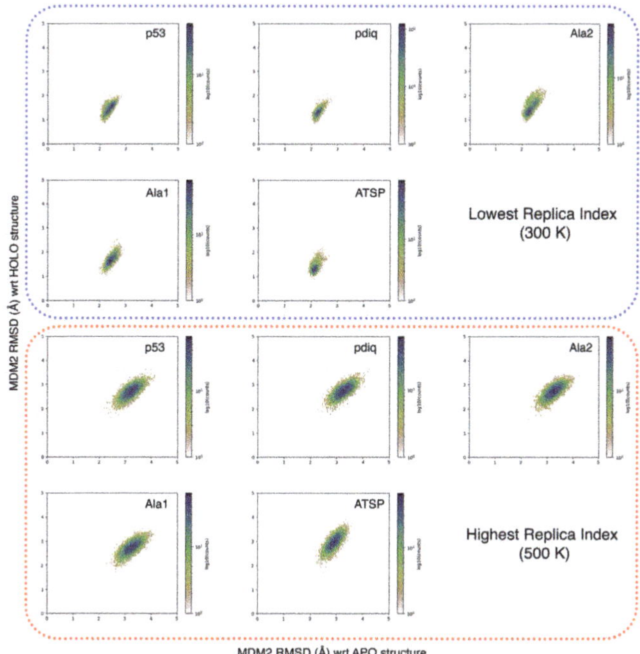

Figure A9. MDM2 protein ensembles with respect to holo and apo experimental structures at the lowest and highest index replicas. Simulations used a flat-bottom harmonic restraint on C_α positions with a 3.5 flat-bottom region.

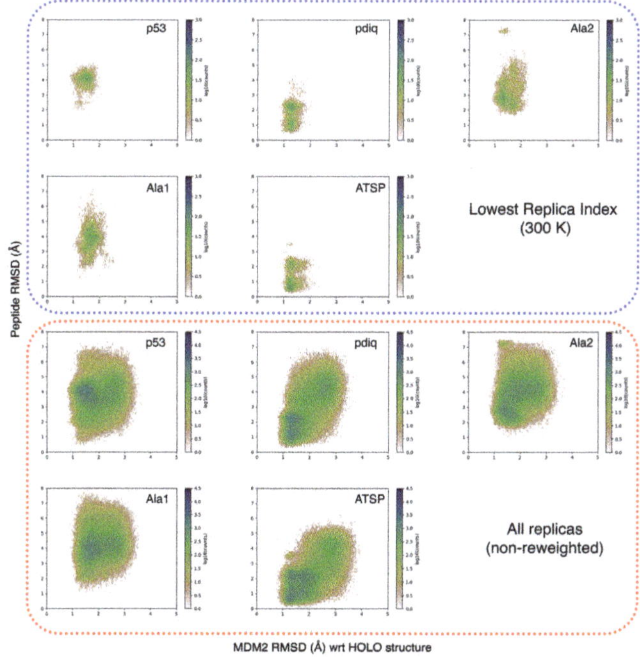

Figure A10. MDM2 versus peptide RMSD ensembles for the five peptides.

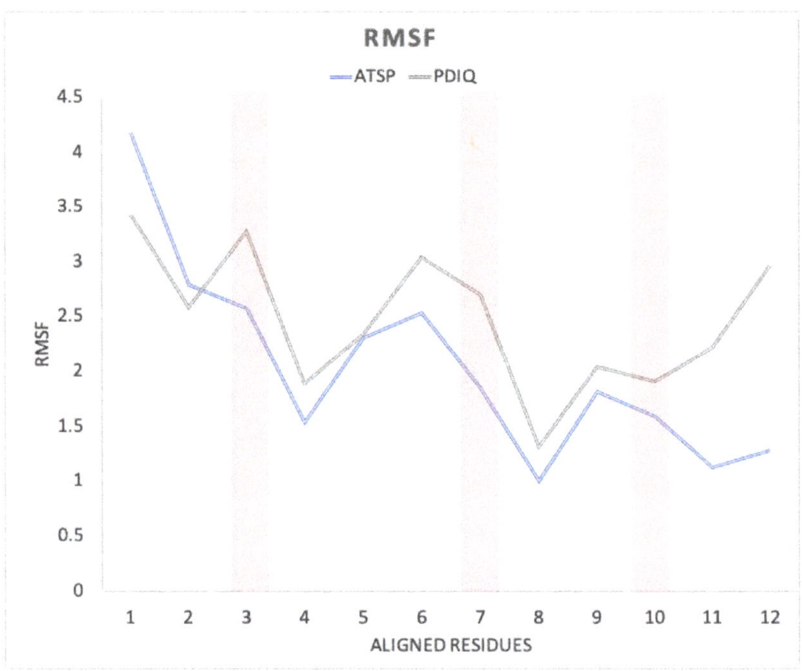

Figure A11. RMSF (Root Mean Square Fluctuation)comparison of the two best peptide binders. The red bars indicate the locations of the anchoring residues. Both peptides have been aligned to match the anchoring residues and end residues removed from *ATSP-7041* for this analysis.

References

1. Verdine, G.L.; Walensky, L.D. The Challenge of Drugging Undruggable Targets in Cancer: Lessons Learned from Targeting BCL-2 Family Members. *Clin. Cancer Res.* **2007**, *13*, 7264–7270, doi:10.1158/1078-0432.ccr-07-2184.
2. Uhlig, T.; Kyprianou, T.; Martinelli, F.G.; Oppici, C.A.; Heiligers, D.; Hills, D.; Calvo, X.R.; Verhaert, P. The emergence of peptides in the pharmaceutical business: From exploration to exploitation. *EuPA Open Proteom.* **2014**, *4*, 58–69, doi:10.1016/j.euprot.2014.05.003.
3. Weng, G.; Gao, J.; Wang, Z.; Wang, E.; Hu, X.; Yao, X.; Cao, D.; Hou, T. Comprehensive Evaluation of Fourteen Docking Programs on Protein–Peptide Complexes. *J. Chem. Theory Comput.* **2020**, *16*, 3959–3969, doi:10.1021/acs.jctc.9b01208.
4. Ciemny, M.; Kurcinski, M.; Kamel, K.; Kolinski, A.; Alam, N.; Schueler-Furman, O.; Kmiecik, S. Protein–peptide docking: Opportunities and challenges. *Drug Discov. Today* **2018**, *23*, 1530–1537, doi:10.1016/j.drudis.2018.05.006.
5. Lee, H.; Heo, L.; Lee, M.S.; Seok, C. GalaxyPepDock: A protein–peptide docking tool based on interaction similarity and energy optimization. *Nucleic Acids Res.* **2015**, *43*, W431–W435, doi:10.1093/nar/gkv495.
6. Porter, K.A.; Xia, B.; Beglov, D.; Bohnuud, T.; Alam, N.; Schueler-Furman, O.; Kozakov, D. ClusPro PeptiDock: Efficient global docking of peptide recognition motifs using FFT. *Bioinform.* **2017**, *33*, 3299–3301, doi:10.1093/bioinformatics/btx216.
7. Trellet, M.; Melquiond, A.S.J.; Bonvin, A.M.J.J. A Unified Conformational Selection and Induced Fit Approach to Protein-Peptide Docking. *PLoS ONE* **2013**, *8*, e58769, doi:10.1371/journal.pone.0058769.
8. Antunes, D.A.; Moll, M.; Devaurs, D.; Jackson, K.R.; Lizée, G.; Kavraki, L.E. DINC 2.0: A New Protein–Peptide Docking Webserver Using an Incremental Approach. *Cancer Res.* **2017**, *77*, e55–e57, doi:10.1158/0008-5472.can-17-0511.
9. Kurcinski, M.; Jamroz, M.; Blaszczyk, M.; Kolinski, A.; Kmiecik, S. CABS-dock web server for the flexible docking of peptides to proteins without prior knowledge of the binding site. *Nucleic Acids Res.* **2015**, *43*, W419–W424, doi:10.1093/nar/gkv456.
10. Alam, N.; Goldstein, O.; Xia, B.; Porter, K.A.; Kozakov, D.; Schueler-Furman, O. High-resolution global peptide-protein docking using fragments-based PIPER-FlexPepDock. *PLoS Comput. Biol.* **2017**, *13*, e1005905, doi:10.1371/journal.pcbi.1005905.
11. De Vries, S.J.; Rey, J.; Schindler, C.; Zacharias, M.; Tuffery, P. The pepATTRACT web server for blind, large-scale peptide–protein docking. *Nucleic Acids Res.* **2017**, *45*, gkx335, doi:10.1093/nar/gkx335.
12. Xu, X.; Yan, C.; Zou, X. MDockPeP: An ab-initio protein–peptide docking server. *J. Comput. Chem.* **2018**, *39*, 2409–2413, doi:10.1002/jcc.25555.
13. Zhou, P.; Jin, B.; Li, H.; Huang, S.Y. HPEPDOCK: A web server for blind peptide–protein docking based on a hierarchical algorithm. *Nucleic Acids Res.* **2018**, *46*, gky357, doi:10.1093/nar/gky357.

14. Zhang, Y.; Sanner, M.F. AutoDock CrankPep: Combining folding and docking to predict protein–peptide complexes. *Bioinformatics* **2019**, *35*, 5121–5127, doi:10.1093/bioinformatics/btz459.
15. Shan, Y.; Kim, E.T.; Eastwood, M.P.; Dror, R.O.; Seeliger, M.A.; Shaw, D.E. How does a drug molecule find its target binding site? *J. Am. Chem. Soc.* **2011**, *133*, 9181-3, doi:10.1021/ja202726y.
16. Migliorini, D.; Denchi, E.L.; Danovi, D.; Jochemsen, A.; Capillo, M.; Gobbi, A.; Helin, K.; Pelicci, P.G.; Marine, J.C. Mdm4 (Mdmx) Regulates *p53*-Induced Growth Arrest and Neuronal Cell Death during Early Embryonic Mouse Development. *Mol. Cell Biol.* **2002**, *22*, 5527–5538, doi:10.1128/mcb.22.15.5527-5538.2002.
17. Marine, J.C. Chapter 3 MDM2 and MDMX in Cancer and Development. *Curr. Top. Dev. Biol.* **2011**, *94*, 45–75, doi:10.1016/b978-0-12-380916-2.00003-6.
18. Wade, M.; Wahl, G.M. Targeting Mdm2 and Mdmx in Cancer Therapy: Better Living through Medicinal Chemistry? *Mol. Cancer Res.* **2008**, *7*, 1–11, doi:10.1158/1541-7786.mcr-08-0423.
19. Schon, O.; Friedler, A.; Bycroft, M.; Freund, S.M.V.; Fersht, A.R. Molecular mechanism of the interaction between MDM2 and p53. *J. Mol. Biol.* **2002**, *323*, 491–501.
20. Popowicz, G.M.; Czarna, A.; Rothweiler, U.; Szwagierczak, A.; Krajewski, M.; Weber, L.; Holak, T.A. Molecular basis for the inhibition of p53 by Mdmx. *Cell Cycle (Georget. Tex.)* **2007**, *6*, 2386–2392, doi:10.4161/cc.6.19.4740.
21. Moll, U.M.; Petrenko, O. The MDM2-p53 interaction. *Mol. Cancer Res.* **2003**, *1*, 1001–1008.
22. Chen, H.F.; Luo, R. Binding induced folding in p53-MDM2 complex. *J. Am. Chem. Soc.* **2007**, *129*, 2930–2937, doi:10.1021/ja0678774.
23. Carvajal, L.A.; Neriah, D.B.; Senecal, A.; Benard, L.; Thiruthuvanathan, V.; Yatsenko, T.; Narayanagari, S.R.; Wheat, J.C.; Todorova, T.I.; Mitchell, K.; et al. Dual inhibition of MDMX and MDM2 as a therapeutic strategy in leukemia. *Sci. Transl. Med.* **2018**, *10*, eaao3003, doi:10.1126/scitranslmed.aao3003.
24. Bernal, F.; Wade, M.; Godes, M.; Davis, T.N.; Whitehead, D.G.; Kung, A.L.; Wahl, G.M.; Walensky, L.D. A Stapled p53 Helix Overcomes HDMX-Mediated Suppression of p53. *Cancer Cell* **2010**, *18*, 411–422, doi:10.1016/j.ccr.2010.10.024.
25. Chang, Y.S.; Graves, B.; Guerlavais, V.; Tovar, C.; Packman, K.; To, K.H.; Olson, K.A.; Kesavan, K.; Gangurde, P.; Mukherjee, A.; et al. Stapled α-helical peptide drug development: A potent dual inhibitor of MDM2 and MDMX for p53-dependent cancer therapy. *Proc. Natl. Acad. Sci. USA* **2013**, *110*, E3445–E3454, doi:10.1073/pnas.1303002110.
26. MacCallum, J.L.; Perez, A.; Dill, K. Determining protein structures by combining semireliable data with atomistic physical models by Bayesian inference. *Proc. Natl. Acad. Sci. USA* **2015**, *112*, 6985–6990, doi:10.1073/pnas.1506788112.
27. Zimmerman, M.I.; Bowman, G.R. FAST Conformational Searches by Balancing Exploration/Exploitation Trade-Offs. *J. Chem. Theory Comput.* **2015**, *11*, 5747–5757, doi:10.1021/acs.jctc.5b00737.
28. Berger-Tal, O.; Nathan, J.; Meron, E.; Saltz, D. The exploration-exploitation dilemma: A multidisciplinary framework. *PLoS ONE* **2014**, *9*, e95693, doi:10.1371/journal.pone.0095693.
29. Uhrinova, S.; Uhrin, D.; Powers, H.; Watt, K.; Zheleva, D.; Fischer, P.; McInnes, C.; Barlow, P.N. Structure of free MDM2 N-terminal domain reveals conformational adjustments that accompany p53-binding. *J. Mol. Biol.* **2005**, *350*, 587–598, doi:10.1016/j.jmb.2005.05.010.
30. Kussie, P.H.; Gorina, S.; Marechal, V.; Elenbaas, B.; Moreau, J.; Levine, A.J.; Pavletich, N.P. Structure of the MDM2 oncoprotein bound to the p53 tumor suppressor transactivation domain. *Science* **1996**, *274*, 948.
31. Feng, H.; Jenkins, L.M.M.; Durell, S.R.; Hayashi, R.; Mazur, S.J.; Cherry, S.; Tropea, J.E.; Miller, M.; Wlodawer, A.; Appella, E.; et al. Structural Basis for p300 Taz2-p53 TAD1 Binding and Modulation by Phosphorylation. *Structure* **2009**, *17*, 202–210, doi:10.1016/j.str.2008.12.009.
32. Krois, A.S.; Ferreon, J.C.; Martinez-Yamout, M.A.; Dyson, H.J.; Wright, P.E. Recognition of the disordered p53 transactivation domain by the transcriptional adapter zinc finger domains of CREB-binding protein. *Proc. Natl. Acad. Sci. USA* **2016**, *113*, E1853–E1862, doi:10.1073/pnas.1602487113.
33. Zhou, H.X.; Rivas, G.; Minton, A.P. Macromolecular Crowding and Confinement: Biochemical, Biophysical, and Potential Physiological Consequences. *Annu. Rev. Biophys.* **2008**, *37*, 375–397, doi:10.1146/annurev.biophys.37.032807.125817.
34. Frishman, D.; Argos, P. Knowledge-based protein secondary structure assignment. *Proteins Struct. Funct. Genet.* **1995**, *23*, 566–579, doi:10.1002/prot.340230412.
35. Humphrey, W.; Dalke, A.; Schulten, K. VMD: Visual molecular dynamics. *J. Mol. Graph.* **1996**, *14*, 33–38, doi:10.1016/0263-7855(96)00018-5.
36. Plattner, N.; Doerr, S.; Fabritiis, G.D.; Noé, F. Complete protein-protein association kinetics in atomic detail revealed by molecular dynamics simulations and Markov modelling. *Nat. Chem.* **2017**, *9*, 1005–1011, doi:10.1038/nchem.2785.
37. Ferruz, N.; Fabritiis, G.D. Binding Kinetics in Drug Discovery. *Mol. Inform.* **2016**, *35*, 216–226, doi:10.1002/minf.201501018.
38. Perez, A.; Sittel, F.; Stock, G.; Dill, K. MELD-Path Efficiently Computes Conformational Transitions, Including Multiple and Diverse Paths. *J. Chem. Theory Comput.* **2018**, *14*, 2109–2116, doi:10.1021/acs.jctc.7b01294.
39. Phan, J.; Li, Z.; Kasprzak, A.; Li, B.; Sebti, S.; Guida, W.; Schönbrunn, E.; Chen, J. Structure-based design of high affinity peptides inhibiting the interaction of p53 with MDM2 and MDMX. *J. Biol. Chem.* **2010**, *285*, 2174 –2183, doi:10.1074/jbc.m109.073056.
40. Wang, J.; Wolf, R.M.; Caldwell, J.W.; Kollman, P.A.; Case, D.A. Development and testing of a general amber force field. *J. Comput. Chem.* **2004**, *25*, 1157–1174, doi:10.1002/jcc.20035.
41. Jakalian, A.; Jack, D.B.; Bayly, C.I. Fast, efficient generation of high-quality atomic charges. AM1-BCC model: II. Parameterization and validation. *J. Comput. Chem.* **2002**, *23*, 1623–1641, doi:10.1002/jcc.10128.

42. Maier, J.A.; Martinez, C.; Kasavajhala, K.; Wickstrom, L.; Hauser, K.E.; Simmerling, C. ff14SB: Improving the Accuracy of Protein Side Chain and Backbone Parameters from ff99SB. *J. Chem. Theory Comput.* **2015**, *11*, 3696–3713, doi:10.1021/acs.jctc.5b00255.
43. Hornak, V.; Abel, R.; Okur, A.; Strockbine, B.; Roitberg, A.; Simmerling, C. Comparison of multiple Amber force fields and development of improved protein backbone parameters. *Proteins* **2006**, *65*, 712–725, doi:10.1002/prot.21123.
44. Nguyen, H.; Roe, D.R.; Simmerling, C. Improved Generalized Born Solvent Model Parameters for Protein Simulations. *J. Chem. Theory Comput.* **2013**, *9*, 2020–2034, doi:10.1021/ct3010485.
45. Hopkins, C.W.; Grand, S.L.; Walker, R.C.; Roitberg, A.E. Long-Time-Step Molecular Dynamics through Hydrogen Mass Repartitioning. *J. Chem. Theory Comput.* **2015**, *11*, 1864–1874, doi:10.1021/ct5010406.
46. Case, D.; Belfon, K.; Ben-Shalom, I.; Brozell, S.; Cerutti, D.; Cheatham, T.I.; Cruzeiro, V.; Darden, T.; Duke, R.; Giambasu, G.; et al. *AMBER 2020*; University of California: San Francisco, CA, USA, 2020.
47. Perez, A.; MacCallum, J.L.; Brini, E.; Simmerling, C.; Dill, K. Grid-based backbone correction to the ff12SB protein force field for implicit-solvent simulations. *J. Chem. Theory Comput.* **2015**, *11*, 4770–4779, doi:10.1021/acs.jctc.5b00662.
48. Nguyen, H.; Maier, J.; Huang, H.; Perrone, V.; Simmerling, C. Folding simulations for proteins with diverse topologies are accessible in days with a physics-based force field and implicit solvent. *J. Am. Chem. Soc.* **2014**, *136*, 13959–13962, doi:10.1021/ja5032776.
49. Robertson, J.C.; Perez, A.; Dill, K. MELD × MD Folds Nonthreadables, Giving Native Structures and Populations. *J. Chem. Theory Comput.* **2018**, *14*, 6734–6740, doi:10.1021/acs.jctc.8b00886.
50. Eastman, P.; Swails, J.; Chodera, J.D.; McGibbon, R.T.; Zhao, Y.; Beauchamp, K.A.; Wang, L.P.; Simmonett, A.C.; Harrigan, M.P.; Stern, C.D.; Wiewiora, R.P.; Brooks, B.R.; Pande, V.S. OpenMM 7: Rapid development of high performance algorithms for molecular dynamics. *PLoS Comput. Biol.* **2017**, *13*, e1005659, doi:10.1371/journal.pcbi.1005659.
51. Perez, A.; MacCallum, J.L.; Dill, K.A. Accelerating molecular simulations of proteins using Bayesian inference on weak information. *Proc. Natl. Acad. Sci. USA* **2015**, *112*, 11846–51, doi:10.1073/pnas.1515561112.
52. Morrone, J.A.; Perez, A.; MacCallum, J.; Dill, K. Computed Binding of Peptides to Proteins with MELD-Accelerated Molecular Dynamics. *J. Chem. Theory Comput.* **2017**, *13*, 870–876, doi:10.1021/acs.jctc.6b00977.
53. Roe, D.R.; Cheatham, T.E., III. PTRAJ and CPPTRAJ: Software for Processing and Analysis of Molecular Dynamics Trajectory Data. *J. Chem. Theory Comput.* **2013**, *9*, 3084–3095, doi:10.1021/ct400341p.
54. Scherer, M.K.; Trendelkamp-Schroer, B.; Paul, F.; Pérez-Hernández, G.; Hoffmann, M.; Plattner, N.; Wehmeyer, C.; Prinz, J.H.; Noé, F. PyEMMA 2: A Software Package for Estimation, Validation, and Analysis of Markov Models. *J. Chem. Theory Comput.* **2015**, *11*, 5525–5542, doi:10.1021/acs.jctc.5b00743.
55. Sittel, F.; Filk, T.; Stock, G. Principal component analysis on a torus: Theory and application to protein dynamics. *J. Chem. Phys.* **2017**, *147*, 244101, doi:10.1063/1.4998259.
56. Pérez-Hernández, G.; Paul, F.; Giorgino, T.; Fabritiis, G.D.; Noé, F. Identification of slow molecular order parameters for Markov model construction. *J. Chem. Phys.* **2013**, *139*, 015102, doi:10.1063/1.4811489.

Article

Possible Transmission Flow of SARS-CoV-2 Based on ACE2 Features

Sk. Sarif Hassan [1], Shinjini Ghosh [2], Diksha Attrish [3], Pabitra Pal Choudhury [4], Alaa A. A. Aljabali [5], Bruce D. Uhal [6], Kenneth Lundstrom [7], Nima Rezaei [8,9], Vladimir N. Uversky [10,*], Murat Seyran [11], Damiano Pizzol [12], Parise Adadi [13], Antonio Soares [14], Tarek Mohamed Abd El-Aziz [14,15], Ramesh Kandimalla [16,17], Murtaza M. Tambuwala [18], Gajendra Kumar Azad [19], Samendra P. Sherchan [20], Wagner Baetas-da-Cruz [21], Kazuo Takayama [22], Ángel Serrano-Aroca [23], Gaurav Chauhan [24], Giorgio Palu [25] and Adam M. Brufsky [26]

1 Department of Mathematics, Pingla Thana Mahavidyalaya, Maligram 721140, India; sarimif@gmail.com
2 Department of Biophysics, Molecular Biology and Bioinformatics, University of Calcutta, Kolkata 700009, West Bengal, India; shinjinighosh2014@gmail.com
3 Dr. B. R. Ambedkar Centre For Biomedical Research (ACBR), University of Delhi (North Campus), Delhi 110007, India; dikshaattrish@gmail.com
4 Applied Statistics Unit, Indian Statistical Institute, Kolkata 700108, West Bengal, India; pabitra@isical.ac.in
5 Department of Pharmaceutics and Pharmaceutical Technology, Yarmouk University-Faculty of Pharmacy, Irbid 566, Jordan; alaaj@yu.edu.jo
6 Department of Physiology, Michigan State University, East Lansing, MI 48824, USA; bduhal@gmail.com
7 PanTherapeutics, Rte de Lavaux 49, CH1095 Lutry, Switzerland; lundstromkenneth@gmail.com
8 Research Center for Immunodeficiencies, Pediatrics Center of Excellence, Children's Medical Center, Tehran University of Medical Sciences, Tehran 1416753955, Iran; rezaei_nima@tums.ac.ir
9 Network of Immunity in Infection, Malignancy and Autoimmunity (NIIMA), Universal Scientific Education and Research Network (USERN), SE-123 Stockholm, Sweden
10 Department of Molecular Medicine, Morsani College of Medicine, University of South Florida, Tampa, FL 33612, USA
11 Doctoral studies in natural and technical sciences (SPL 44), University of Vienna, 1010 Wien, Austria; muratseyran@gmail.com
12 Italian Agency for Development Cooperation—Khartoum, Sudan Street 33, Al Amarat, Khartoum 825109, Sudan; damianopizzol8@gmail.com
13 Department of Food Science, University of Otago, Dunedin 9054, New Zealand; parise.adadi@postgrad.otago.ac.nz
14 Department of Cellular and Integrative Physiology, University of Texas Health Science Center at San Antonio, 7703 Floyd Curl Dr, San Antonio, TX 77030, USA; soaresa@uthscsa.edu (A.S.); mohamedt1@uthscsa.edu (T.M.A.E.-A.)
15 Zoology Department, Faculty of Science, Minia University, El-Minia 61519, Egypt
16 Applied Biology, CSIR-Indian Institute of Chemical Technology Uppal Road, Tarnaka, Hyderabad 500007, Telangana State, India; ramesh.kandimalla@iict.res.in
17 Department of Biochemistry, Kakatiya Medical College, Warangal, Telangana 500022, India
18 School of Pharmacy and Pharmaceutical Science, Ulster University, Coleraine BT52 1SA, Northern Ireland, UK; m.tambuwala@ulster.ac.uk
19 Department of Zoology, Patna University, Patna, Bihar 800005, India; gkazad@patnauniversity.ac.in
20 Department of Environmental Health Sciences, Tulane University, New Orleans, LA 70112, USA; sshercha@tulane.edu
21 Translational Laboratory in Molecular Physiology, Centre for Experimental Surgery, College of Medicine, Federal University of Rio de Janeiro (UFRJ), Rio de Janeiro 21941901, Brazil; wagner.baetas@gmail.com
22 Center for iPS Cell Research and Application, Kyoto University, Kyoto 606-8501, Japan; kazuo.takayama@cira.kyoto-u.ac.jp
23 Biomaterials and Bioengineering Lab, Translational Research Centre San Alberto Magno, Catholic University of Valencia San Vicente Mártir, c/Guillem de Castro 94, 46001 Valencia, Spain; angel.serrano@ucv.es

[24] School of Engineering and Sciences, Tecnologico de Monterrey, Av. Eugenio Garza Sada 2501 Sur, Monterrey 64849, Nuevo León, Mexico; gchauhan@tec.mx
[25] Department of Molecular Medicine, University of Padova, Via Gabelli 63, 35121 Padova, Italy; giorgio.palu@unipd.it
[26] Division of Hematology/Oncology, Department of Medicine, UPMC Hillman Cancer Center, University of Pittsburgh School of Medicine, Pittsburgh, PA 15260, USA; brufskyam@upmc.edu
* Correspondence: vuversky@usf.edu

Received: 4 November 2020; Accepted: 10 December 2020; Published: 13 December 2020

Abstract: Angiotensin-converting enzyme 2 (ACE2) is the cellular receptor for the Severe Acute Respiratory Syndrome Coronavirus 2 (SARS-CoV-2) that is engendering the severe coronavirus disease 2019 (COVID-19) pandemic. The spike (S) protein receptor-binding domain (RBD) of SARS-CoV-2 binds to the three sub-domains viz. amino acids (aa) 22–42, aa 79–84, and aa 330–393 of ACE2 on human cells to initiate entry. It was reported earlier that the receptor utilization capacity of ACE2 proteins from different species, such as cats, chimpanzees, dogs, and cattle, are different. A comprehensive analysis of ACE2 receptors of nineteen species was carried out in this study, and the findings propose a possible SARS-CoV-2 transmission flow across these nineteen species.

Keywords: ACE2; viral spike receptor-binding domain; SARS-CoV-2; transmission; bioinformatics

1. Introduction

We had been acquainted with the term beta-coronavirus for about two decades when we first encountered the Severe Acute Respiratory Syndrome Coronavirus (SARS-CoV) outbreak that emerged in 2002, infecting about 8000 people with a 10% mortality rate [1]. It was followed by the emergence of the Middle East Respiratory Syndrome Coronavirus (MERS-CoV) in 2012 with 2300 cases and mortality rate of 35% [2]. The third outbreak, caused by SARS-CoV-2, was first reported in December 2019 in China, Wuhan province, which rapidly took the form of a pandemic [3–5]. To date, this new human coronavirus has affected 65.5 million people worldwide and is held accountable for over 1.5 million deaths [6]. SARS-CoV-2 is an enveloped single-stranded plus sense RNA virus whose genome is about 30 kb in length, and which encodes for 16 non-structural proteins, four structural, and six accessory proteins [7]. The four major structural proteins which play a vital role in viral pathogenesis are Spike protein (S), Nucleocapsid protein (N), Membrane protein (M), and Envelope protein (E) [8,9]. SARS-CoV-2 infection is mainly characterized by pneumonia [10]; however, multi-organ failure involving myocardial infarction, hepatic, and renal damage is also reported in patients infected with this virus [11]. SARS-CoV-2 binds to the Angiotensin-converting enzyme 2 (ACE2) receptor on the host cell surface via its S protein [12,13]. ACE2 plays an essential role in viral attachment and entry [14,15]. The study of the interaction of ACE2 and S protein is of utmost importance [16–18].

The S1 subunit of the S protein has two domains, the C-terminal and the N-terminal domains, which fold independently, and either of the domains can act as Receptor Binding Domain (RBD) for the interaction and binding to the ACE2 receptor widely expressed on the surface of many cell types of the host [19,20]. The human ACE2 protein is 805 amino acids long, containing two functional domains: the extracellular N-terminal claw-like peptidase M2 domain and the C-terminal transmembrane collectrin domain with a cytosolic tail [21]. The RBD of the S protein binds to three different regions of ACE2, which are located at amino acids (aa) 24–42, 79–84, and 330–393 positions present in the claw-like peptidase domain of ACE2 [14]. These binding regions are designated in our study as domains D1, D2, and D3, respectively. ACE2 modulates angiotensin activities, which promote aldosterone release and increase blood pressure and inflammation, thus causing damage to blood vessel linings and various types of tissue injury [22]. ACE2 converts Angiotensin II to other molecules and reduces this effect [23]. However, when SARS-CoV-2 binds to ACE2, the function of ACE2 is inhibited and, in turn, leads to endocytosis of the virus particle into the host cell [24].

Zoonotic transmission of this virus from bat to human and random mutations acquired by SARS-CoV-2 during human to human transmission has also empowered this virus with the ability to undergo interspecies transmission, and, recently, many cases have been reported stating that different species can be infected by this virus [25,26].

In this study, we aim to determine the susceptibility of other species, whether they bear the capability of being a possible host of SARS-CoV-2. We chose nineteen different species (*Bos taurus, Capra hircus, Danio rerio, Equus caballus, Felis catus, Gallus gallus, Homo sapiens, Macaca mulatta, Manis javanica, Mesocricetus auratus, Mustela putorius furo, Pelodiscus sinensis, Pteropus alecto, Pteropus vampyrus, Pan troglodytes, Rattus norvegicus, Rhinolophus ferrumequinum, Salmo salar,* and *Sus scrofa*) and analyzed the ACE2 protein sequence from eighteen non-human species in relation to the human ACE2 sequence and determined the degree of variability by which the sequences differed from each other. We performed a comprehensive bioinformatics analysis in addition to the phylogenetic analysis based on full-length sequence homology, polarity along with individual domain sequence homology and secondary structure prediction of these protein sequences. These findings could have emerged to six distinct clusters of nineteen species based on the collective analysis and thereby provided a prediction of the interspecies SARS-CoV-2 transmission.

2. Results

Based on amino acid homology, secondary structures, bioinformatics, and polarity of the three domains D1, D2, and D3 of ACE2, all nineteen species were clustered. Note that, since ACE2 from Salmo salar is missing 119 N-terminal residues, this protein was not included in the analysis of the sequence conservation of the D1 (residues 24–42) and D2 domains (residues 79–84) involved in the interaction with SARS-CoV spike glycoprotein. Finally, a cumulative set of nineteen species clusters was built, among which the SARS-CoV-2 transmission may occur.

2.1. Phylogeny and Clustering Based on ACE2 Domain-Based Homology

First, we examined all the substitutions with similar properties and similar side chain binding atoms, signifying that the substitutions would not impede the SARS-CoV-2 transmission. Note that all the mutations are considered concerning the human ACE2 domains D1, D2, and D3 (Figure 1).

Sl. No.	SPECIES	D1 (24–42 aa)	D2 (79–84 aa)	D3 (330–393 aa)
1	Pan troglodytes	0	0	0
2	Macaca mulatta	0	0	0
3	Equus caballus	Q24L, D30L, H34S, D38E, Y41H	M82T	0
4	Felis catus	Q24L, D30E, D38E	M82T	0
5	Mesocricetus auratus	H34Q	M82T	0
6	Manis javanica	Q24E, D30E, H34S, D38E	L79I, M82N	G354H
7	Pteropus alecto	Q24L, D30E, H34T	M82A	N330K, R393K
8	Capra hircus	D30E	L79M, M82T	0
9	Sus scrofa	Q24L, D30E, H34L	L79I, M82T	R393K
10	Mustela putorius furo	Q24L, D30E, H34Y, D38E	L79H, M82T	G354R
11	Bos taurus	D30E	L79M, M82T	0
12	Rattus norvegicus	Q24K, T27S, D30N, H34Q	L79I, M82N, Y83F	K353H
13	Pteropus vampyrus	Q24L, D30E, H34T	M82A	N330K, R393K
14	Rhinolophus ferrumequinum	Q24L, T27K, K31D, H34S, D38N, Y41H	M82N, Y83F	0
15	Gallus gallus	DEL24Q, D30A, K31E, H34V, E35R, Q42E	L79N, M82R, Y83F	G354N
16	Pelodiscus sinensis	Q24E, T27N,D30S, K31E, H34V, E35Q, Q42A	L79N, M82K	G354K
17	Danio rerio	Q24R, T27E, D30N, H34E, E37S	L79E, M82A	INSERT353N, K353R, G354K
18	Salmo salar	Domain does not exist	Domain does not exist	N330D, INSERT353N, K353R, G354E

Figure 1. Substitutions in D1, D2, and D3 domains of ACE2 across eighteen species.

In D1 domain: out of eighteen species, eight species were found to possess a substitution at position 30 where D (aspartate) was substituted by E (glutamate), and four species were found to carry the D38E substitution. It was reported that, in the aspartate side chain, the oxygen atom was involved in ionic-ionic interaction and the side-chain oxygen atom was also present in glutamate, so this substitution may not affect the protein–protein interaction properties [27,28]. In the T27S substitution, threonine and serine both possess OH that participates in binding, and in the H34L substitution, both histidine and leucine use the NH group for interaction with another amino acid (backbone HN). Consequently, if we consider only the critical perspective for these substitutions, we can conclude that these changes would not impede the binding between the S and ACE2 protein.

In D2 domain: L79I bears importance across eighteen species since both of these amino acids (leucine and isoleucine) share similar chemical properties. Thus, if we analyze the changes in amino acid residues based on their chemical properties, which is the main contributing factor for protein–protein interaction, we can conclude that it will not significantly affect the binding between ACE-2 and RBD of the S protein.

In D3 domain: out of eleven substitutions, three substitutions (R393K, K353H, and K353R) were observed of the similar type with similar side chain interacting atoms and therefore changed at these positions would not affect the interaction of ACE2 with that of the S protein.

Secondly, across all nineteen species, homology was derived based on amino acid sequences, and, consequently, associated phylogenetic trees were drawn (Figure 2).

Six clusters of the nineteen species were formed using the K-means clustering technique based on sequence homology of the three domains (Figure 3). The clusters of species $\{S1, S2, S3\}$ and $\{S6, S13\}$ stayed together for the ACE2 full-length sequence homology and the combination of three domain-based sequence similarity. The species S16, S17, and S18 also followed the same as observed.

Furthermore, it was observed that sequence homology of the D1, D2, and D3 domains clustered the species S15 into the cluster where S9, S10, and S12 belong, although S15 was similar ACE2 sequence of S8 and S9. Despite S4 being very similar to S9, S10, and S12 for full-length ACE2 homology, it combined with S5 and S11 concerning the three domain-based sequence spatial organizations. In addition, S7 was found to be in the proximity of S6 and S13 although S7 was very much similar to S5 and S11 based on ACE2 homology.

2.2. Clustering Based on Secondary Structures

For each existing domain of ACE2 of the nineteen species, the secondary structure was predicted (Figure 4). For each domain, species are grouped into several subgroups.

Concerning the D1 domain:

- *Bos taurus* and *Capra hircus*
- *Equus caballus* and *Felis catus*
- *Mustela putorius furo* has a structure closer to *Equus caballus* and *Felis catus* as it has only one difference of a coil present at position 42. In addition, *Mesocricetus auratus* has a secondary structure similar to the above two, except an extended helix at position four. Similarly, *Sus scrofa* has an extended sheet instead of a helix at position 42. Thus, *Mustela putorius furo, Mesocricetus auratus,* and *Sus scrofa* can be put in the same cluster as *Equus caballus* and *Felis catus*.
- *Homo sapiens, Macaca mulatta,* and *Pan troglodytes*
- *Manis javanica* and *Rhinolophus ferrumequinum*
- *Pteropus alecto* and *Pteropus Vampyrus*
- *Rattus norvegicus* and *Pelodiscus sinensis* have similar structures differing by the presence of an extra coil at position 39 for *Rattus norvegicus*.
- *Gallus gallus* and *Danio rerio* have a unique secondary structure in comparison to the others.

These individual eight clusters show six different secondary structures in D1 shared by sixteen species, which shows high similarities in their secondary structures, while the remaining two have a unique secondary structure for D1 domain. Thus, these eight clusters have similar secondary structures indicating that the species in the eight clusters are closely related.

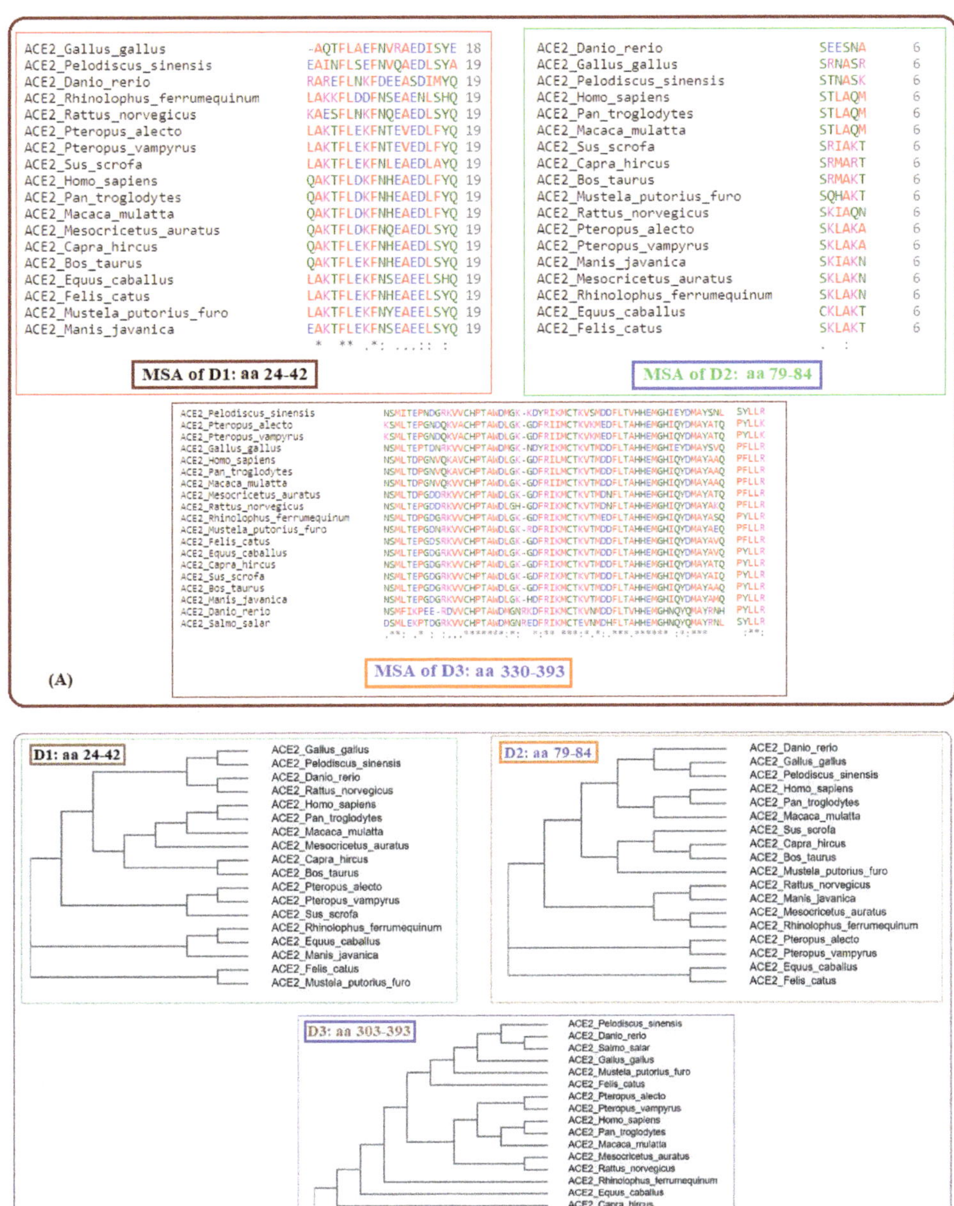

Figure 2. Multiple sequence alignments of D1, D2, and D3 domains of ACE2 of nineteen species (**A**) and respective phylogenies (**B**).

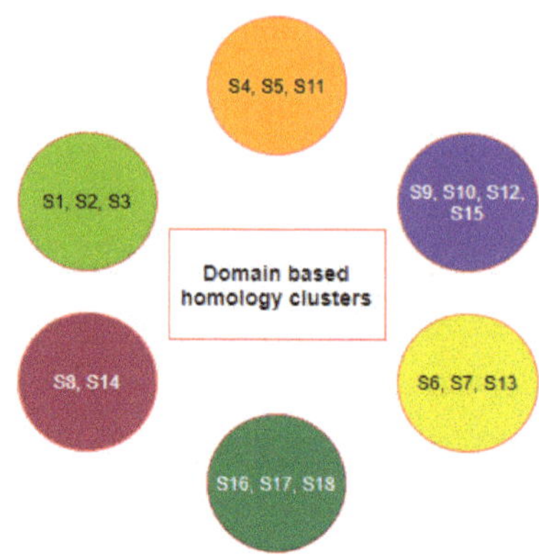

Figure 3. Clusters of species based on domain-based sequence homology.

With respect to the D2 domain:

- *Homo sapiens, Macaca mulatta,* and *Pan troglodytes*
- *Bos taurus, Mustela putorius furo, Pteropus alecto,* and *Pteropus vampyrus*
- *Equus caballus, Felis catus, Manis javanica, Pelodiscus sinensis,* and *Rhinolophus ferrumequinum*
- *Danio rerio* and *Gallus gallus*

Similarly, for the D2 domain, we found four clusters with the same secondary structure, indicating that they are closely related.

With respect to the D3 domain:

- *Homo sapiens* and *Pan troglodytes*
- *Bos Taurus, Rhinolophus ferrumequinum, Sus scrofa,* and *Capra hircus*
- *Equus caballus* and *Felis catus*
- *Pteropus alecto* and *Pteropus vampyrus*

Again for the D3 domain, four different clusters were bearing similar secondary structures; therefore, these species are also closely related.

Based on the similarity among the three domains, all eighteen species were clustered (Figure 5).

From the clusters (Figure 5) based on the secondary structure of the three domains of ACE2, it was observed that the species S4 was clustered uniquely, though S4 is clustered with S9 and S19 based on ACE2 full-length sequence homology. Furthermore, S6 and S13 were found to be similar based on ACE2 homology, but they got clustered into two different clusters when the secondary structure of three domains was concerned. In contrast, the group of species $\{S1, S2, S3\}$, $\{S9, S12\}$, $\{S16, S18\}$, and $\{S5, S7, S11\}$ remained in the same clusters concerning ACE2 homology as well as individual secondary structures of the domains.

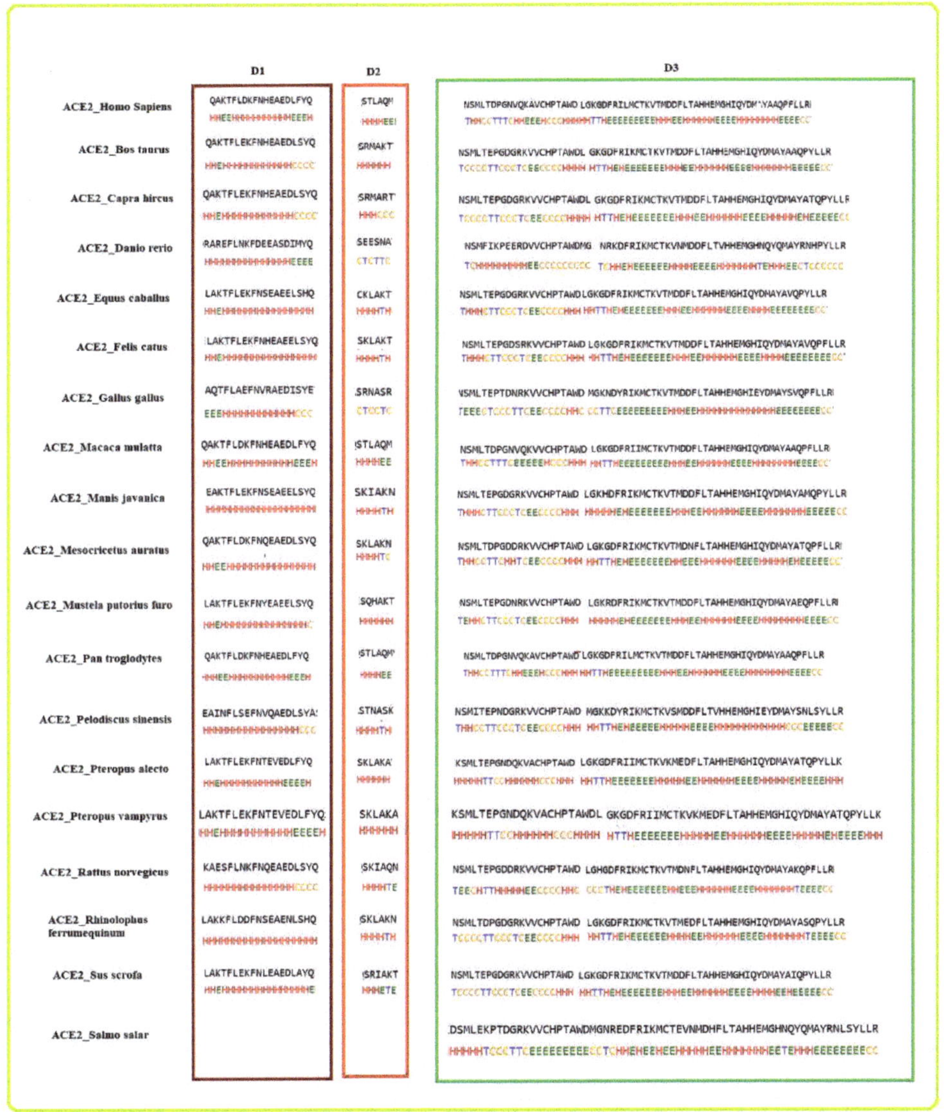

Figure 4. Predicted secondary structures of D1, D2, and D3 domains for 18 species and only D3 domain for *Salmo salar*.

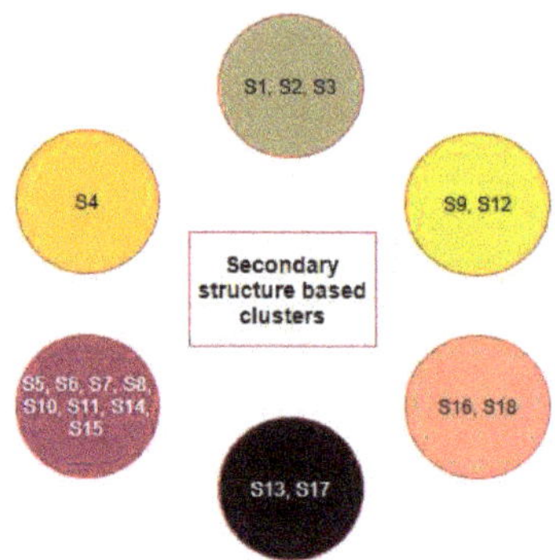

Figure 5. Clusters of species based on the secondary structures of the D1, D2, and D3 domains.

2.3. Clustering Based on Bioinformatics

Twelve bioinformatics features viz. Shannon entropy, instability index, aliphatic index, charged residues, half-life, melting temperature, N-terminal of the sequence, molecular weight, extinction coefficient, net charge at pH7, and isoelectric point of the D1, D2, and D3 domains of ACE2 for all nineteen species were determined (Figure 6).

For each species, a twelve-dimensional feature vector was found (Figure 6). For each domain D1, D2, and D3 domain, a distance matrix was determined using the Euclidean distance

$$d(S,T) = \sqrt{\sum_{i=1}^{12}(f_i - g_i)^2}$$

Note that here f_i and g_i denote the ith feature for the species S and T, respectively. These distance matrices with heatmap representation for all three domains are presented in Figures 7–9. In addition, by inputting the distance matrix, using the K-means clustering technique, several clusters of species were formed for D1 and D2 domains in eighteen species (Figures 7 and 8) and D3 domain in all nineteen species (Figure 9).

A final set of six clusters was formed using the K-means clustering method to have all three domains for eighteen different species (Figure 10). Although the species S7 was clustered with the species S5 and S11 as per full-length ACE2 sequence homology, S7 formed a unique singleton cluster when the bioinformatics features were taken into consideration. Similarly, the species S16 formed a singleton cluster though it was clustered with S17, S18, and S19 as per the amino acid homology of ACE2. The sequence homology of ACE2 made the four species S16, S17, S18, and S19 into a single cluster, but bioinformatics features placed the species S18 in a cluster where the other three species S1, S2, and S3 belonged. Based on bioinformatics features, S4 clustered together with S15 though the ACE2 receptor of S4 was sequentially similar to ACE2 of S9, S10, and S12.

The clusters $\{S1, S2, S3\}$, $\{S6, S13\}$, and $\{S9, S10, S12\}$ were unaltered with respect to the full length ACE2 homology and bioinformatics features.

Sl No	SPECIES	Domain	Shannon Entropy	Instability index:	Aliphatic index	Negatively charged residues	Positively charged residues	N-terminal of the sequence	The estimated half-life	Melting Temperature Prediction	MW (g/mol)	Extinction coefficient ε(M-1 cm-1)	Net Charge at pH7	Estimated Solubility in Water	pI
1	Homo sapiens	D1	3.366	38.79 (Stable)	51.58	4	2	Q	0.8 hours	55℃~65℃	2344.53	1280	-1.9	Good	4.42
2	Pan troglodytes	D1	3.366	38.79 (Stable)	51.58	4	2	Q	0.8 hours	55℃~65℃	2344.53	1280	-1.9	Good	4.42
3	Macaca mulatta	D1	3.366	38.79 (Stable)	51.58	4	2	Q	0.8 hours	55℃~65℃	2344.53	1280	-1.9	Good	4.42
4	Equus caballus	D1	3.156	43.26 (Unstable)	72.11	4	2	L	5.5 hours	lower than 55℃	2221.42	0	-1.9	Good	0.61
5	Felis catus	D1	3.261	33.13 (Stable)	72.11	4	2	L	5.5 hours	55℃~65℃	2297.52	1280	-1.9	Good	3.93
6	Mesocricetus auratus	D1	3.366	27.81 (Stable)	51.58	4	2	Q	0.8 hours	higher than 65℃	2275.43	1280	-2	Good	3.93
7	Manis javanica	D1	3.111	43.26 (Unstable)	51.58	5	2	E	1 hour	lower than 55℃	2263.41	1280	-3	Good	3.97
8	Pteropus alecto	D1	3.287	48.93 (Unstable)	82.11	4	2	L	5.5 hours	higher than 65℃	2335.6	1280	-2	Good	4.04
9	Capra hircus	D1	3.471	26.11 (Stable)	51.58	4	2	Q	0.8 hours	higher than 65℃	2298.46	1280	-1.9	Good	4.52
10	Sus scrofa	D1	3.116	26.11 (Stable)	97.89	4	2	L	5.5 hours	higher than 65℃	2243.5	1280	-2	Good	4.04
11	Mustela putorius furo	D1	3.156	29.16 (Stable)	72.11	4	2	L	5.5 hours	55℃~65℃	2323.55	2560	-2	Good	4.15
12	Bos taurus	D1	3.471	26.11 (Stable)	51.58	4	2	Q	0.8 hours	higher than 65℃	2298.46	1280	-1.9	Good	4.52
13	Rattus norvegicus	D1	3.261	48.38 (Unstable)	51.58	4	2	K	1.3 hours	lower than 55℃	2261.4	1280	-2	Good	4.04
14	Pteropus vampyrus	D1	3.287	48.93 (Unstable)	82.11	4	2	L	5.5 hours	higher than 65℃	2335.6	1280	-2	Good	4.04
15	Rhinolophus ferrumequinum	D1	3.261	15.64 (Stable)	72.11	4	2	L	5.5 hours	lower than 55℃	2206.37	0	-2	Good	4.42
16	Gallus gallus	D1	3.53	22.81 (Stable)	76.11	4	1	A	4.4 hours	lower than 55℃	2103.24	1280	-3	Good	3.54
17	Pelodiscus sinensis	D1	3.326	34.3 (Stable)	92.63	4	0	E	1 hour	lower than 55℃	2160.29	1280	-4	Good	0.53
18	Danio rerio	D1	3.577	58.05 (Unstable)	51.58	5	3	R	1 hour	lower than 55℃	2362.57	1280	-2	Good	4.16
1	Homo sapiens	D2	2.585	8.33 (Stable)	81.67	0	0	S	1.9 hours	lower than 55℃	649.76	0	0	Poor	3.44
2	Pan troglodytes	D2	2.585	8.33 (Stable)	81.67	0	0	S	1.9 hours	lower than 55℃	649.76	0	0	Poor	3.44
3	Macaca mulatta	D2	2.585	8.33 (Stable)	81.67	0	0	S	1.9 hours	lower than 55℃	649.76	0	0	Poor	3.44
4	Equus caballus	D2	2.252	-5.82 (Stable)	81.67	0	2	C	1.2 hours	lower than 55℃	662.84	0	1.9	Good	10.23
5	Felis catus	D2	2.252	-5.82 (Stable)	81.67	0	2	S	1.9 hours	lower than 55℃	646.78	0	2	Good	10.69
6	Mesocricetus auratus	D2	2.252	-5.82 (Stable)	81.67	0	2	S	1.9 hours	lower than 55℃	659.78	0	2	Good	10.69
7	Manis javanica	D2	2.252	-5.82 (Stable)	81.67	0	2	S	1.9 hours	higher than 65℃	659.78	0	2	Good	10.69
8	Pteropus alecto	D2	1.918	-5.82 (Stable)	98.33	0	2	S	1.9 hours	lower than 55℃	616.75	0	2	Good	10.69
9	Capra hircus	D2	2.252	61 (Unstable)	16.67	0	2	S	1.9 hours	lower than 55℃	720.85	0	2	Good	12.1
10	Sus scrofa	D2	2.585	40.43 (Unstable)	81.67	0	2	S	1.9 hours	lower than 55℃	674.79	0	2	Good	11.39
11	Mustela putorius furo	D2	2.585	40.43 (Unstable)	16.67	0	1	S	1.9 hours	higher than 65℃	670.72	0	1.1	Good	9.86
12	Bos taurus	D2	2.585	61 (Unstable)	16.67	0	2	S	1.9 hours	higher than 65℃	692.83	0	2	Good	11.39
13	Rattus norvegicus	D2	2.585	-5.82 (Stable)	81.67	0	1	S	1.9 hours	lower than 55℃	659.73	0	1	Good	9.86
14	Pteropus vampyrus	D2	1.918	-5.82 (Stable)	98.33	0	2	S	1.9 hours	lower than 55℃	616.75	0	2	Good	10.69
15	Rhinolophus ferrumequinum	D2	2.252	-5.82 (Stable)	81.67	0	2	S	1.9 hours	lower than 55℃	659.78	0	2	Good	10.69
16	Gallus gallus	D2	1.918	93.1 (Unstable)	16.67	0	2	S	1.9 hours	lower than 55℃	689.72	0	2	Good	12.1
17	Pelodiscus sinensis	D2	2.252	-16.72 (Stable)	16.67	0	1	S	1.9 hours	lower than 55℃	606.63	0	1	Good	9.86
18	Danio rerio	D2	1.918	126.87 (Unstable)	16.67	2	0	S	1.9 hours	lower than 55℃	635.58	0	-2	Good	0.76
1	Homo sapiens	D3	4.132	36.32 (Stable)	71.72	7	5	N	1.4 hours	55℃~65℃	7253.32	8250	-1.8	Poor	6.05
2	Pan troglodytes	D3	4.132	36.32 (Stable)	71.72	7	5	N	1.4 hours	55℃~65℃	7253.32	8250	-1.8	Poor	6.05
3	Macaca mulatta	D3	4.161	31.98 (Stable)	74.69	7	5	N	1.4 hours	higher than 65℃	7281.37	8250	-1.8	Poor	6.05
4	Equus caballus	D3	4.153	40.09 (Unstable)	67.03	8	7	N	1.4 hours	55℃~65℃	7341.43	9530	-0.8	Good	6.5
5	Felis catus	D3	4.178	49.11 (Unstable)	67.03	8	7	N	1.4 hours	higher than 65℃	7355.46	8250	-0.8	Good	6.5
6	Mesocricetus auratus	D3	4.134	33.55 (Stable)	62.5	8	7	N	1.4 hours	55℃~65℃	7370.43	8250	-0.8	Good	6.5
7	Manis javanica	D3	4.143	40.23 (Unstable)	62.5	8	7	N	1.4 hours	55℃~65℃	7453.59	9530	-0.7	Good	6.61
8	Pteropus alecto	D3	4.135	45.32 (Unstable)	65.62	8	7	K	1.3 hours	55℃~65℃	7356.48	9530	-0.7	Good	6.5
9	Capra hircus	D3	4.143	38.91 (Stable)	62.5	8	7	N	1.4 hours	55℃~65℃	7343.4	9530	-0.8	Good	6.5
10	Sus scrofa	D3	4.161	40.09 (Unstable)	68.59	8	7	N	1.4 hours	55℃~65℃	7355.46	9530	-0.8	Good	6.5
11	Mustela putorius furo	D3	4.186	52.52 (Unstable)	62.5	9	8	N	1.4 hours	55℃~65℃	7511.6	8250	-0.8	Good	6.5
12	Bos taurus	D3	4.147	40.09 (Unstable)	64.6	8	7	N	1.4 hours	55℃~65℃	7313.38	9530	-0.8	Good	6.5
13	Rattus norvegicus	D3	4.172	45.48 (Unstable)	62.5	8	7	N	1.4 hours	55℃~65℃	7420.49	8250	-0.7	Good	6.61
14	Pteropus vampyrus	D3	4.135	45.32 (Unstable)	65.62	8	7	K	1.3 hours	55℃~65℃	7356.48	9530	-0.7	Good	6.5
15	Rhinolophus ferrumequinum	D3	4.172	40.09 (Unstable)	62.5	8	7	N	1.4 hours	55℃~65℃	7329.38	9530	-0.8	Good	6.5
16	Gallus gallus	D3	4.171	47.3 (Unstable)	59.38	9	7	N	1.4 hours	55℃~65℃	7534.61	9530	-1.8	Good	6.06
17	Pelodiscus sinensis	D3	4.119	43.94 (Unstable)	63.91	9	8	N	1.4 hours	55℃~65℃	7524.62	10810	-0.8	Good	6.5
18	Danio rerio	D3	4.18	28.67 (Stable)	51.72	8	9	N	1.4 hours	higher than 65℃	7797.92	9530	8.02	Good	1.3
19	Salmo salar	D3	4.18	29.68 (Stable)	54	9	8	D	1.1 hours	higher than 65℃	7770.81	9530	-0.6	Good	6.62

Figure 6. Bioinformatics of the D1, D2, and D3 domains of ACE2 from eighteen species. For *Salmo salar*, only D3 bioinformatics was presented.

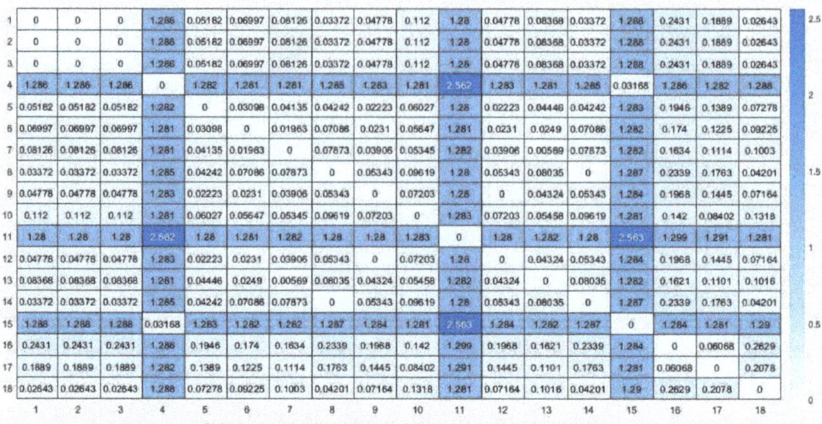

	1	2	3	4	5	6	7	8	9	10	11	12	13	14	15	16	17	18
1	0	0	0	1.286	0.05182	0.06997	0.08126	0.03372	0.04778	0.112	1.28	0.04778	0.08368	0.03372	1.288	0.2431	0.1889	0.02643
2	0	0	0	1.286	0.05182	0.06997	0.08126	0.03372	0.04778	0.112	1.28	0.04778	0.08368	0.03372	1.288	0.2431	0.1889	0.02643
3	0	0	0	1.286	0.05182	0.06997	0.08126	0.03372	0.04778	0.112	1.28	0.04778	0.08368	0.03372	1.288	0.2431	0.1889	0.02643
4	1.286	1.286	1.286	0	1.282	1.281	1.281	1.285	1.283	1.281	2.562	1.283	1.281	1.285	0.03168	1.286	1.282	1.288
5	0.05182	0.05182	0.05182	1.282	0	0.03098	0.04135	0.04242	0.02223	0.06027	1.28	0.02223	0.04446	0.04242	1.283	0.1946	0.1389	0.07278
6	0.06997	0.06997	0.06997	1.281	0.03098	0	0.01963	0.07086	0.0231	0.05647	1.281	0.0231	0.0249	0.07086	1.282	0.174	0.1225	0.09225
7	0.08126	0.08126	0.08126	1.281	0.04135	0.01963	0	0.07873	0.03906	0.05345	1.282	0.03906	0.00569	0.07873	1.282	0.1634	0.1114	0.1003
8	0.03372	0.03372	0.03372	1.285	0.04242	0.07086	0.07873	0	0.05343	0.09619	1.28	0.05343	0.08035	0	1.287	0.2339	0.1763	0.04201
9	0.04778	0.04778	0.04778	1.283	0.02223	0.0231	0.03906	0.05343	0	0.07203	1.28	0	0.04324	0.05343	1.284	0.1968	0.1445	0.07164
10	0.112	0.112	0.112	1.281	0.06027	0.05647	0.05345	0.09619	0.07203	0	1.283	0.07203	0.05458	0.09619	1.281	0.142	0.08402	0.1318
11	1.28	1.28	1.28	2.562	1.28	1.281	1.282	1.28	1.28	1.283	0	1.28	1.282	1.28	2.563	1.299	1.291	1.281
12	0.04778	0.04778	0.04778	1.283	0.02223	0.0231	0.03906	0.05343	0	0.07203	1.28	0	0.04324	0.05343	1.284	0.1968	0.1445	0.07164
13	0.08368	0.08368	0.08368	1.281	0.04446	0.0249	0.00569	0.08035	0.04324	0.05458	1.282	0.04324	0	0.08035	1.282	0.1621	0.1101	0.1016
14	0.03372	0.03372	0.03372	1.285	0.04242	0.07086	0.07873	0	0.05343	0.09619	1.28	0.05343	0.08035	0	1.287	0.2339	0.1763	0.04201
15	1.288	1.288	1.288	0.03168	1.283	1.282	1.282	1.287	1.284	1.281	2.563	1.284	1.282	1.287	0	1.284	1.281	1.29
16	0.2431	0.2431	0.2431	1.286	0.1946	0.174	0.1634	0.2339	0.1968	0.142	1.299	0.1968	0.1621	0.2339	1.284	0	0.06068	0.2629
17	0.1889	0.1889	0.1889	1.282	0.1389	0.1225	0.1114	0.1763	0.1445	0.08402	1.291	0.1445	0.1101	0.1763	1.281	0.06068	0	0.2078
18	0.02643	0.02643	0.02643	1.288	0.07278	0.09225	0.1003	0.04201	0.07164	0.1318	1.281	0.07164	0.1016	0.04201	1.29	0.2629	0.2078	0

Distance matrix derived from bioinformatics feature vectors of D1

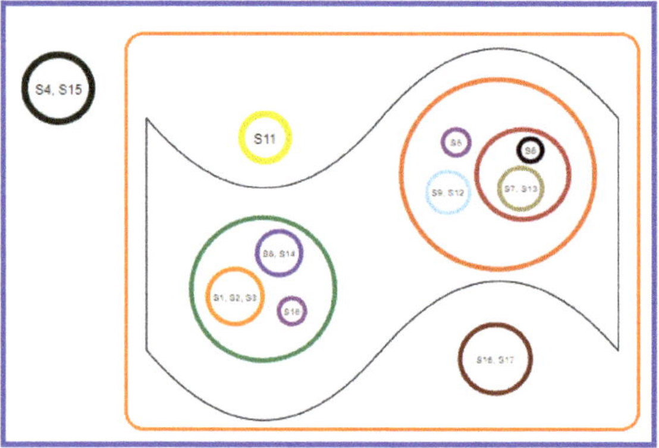

Clusters based on bioinformatics of D1

Figure 7. Distance matrix based on the bioinformatics feature vectors of D1 of ACE2 across eighteen eighteen species and associated clusters.

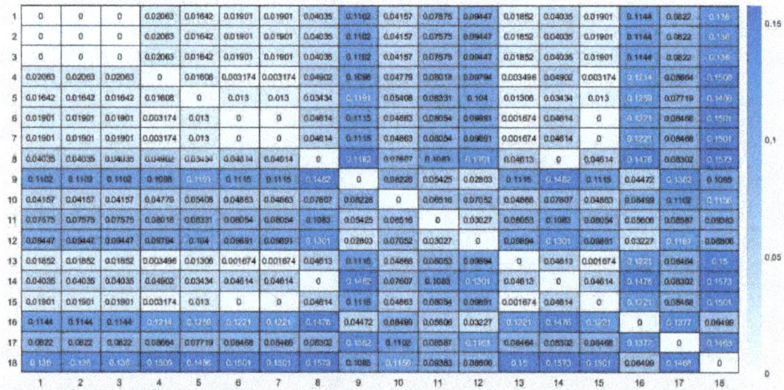

	1	2	3	4	5	6	7	8	9	10	11	12	13	14	15	16	17	18
1	0	0	0	0.02063	0.01642	0.01901	0.01901	0.04035	0.1102	0.04157	0.07575	0.09447	0.01852	0.04035	0.01901	0.1144	0.0822	0.136
2	0	0	0	0.02063	0.01642	0.01901	0.01901	0.04035	0.1102	0.04157	0.07575	0.09447	0.01852	0.04035	0.01901	0.1144	0.0822	0.136
3	0	0	0	0.02063	0.01642	0.01901	0.01901	0.04035	0.1102	0.04157	0.07575	0.09447	0.01852	0.04035	0.01901	0.1144	0.0822	0.136
4	0.02063	0.02063	0.02063	0	0.01608	0.003174	0.003174	0.04902	0.1098	0.04779	0.08018	0.09794	0.003496	0.04902	0.003174	0.1214	0.08664	0.1506
5	0.01642	0.01642	0.01642	0.01608	0	0.013	0.013	0.03434	0.1191	0.05408	0.08331	0.104	0.01306	0.03434	0.013	0.1259	0.07719	0.1486
6	0.01901	0.01901	0.01901	0.003174	0.013	0	0	0.04614	0.1115	0.04863	0.08054	0.09891	0.001674	0.04614	0	0.1221	0.08468	0.1501
7	0.01901	0.01901	0.01901	0.003174	0.013	0	0	0.04614	0.1115	0.04863	0.08054	0.09891	0.001674	0.04614	0	0.1221	0.08468	0.1501
8	0.04035	0.04035	0.04035	0.04902	0.03434	0.04614	0.04614	0	0.1462	0.07607	0.1083	0.1301	0.04613	0	0.04614	0.1476	0.08302	0.1573
9	0.1102	0.1102	0.1102	0.1098	0.1191	0.1115	0.1115	0.1462	0	0.08228	0.05425	0.02803	0.1116	0.1462	0.1116	0.04472	0.1362	0.1088
10	0.04157	0.04157	0.04157	0.04779	0.05408	0.04863	0.04863	0.07607	0.08228	0	0.06516	0.07052	0.04866	0.07607	0.04863	0.08499	0.1102	0.1156
11	0.07575	0.07575	0.07575	0.08018	0.08331	0.08054	0.08054	0.1083	0.05425	0.06516	0	0.03027	0.08053	0.1083	0.08054	0.05606	0.08587	0.09383
12	0.09447	0.09447	0.09447	0.09794	0.104	0.09891	0.09891	0.1301	0.02803	0.07052	0.03027	0	0.09894	0.1301	0.09891	0.03227	0.1161	0.08806
13	0.01852	0.01852	0.01852	0.003496	0.01306	0.001674	0.001674	0.04613	0.1116	0.04866	0.08053	0.09894	0	0.04613	0.001674	0.1221	0.08464	0.15
14	0.04035	0.04035	0.04035	0.04902	0.03434	0.04614	0.04614	0	0.1462	0.07607	0.1083	0.1301	0.04613	0	0.04614	0.1476	0.08302	0.1573
15	0.01901	0.01901	0.01901	0.003174	0.013	0	0	0.04614	0.1116	0.04863	0.08054	0.09891	0.001674	0.04614	0	0.1221	0.08468	0.1501
16	0.1144	0.1144	0.1144	0.1214	0.1259	0.1221	0.1221	0.1476	0.04472	0.08499	0.05606	0.03227	0.1221	0.1476	0.1221	0	0.1377	0.06499
17	0.0822	0.0822	0.0822	0.08664	0.07719	0.08468	0.08468	0.08302	0.1362	0.1102	0.08587	0.1161	0.08464	0.08302	0.08468	0.1377	0	0.1468
18	0.136	0.136	0.136	0.1506	0.1486	0.1501	0.1501	0.1573	0.1088	0.1156	0.09383	0.08806	0.15	0.1573	0.1501	0.06499	0.1468	0

Distance matrix derived from bioinformatics feature vectors of D2

Figure 8. *Cont.*

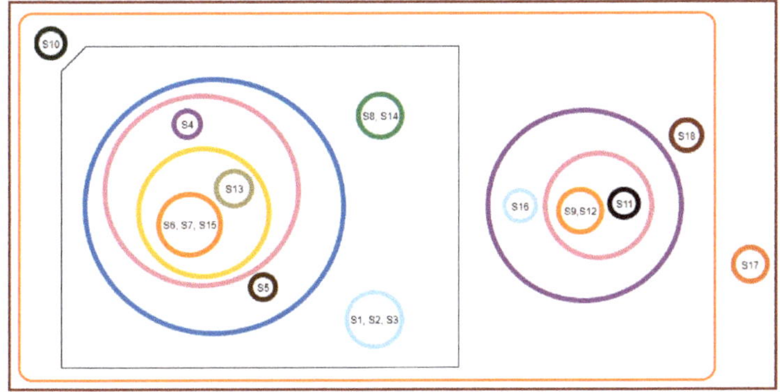

Figure 8. Distance matrix based on the bioinformatics feature vectors of D2 of ACE2 across eighteen species and associated clusters.

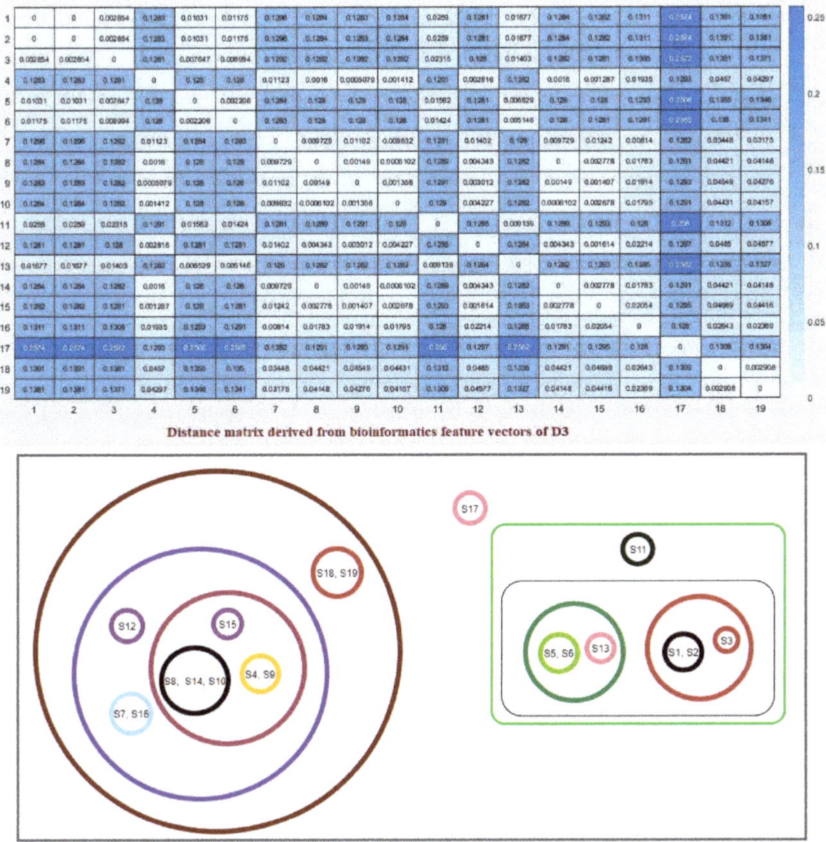

Figure 9. Distance matrix based on the bioinformatics feature vectors of D3 of ACE2 across nineteen species and associated clusters.

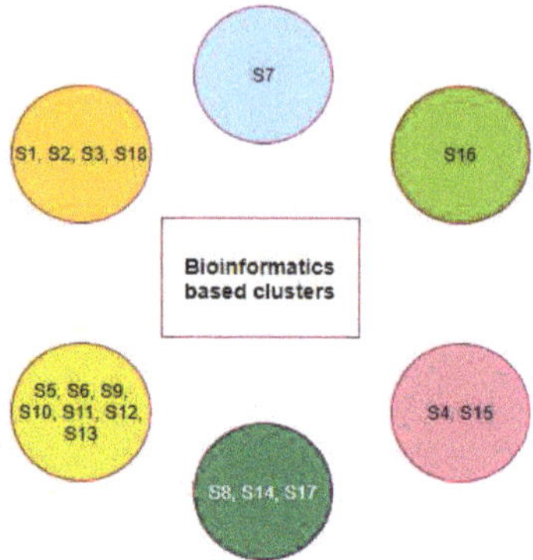

Figure 10. Clusters of species based on the bioinformatics of the D1, D2, and D3 domains.

2.4. Phylogeny and Clustering Based on Polarity

In the D1 domain, it was observed that the polarity of thirteen amino acids among nineteen (24–42 aa) amino acids were found to be conserved across eighteen species. Based on the amino acids' polarity and non-polarity nature, the species were arranged in a phylogenetic tree (Figure 11).

It was found that *Homo sapiens, Pan troglodyte, Macaca mulatta*, and *Danio rerio* closer according to this analysis. *Pteropus alecto, Pteropus vampyrus*, and *Sus scrofa* occurred in parallel along with the above three and formed a different clade indicating the closeness based on polarity. Again, the case for *Gallus gallus, Rhinolophus ferrumequinum, Mustela putorius furo, Equus caballus*, and *Felis catus* is similar. Two separate groups, *Mesocricetus auratus, Manis javanica*, and *Capra hircus, Bos taurus*, were similarly placed nearby, indicating that the polarity of amino acids of the proteins for these species was similar. *Pelodiscus sinensis* and *Rattus norvegicus* occurred separately and were not grouped with any other species but bears similarity with both the groups containing species *Mesocricetus auratus, Manis javanica, Pteropus alecto, Pteropus vampyrus, and Sus scrofa*.

In the D2 domain, out of six amino acid long sequences, the polarity of three amino acids was conserved across eighteen species, and among them, one amino acid was a binding residue. *Homo sapiens, Pan troglodytes, Macaca mulatta, Pteropus vampyrus*, and *Pteropus alecto* were grouped together since the overall polarity of their amino acid chain was found to be similar, and, simultaneously, *Danio rerio, Mustela putorius furo, Gallus gallus*, and *Pelodiscus sinensis* were placed together. In addition, three groups comprising *Manis javanica, Capra hircus, Bos taurus, Sus scrofa, Rattus norvegicus, and Rhinolophus ferrumequinum*, respectively, were placed in close proximity based on their polarity and non-polarity of the amino acids in the protein sequence. However, *Equus caballus, Felis cattus*, and *Mesocricetus auratus* were placed separately since they did not show much resemblance based on polarity.

In the D3 domain of ACE2 sequences of *Salmo salar* and *Danio rerio*, there was an insertion of a polar amino acid into one of the binding residue positions that may affect the binding of ACE2 to that of RBD of SARS-CoV2 negatively. A total of three binding residues were already reported in the D3 domain, of which one of them remained conserved concerning polarity across the nineteen species. *Rattus norvegicus, Mustela putorius furo, Mesocricetus auratus*, and *Felis catus* were grouped under a single clade based on the polarity of their protein sequence. It was a similar case for *Danio rerio, Pelodiscus sinensis, Salmo salar*, and *Gallus gallus*. Due to the sequence similarity between *Pteropus vampyrus* and *Pteropus alecto*, their polarity of the protein sequence was also similar and thus grouped. Sequence similarity was also observed for *Homo sapiens, Pan troglodytes*, and *Macaca mulatta*, so again these were categorized together. Two groups comprised of *Rhinolophus ferrumequinum, Capra hircus*, and *Bos taurus, Sus scrofa*, respectively, were sorted together indicating their similar nature of polarity and non-polarity of protein sequence. Lastly, *Manis javanica* and *Equus caballus* were placed separately signifying that the sequences of both species were quite distinct.

The individual groups of species based on the polarity of individual D1, D2, and D3 domains have emerged into six disjoint clusters (Figure 12).

Here, the clusters $\{S1, S2, S3\}$, $\{S16, S17, S18\}$, $\{S8, S14\}$, $\{S6, S13\}$, and $\{S10, S12\}$ remained invariant with regard to the homology of full length ACE2 as well as polarity sequence of the D1, D2, and D3 domains.

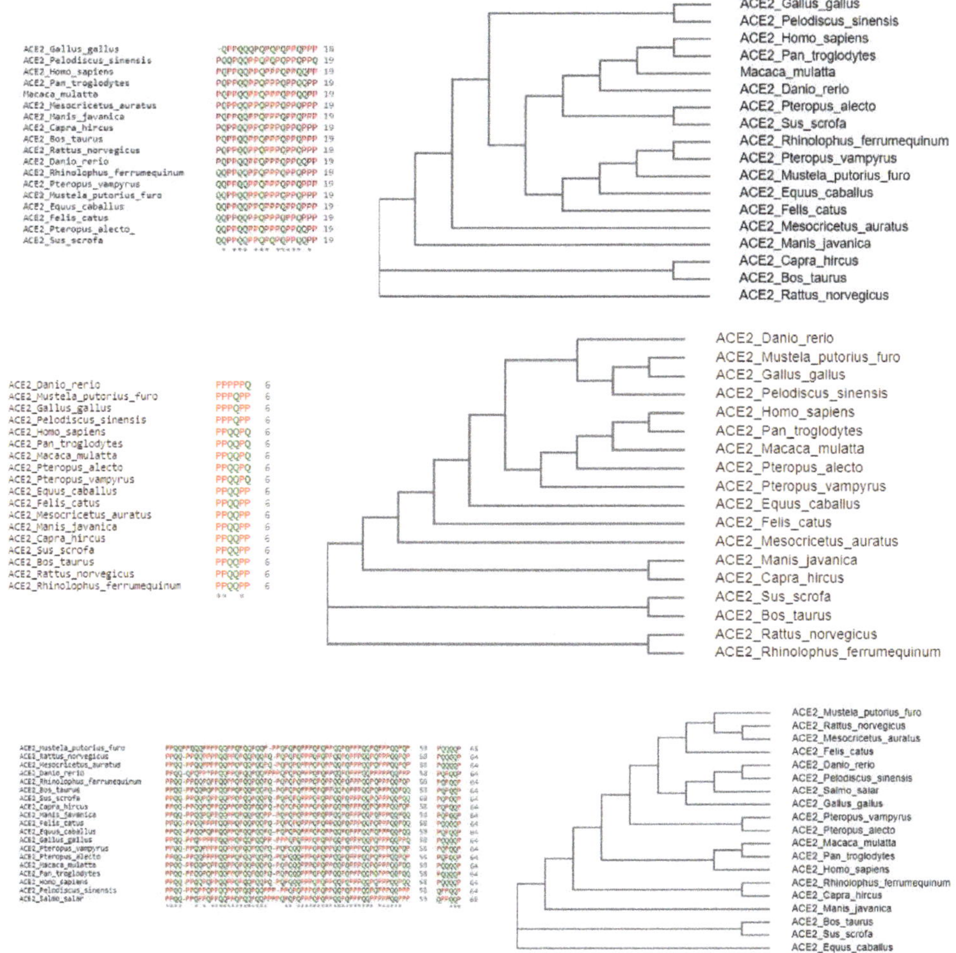

Figure 11. Polarity sequence of the D1, D2, and D3 domains across all species alignment and associated phylogenetic relationships.

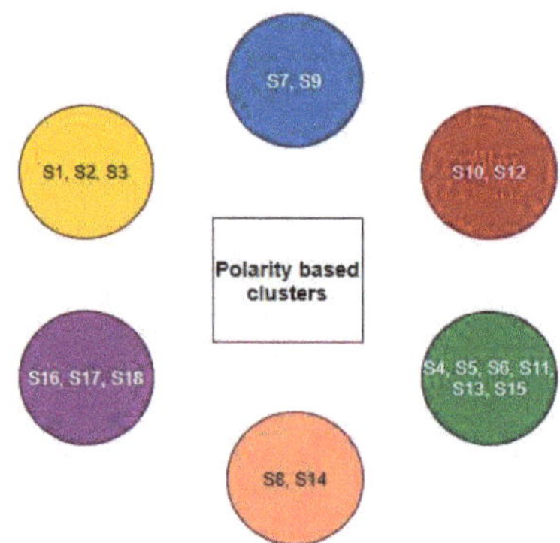

Figure 12. Clusters based on groups of species based on domain-wise polarity.

2.5. Possible Clusters of Transmission of SARS-CoV-2

Based on all the different clusters formed on the basis of amino acid homology, secondary structures, bioinformatics, and polarity of the D1, D2, and D3 domains of ACE2, final clusters of all nineteen species were devised using the K-means clustering method Figure 13.

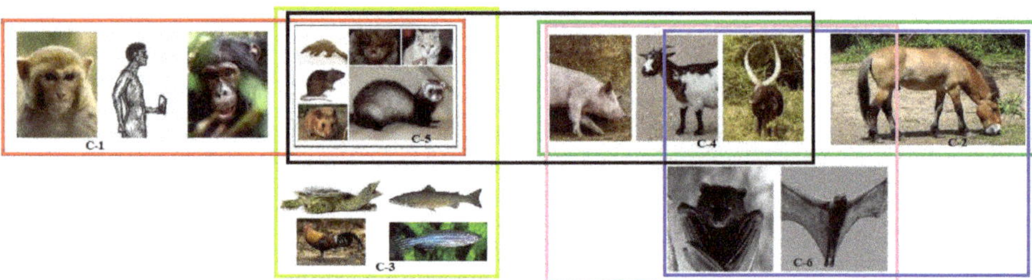

Figure 13. Schematic representation of a possible set of clusters of transmission of SARS-CoV-2.

In Figure 13, it was found that the cluster-1 (C-1) comprising of *Homo sapiens, Pan troglodyte*, and *Macaca mulatta* were close to cluster-5 (C-5) comprising of *Felis catus* (Cat), *Mesocricetus auratus* (Golden Hamster), *Manis javanica* (Sunda pangolin), *Mustela putorius furo* (Ferret), *Rattus norvegicus* (Rat), and *Rhinolphus ferrumequinum* (Greater horseshoe bat) (Figure 13). This C-5 is also close to cluster-3 (C-3) [*Gallus gallus* (red jungle fowl), *Pelodiscus sinensis* (Chinese shell turtle), *Danio rerio* (zebrafish) and *Salmo salar*], and cluster-4 (C-4) [*Capra hircus* (Goat), *Bos taurus* (Cattle), and *Sus scrofa* (pig)]. C-4 also showed resemblance with cluster-2 (C-2) [*Pteropus alecto, Pteropus vampyrus*], and cluster-6 (C 6) that is comprised of *Equus caballus* (horse) only. However, both C-2 and C-6 were also close to each other.

Furthermore, pooled analyses based on the two types of substitutions (one is affecting SARS-CoV-2 transmission (M1), and the other one is SARS-CoV-2 non-affecting transmission(M2)) for all six of the final clusters, which are presented in Table 1.

Based on Table 1, information regarding the number of M1 and M2 substitutions and the intra-species transmission of SARS-CoV-2 were presented as follows:

- C-1: None of the species bear any mutation in the binding residues and are conserved, so viral transmission is immaculate.
- C-2: This cluster has an equal number of transmission affecting and transmission non-affecting types of substitutions. Therefore, both have an equal probability of getting infected from each other.

- C-3: Here, again, *Gallus gallus*, *Pelodiscus sinensis*, and *Danio rerio* have a similar ratio of S1 to S2, signifying possible flow of viral transmission within these three species. However, *Salmo salar* is unique and distant, and therefore, the probability of viral transmission is unlikely.
- C-4: The species in this cluster have a similar number of transmission-affecting and transmission non-affecting types of substitutions show that the flow of viral transmission would be continuous among these three species.
- C-5: Transmission between *Felis catus* and *Mesocrietus auratus* is highly likely, which is the same for *Manis javanica*, *Mustela putorius furo*, and *Rattus norvegicus* as indicated by their similar number of substitutions. Therefore, the inter-transmission between these species is highly plausible. While *Rhinolophus ferrumequinum* has a relatively high value of transmission affecting substitutions from all of the above, its susceptibility to getting infected from other species is uncertain.
- C-6: A total of five transmission affecting substitutions in the three domains for *Homo sapiens* were observed.

Table 1. M1 and M2 substitutions across nineteen ACE2 receptors.

Cluster	Species	DI (M1, M2)	D2 (M1, M2)	D3 (M1, M2)	Total (M1, M2)
1	*Human ACE2*	0	0	0	0
1	*Pan troglodytes*	0	0	0	0
1	*Macaca mulatta*	0	0	0	0
6	*Equus caballus*	(4,1)	(1,0)	(0,0)	(5,1)
5	*Felis catus*	(1,2)	(1,0)	(0,0)	(2,2)
5	*Mesocrietus auratus*	(1,0)	(1,0)	(0,0)	(2,0)
5	*Manis javanica*	(2,2)	(1,1)	(1,0)	(4,3)
5	*Mustela putorius furo*	(2,2)	(2,0)	(1,0)	(5,2)
5	*Rattus norvegicus*	(3,1)	(3,1)	(0,1)	(6,3)
5	*Rhinolophus ferrumequinum*	(6,0)	(2,0)	(0,0)	(8,0)
4	*Capra hircus*	(0,1)	(2,0)	(0,0)	(2,1)
4	*Bos taurus*	(0,1)	(2,0)	(0,0)	(2,1)
4	*Sus scrofa*	(1,2)	(1,1)	(0,1)	(2,4)
2	*Pteropus vampyrus*	(2,1)	(1,0)	(1,1)	(4,2)
2	*Pteropus alecto*	(2,1)	(1,0)	(1,1)	(4,2)
3	*Gallus gallus*	(6,0)	(3,0)	(1,0)	(10,0)
3	*Pelodiscus sinensis*	(7,0)	(2,0)	(1,0)	(10,0)
3	*Danio rerio*	(5,0)	(2,0)	(2,1)	(9,1)
3	*Salmo salar*	NA	NA	(3,1)	(3,1)

3. Discussion

In this study, we amassed the ACE2 protein sequences of nineteen species to investigate the possible transmission of SARS-CoV-2 among these species in relation to human ACE2 protein. Multiple sequence alignments of these ACE2 receptors enabled us to estimate the similarity concerning amino acids and, from that, we observed that *Salmo salar* (Salmon fish) was quite distant. It also gave us the idea that some of the amino acid substitutions in the binding residues occurring across the species with respect to human ACE2 resulted in amino acids have similar binding properties, indicating that their interactions with RBD of the S protein will be similar to that of humans, thus making transmission across these species feasible. It was observed that ACE2 sequences from *Homo sapiens* and *Pan troglodytes* (Chimpanzee) were almost identical (showing 99.01% sequence identity). Although ACE2 from *Macaca mulatta* (Rhesus macaque) also shared a high percentage of sequence identity with human protein (95.16%), it possesses substitutions at 39 positions. However, no substitutions were observed in the amino acid residues involved in the interaction with the RBD of the S protein, making the viral transmission across these species highly likely. Again, *Pteropus vampyrus* (Large flying fox) and *Pteropus alecto* (Black flying fox) have precisely the same ACE2 sequence, and thus signifying high viral transmission and that both of them have an equal chance of getting infected by each other.

Further analysis led us to present a possible transmission flow among the nineteen species, as illustrated in Figure 13. The multifaceted examination of the ACE2 protein indicated that interspecies SARS-CoV-2 transmission is quite possible, and we have tried to provide a better insight into it by predicting the possible transmission among species within the same cluster and between clusters too. However, further in-depth analysis is necessary in the future for the identification of new hosts of SARS-CoV-2 as well as for determination of possible ways to prevent inter-species transmission.

The results reported in this study allow us to propose possible routes of the SARS-CoV-2 transmission flow among species. Unsurprisingly, our results indicate that, among the species studied, it is the members of primates that are the most at risk, followed by those of carnivores, cetartiodactyls, and finally bats. It is settling to see that the predicted transmission flow based on the results of our analyses is in line with the conventional evolutionary knowledge and reported infection cases. One should keep in mind though that the major goal of this study was to provide formally comprehensive structural evidence that could help in clarifying why some hosts are more susceptible than others to SARS-CoV-2 and could constitute a reservoir for further virus spillover. Obviously, more detailed studies are needed in the future to take into account structural properties of ACE2 and peculiarities of its interaction with the RBD of the S protein [15–18], and the presence of different ACE2 isoforms in individual animal species (e.g., humans have at least five ACE2 isoforms [26]). Moreover, one should consider the epigenetic regulation and expression determination of ACE2 (e.g., despite having the same protein sequence, ACE2 is differently expressed in different human cells, and different levels of expression of ACE2 are found in the same type of nasal epithelial cells or pneumocytes from humans and mice). It will also be necessary to analyze more ACE2 sequences from other species and to investigate the possibility for these different species of transforming themselves, in the long term, into healthy carriers of the virus or even into transmitters and diffusers of the disease.

It is well known that protein pairs with a sequence identity greater than 40% are very likely to be structurally similar, whereas protein pairs with a sequence identity of 20–35% represent a 'twilight zone', where structural similarity in pairs is considerably less common, with less than 10% of protein pairs with sequence identity below 25% have similar structures [29–31]. Sequence identity of the ACE2 proteins from nineteen species analyzed in this study ranges from 99.01% (*Homo sapiens* vs. *Pan troglodytes*) to 58.0% (*Homo sapiens* vs. *Danio rerio*), with the lowest identity of 57.13% being between the proteins from *Danio rerio* and *Rhinolophus ferrumequinum*. Therefore, one might expect rather close overall structural organization of all these proteins, even the most distant ones. In fact, even the lowest level of sequence identity for the pair of ACE2 proteins is still well above the sequence identity of 20–35% characteristics for the 'twilight zone'. On the other hand, fold-level, global structural similarity does not exclude the presence of local structural variability that might define, for example, the peculiarities of protein–protein interactions. Structural information is currently available only for the ACE2 from *Homo sapiens* and *Felis catus*. Therefore, previous studies that analyzed the peculiarities of interactions between the viral spike protein and host ACE2 from many household and other animals, such as *Pan troglodytes* (chimpanzee), *Macaca mulatta* (Rhesus monkey), *Felis catus* (domestic cat), *Equus caballus* (horse), *Oryctolagus cuniculus* (rabbit), *Canis lupus familiaris* (dog), *Sus scrofa* (pig), *Avis aries* (sheep), *Bos taurus* (cattle), *Mus musculus* (house mouse), and *Mustela putorius furo* (ferret) [32–34] were focused on the structural part of these interactions and utilized a typical set of structural biology approaches, such as homology modelling and docking. Therefore, in line with our previous study [25], we decided to compare the peculiarities of the per-residue intrinsic disorder predispositions of the ACE2 proteins from nineteen species analyzed in this study rather than building their homology models. Figure 14 summarizes the results of this analysis and shows that, although these proteins have rather similar intrinsic disorder predispositions, their disorder profiles are not identical.

Furthermore, such differences in the intrinsic disorder predisposition are not equally spread through the protein sequences, with some regions (e.g., the N-terminal 150 residues and residues 500–700) of the disorder profiles showing rather noticeable variability. Figure 14 also shows that the S protein binding domains D1 and D2 of ACE2 proteins are characterized by high variability of their intrinsic disorder predispositions, whereas D3 domains are more conserved. We also looked at the peculiarities of intrinsic disorder profiles of ACE2 proteins in six clusters with the major focus at the S protein binding domains D1, D2, and D3 (see Figure 15).

This comparison revealed that the in-cluster variability of intrinsic disorder propensity was noticeably lower than the diversity between the clusters as a rule. These observations support the notion that the capability of ACE2 to interact with SARS-CoV-2 protein S can be dependent on the peculiarities of the ACE2 local intrinsic disorder predisposition [25].

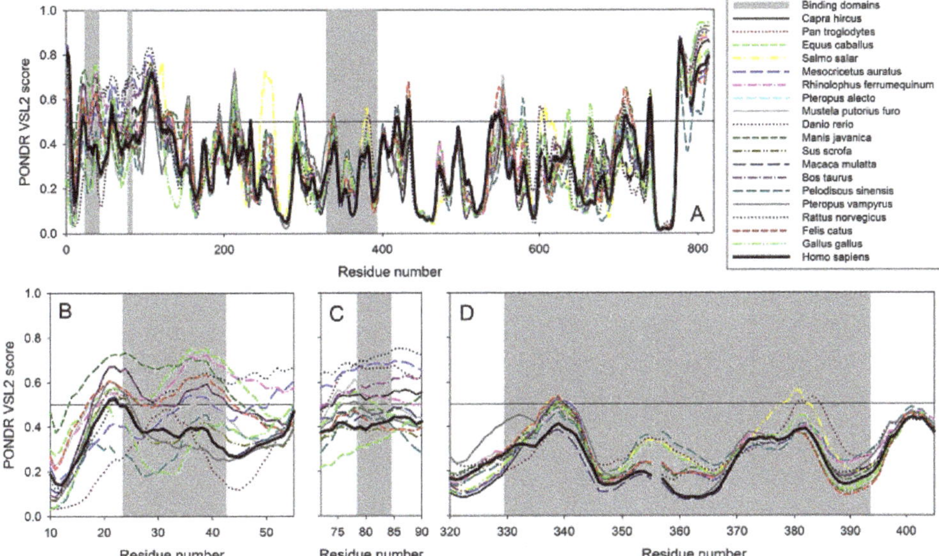

Figure 14. Per-residue intrinsic disorder predisposition of ACE2 proteins. (**A**) Peculiarities of the intrinsic disorder distribution within the amino sequences of ACE2 protein from nineteen species analyzed in this study. Light gray vertical bars show the location of the ACE2 regions responsible for interaction with SARS-CoV-2 S protein, domains D1 (residues 24–42), D2 (residues 79–84), and D3 (residues 330–393). (**B**–**D**). Zoomed-in disorder profiles focusing at the domains D1 (**B**), D2 (**C**), and D3 (**D**) responsible for the ACE2-S interaction. Disorder predispositions were evaluated using the PONDR® VSL2 algorithm.

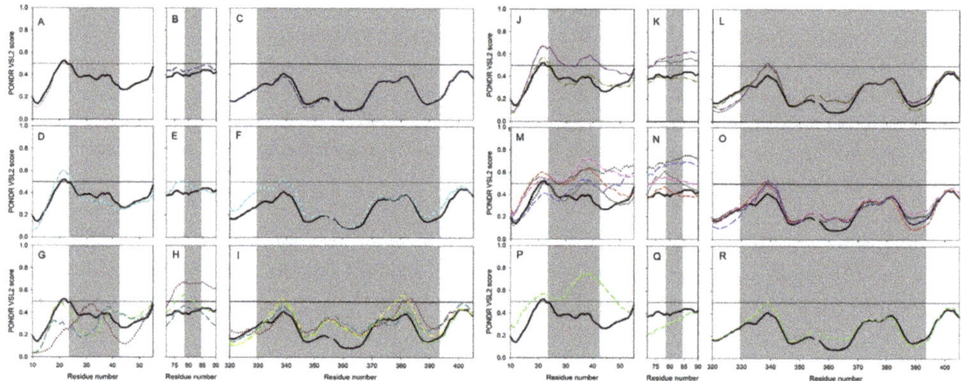

Figure 15. Peculiarities of intrinsic disorder predisposition within the D1 (**A,D,G,J,M**), and (**P**), D2 (**B,E,H,K,N**), and (**Q**), and D3 domains (**C,F,I,L,O**), and (**R**) of ACE2 proteins from cluster 1 (**A,B**), and (**C**), cluster 2 (**D,E**), and (**F**), cluster 3 (**G,H**), and (**I**)), cluster 4 (**J,K**), and (**L**), cluster 5 (**M,N**), and (**O**), and cluster 6 (**P,Q**), and (**R**). For keys, see Figure 14.

4. Materials and Methods

4.1. Data Acquisition and Findings

The ACE2 protein receptor sequences from nineteen species *Homo sapiens* (Human), *Capra hircus* (Domestic goat), *Pan troglodytes* (Chimpanzee), *Equus caballus* (Horse), *Salmo salar* (Atlantic salmon), *Mesocricetus auratus* (Golden hamster), *Rhinolophus ferrumequinum* (Greater horseshoe bat), *Pteropus alecto* (Black flying fox), *Mustela putorius furo* (Domestic ferret), *Danio rerio* (Zebrafish), *Manis javanica* (Sunda pangolin), *Sus scrofa* (Domestic pig), *Macaca mulatta* (Rhesus macaque), *Bos taurus* (Aurochs), *Pelodiscus sinensis* (Chinese soft-shelled turtle), *Pteropus vampyrus* (Large flying fox), *Rattus norvegicus* (Brown rat), *Felis catus* (Domestic cat), and *Gallus gallus* (Red jungle fowl) were derived from the NCBI database [35]. Nineteen species and their respective ACE2 protein accession IDs with length are presented in Table 2.

Table 2. Nineteen species and their associated ACE2 sequences.

Name	Species	ACE2 Accession ID	Length
S1	*Homo sapiens*	NP_001358344.1	805
S2	*Pan troglodytes*	PNI38577.1	805
S3	*Macaca mulatta*	XP_028697658.1	805
S4	*Equus caballus*	XP_001490241.1	805
S5	*Felis catus*	NP_001034545.1	805
S6	*Mesocricetus auratus*	XP_005074266.1	805
S7	*Manis javanica*	XP_017505752.1	805
S8	*Pteropus alecto*	XP_006911709.1	805
S9	*Capra hircus*	AHI85757.1	804
S10	*Sus scrofa*	NP_001116542.1	805
S11	*Mustela putorius furo*	XP_004758943.1	805
S12	*Bos taurus*	NP_001019673.2	804
S13	*Rattus norvegicus*	NP_001012006.1	805
S14	*Pteropus vampyrus*	XP_011361275.1	804
S15	*Rhinolophus ferrumequinum*	XP_032963186.1	805
S16	*Gallus gallus*	XP_416822.2	808
S17	*Pelodiscus sinensis*	XP_006122891.1	808
S18	*Danio rerio*	XP_005169417.1	807
S19	*Salmo salar*	XP_014062928.1	695

The nearest neighborhood phylogeny of the nineteen species derived from the NCBI public server based on ACE2 protein sequence similarity is shown in Figure 16A [36].

Figure 16. ACE2 full-length sequence-based phylogeny among nineteen species (**A**) and its derived clusters (**B**).

ACE2 sequence similarity among the species derives six clusters as shown in (Figure 16B). The contact residues of the receptor-binding domain (RBD) of the spike protein (YP_009724390.1) of SARS-CoV-2 with the homo sapiens ACE2 interface are presented in Table 3 [14].

Table 3. Contact residues of RBD spike protein of SARS-CoV-2 and Homo sapiens ACE2.

SARS-CoV-2 RBD	ACE2 (Homo Sapiens)	SARS-CoV-2 RBD	ACE2 (Homo Sapiens)
K417	Q24	Q493	Q42
G446	T27	G496	L79
Y449	F28	Q498	M82
Y453	D30	T500	Y83
L455	K31	N501	N330
F456	H34	G502	K353
A475	E35	Y505	G354
F486	E37		D355
N487	D38		R357
Y489	Y41		R393

The three designated domains, D1 (aa 24–42), D2 (aa 79–84), and D3 (aa 330–393) respectively contain the residues which bind to the RBD of the S protein.

4.2. Methods

Examining amino acid substitutions: For human ACE2 receptor, substitutions were examined for all species, and only those substitutions are accounted for, which occurred in the binding residues in the mentioned three domains D1, D2, and D3 [14]. Based on the character of the substitutions which interfered with the binding residues of the ACE2 across various species, two types were defined: substitutions affected transmission (M1) and substitutions which did not affect transmission (M2).

Multiple sequence alignments and associated phylogenetic trees were developed using the NCBI web-suite across all individual binding domains D1, D2, and D3 in eighteen species and D3 in *Salmo salar* [37,38].

K-means clustering: The algorithmic clustering technique derives homogeneous subclasses within the data such that data points in each cluster are as similar as possible according to a widely used distance measure viz. Euclidean distance. One of the most commonly used simple clustering techniques is the *K-means clustering* [39,40]. The algorithm is described below in brief:

Algorithm: K-means algorithm is an iterative algorithm that tries to form equivalence classes from the feature vectors into K (pre-defined) clusters where each data point belongs to only one cluster [39].

- Assign the number of desired clusters (K) (in the present study, $K = 6$).
- Find centroids by first shuffling the dataset and then randomly selecting K data points for the centroids without replacement.
- Keep iterating until there is no change to the centroids.
- Find the sum of the squared distance between data points and all centroids.
- Assign each data point to the closest cluster (centroid).
- Compute the centroids for the clusters by taking the average of the all data points that belong to each cluster.

In this present study, nineteen species were clustered using *Matlab* by inputting the distance matrix derived from the feature vectors associated with the three domains of ACE2 across all species.

Secondary structure predictions: The secondary structure of full-length ACE2 sequence of all species were predicted using the web-server CFSSP (Chou and Fasman Secondary Structure Prediction Server) [41]. This server predicts secondary structure regions from the protein sequence such as alpha-helix, beta-sheet, and turns from the amino acid sequence [41]. On obtaining the full-length ACE2 secondary structures, individual domains D1, D2, and D3 were cropped for each species.

Bioinformatics features: Several bioinformatics features viz. Shannon entropy, instability index, aliphatic index, charged residues, half-life, melting temperature, N-terminal of the sequence, molecular weight, extinction coefficient, net charge at pH7, and isoelectric point of D1, D2, and D3 domains of ACE2 for all nineteen species were determined using the web-servers *Pfeature and ProtParam* [42,43].

Computational analysis of the intrinsic disorder predisposition: Per-residue propensity of the ACE2 proteins from nineteen species for the intrinsic disorder were evaluated by the PONDR® VSL2 algorithm [44,45], which is one of the more accurate stand-alone per-residue disorder predictors [46,47]. In these analyses, residues with the disorder scores exceeding the threshold value of 0.5 are considered as intrinsically disordered, whereas residues with the predicted disorder scores between 0.2 and 0.5 are considered as flexible.

Shannon entropy: Shannon entropy measures the amount of complexity in a primary sequence of ACE2. It was determined using the web-server *Pfeature* by the formula

$$SE = -\sum_{i=1}^{20} p_i log_2(p_i),$$

where p_i denotes the frequency probability of a given amino acid in the sequence [42].

Instability index: Instability index is determined using the web-server *ProtParam*, and it estimates the stability of a protein in a test tube. A protein whose instability index is smaller than 40 is predicted as stable. A value above 40 predicts that the protein may be unstable [42].

Aliphatic index: Aliphatic index of a protein is defined as the relative volume gathered by aliphatic side chains (alanine, valine, isoleucine, and leucine). It may be regarded as a positive factor for increasing the thermostability of globular proteins, such as ACE2 [42].

N-terminal: It was reported that the N-terminal of a protein is responsible for its function. For each domain sequence, N-terminal residue was determined using the *Pfeature* [42].

In vivo half-life: The half-life predicts the time it takes for half of the protein amount to degrade after its synthesis in the cell. The N-end rule originated from the observations that the identity of the N-terminal residue of a protein plays an essential role in determining its stability in vivo [48].

Extinction coefficients: The extinction coefficient measures how much light a protein absorbs at a particular wavelength. It is useful to estimate this coefficient when a protein is purified [48].

Polarity sequence: Every amino acid in the domains D1, D2, and D3 of ACE2 were recognized as polar (P) and non-polar (Q) and thus every D1, D2, and D3 for eighteen species and the domain of

Salmo salar turned out to be binary sequences with two symbols P and Q. Then, homology of these sequences for each domain was made and, consequently, a phylogenetic relationship was drawn.

Author Contributions: S.S.H. conceived the problem. S.S.H., V.N.U., D.A., and S.G. carried out the work. S.S.H. and P.P.C., S.S.H. V. N.U. and G.K.A. analyzed the results and wrote the primary draft of the article; K.L., A.A.A.A., M.M.T., and P.A. edited the manuscript. B.D.U., M.S., D.P., A.S., T.M.A.E.-A., R.K., K.T., G.P. and A.M.B. have reviewed critically, and N.R., S.P.S., W.B.-d.-C., Á.S.-A., G.C., and have read the final draft. All authors have agreed to the published version of the manuscript.

Funding: This research received no external funding.

Conflicts of Interest: The authors do not have any conflicts of interest to declare.

References

1. World Health Organization. Severe acute respiratory syndrome (SARS). *Wkly. Epidemiol. Rec.* **2003**, *78*, 89.
2. World Health Organization. Outbreak news: Severe acute respiratory syndrome (SARS). *Wkly. Epidemiol. Rec.* **2003**, *78*, 81–83.
3. Zhang, J.J.; Dong, X.; Cao, Y.Y.; Yuan, Y.D.; Yang, Y.B.; Yan, Y.Q.; Akdis, C.A.; Gao, Y.D. Clinical characteristics of 140 patients infected with SARS-CoV-2 in Wuhan, China. *Allergy* **2020**, *75*, 1730–1741. [CrossRef] [PubMed]
4. Liu, Y.; Ning, Z.; Chen, Y.; Guo, M.; Liu, Y.; Gali, N.K.; Sun, L.; Duan, Y.; Cai, J.; Westerdahl, D.; et al. Aerodynamic analysis of SARS-CoV-2 in two Wuhan hospitals. *Nature* **2020**, *582*, 557–560. [CrossRef]
5. Sun, J.; He, W.T.; Wang, L.; Lai, A.; Ji, X.; Zhai, X.; Li, G.; Suchard, M.A.; Tian, J.; Zhou, J.; et al. COVID-19: epidemiology, evolution, and cross-disciplinary perspectives. *Trends Mol. Med.* **2020**, *26*, 483–495. [CrossRef]
6. World Health Organization. *Coronavirus Disease (COVID-19): Weekly Epidemiological, Update 1*; World Health Organization: Washington, DC, USA, 2020.
7. Zhang, Y.Z.; Holmes, E.C. A genomic perspective on the origin and emergence of SARS-CoV-2. *Cell* **2020**, *181*, 223–227. [CrossRef]
8. Zheng, M.; Song, L. Novel antibody epitopes dominate the antigenicity of spike glycoprotein in SARS-CoV-2 compared to SARS-CoV. *Cell. Mol. Immunol.* **2020**, *17*, 536–538. [CrossRef]
9. Minakshi, R.; Jan, A.T.; Rahman, S.; Kim, J. A testimony of the surgent SARS-CoV-2 in the immunological panorama of the human host. *Front. Cell. Infect. Microbiol.* **2020**, *10*, 539. [CrossRef]
10. Yang, X.; Yu, Y.; Xu, J.; Shu, H.; Liu, H.; Wu, Y.; Zhang, L.; Yu, Z.; Fang, M.; Yu, T.; et al. Clinical course and outcomes of critically ill patients with SARS-CoV-2 pneumonia in Wuhan, China: A single-centered, retrospective, observational study. *Lancet Respir. Med.* **2020**, *8*, 475–481. [CrossRef]
11. Zaim, S.; Chong, J.H.; Sankaranarayanan, V.; Harky, A. COVID-19 and multi-organ response. *Curr. Probl. Cardiol.* **2020**, *45*, 100618. [CrossRef]
12. Saponaro, F.; Rutigliano, G.; Sestito, S.; Bandini, L.; Storti, B.; Bizzarri, R.; Zucchi, R. ACE2 in the era of SARS-CoV-2: Controversies and novel perspectives. *Front. Mol. Biosci.* **2020**, *7*. [CrossRef] [PubMed]
13. Li, S.; Han, J.; Zhang, A.; Han, Y.; Chen, M.; Liu, Z.; Shao, M.; Cao, W. Exploring the Demographics and Clinical Characteristics Related to the Expression of Angiotensin-Converting Enzyme 2, a Receptor of SARS-CoV-2. *Front. Med.* **2020**, *7*, 530. [CrossRef] [PubMed]
14. Lan, J.; Ge, J.; Yu, J.; Shan, S.; Zhou, H.; Fan, S.; Zhang, Q.; Shi, X.; Wang, Q.; Zhang, L.; et al. Structure of the SARS-CoV-2 spike receptor-binding domain bound to the ACE2 receptor. *Nature* **2020**, *581*, 215–220. [CrossRef] [PubMed]
15. Qiu, Y.; Zhao, Y.B.; Wang, Q.; Li, J.Y.; Zhou, Z.J.; Liao, C.H.; Ge, X.Y. Predicting the angiotensin converting enzyme 2 (ACE2) utilizing capability as the receptor of SARS-CoV-2. *Microbes Infect.* **2020**, *22*, 221–225. [CrossRef] [PubMed]
16. Hussain, M.; Jabeen, N.; Raza, F.; Shabbir, S.; Baig, A.A.; Amanullah, A.; Aziz, B. Structural variations in human ACE2 may influence its binding with SARS-CoV-2 spike protein. *J. Med Virol.* **2020**, *92*, 1580–1586. [CrossRef] [PubMed]
17. McMillan, P.; Uhal, B.D. COVID-19–A theory of autoimmunity to ACE-2. *MOJ Immunol.* **2020**, *7*, 17.
18. Luan, J.; Lu, Y.; Jin, X.; Zhang, L. Spike protein recognition of mammalian ACE2 predicts the host range and an optimized ACE2 for SARS-CoV-2 infection. *Biochem. Biophys. Res. Commun.* **2020**, *562*, 165–169. [CrossRef]

19. Samavati, L.; Uhal, B.D. ACE2, Much More Than Just a Receptor for SARS-COV-2. *Front. Cell. Infect. Microbiol.* **2020**, *10*, 317. [CrossRef]
20. Veeramachaneni, G.K.; Thunuguntla, V.; Bobbillapati, J.; Bondili, J.S. Structural and simulation analysis of hotspot residues interactions of SARS-CoV 2 with human ACE2 receptor. *J. Biomol. Struct. Dyn.* **2020**, 1–11. [CrossRef]
21. Li, W.; Zhang, C.; Sui, J.; Kuhn, J.H.; Moore, M.J.; Luo, S.; Wong, S.K.; Huang, I.C.; Xu, K.; Vasilieva, N.; et al. Receptor and viral determinants of SARS-coronavirus adaptation to human ACE2. *EMBO J.* **2005**, *24*, 1634–1643. [CrossRef]
22. Gheblawi, M.; Wang, K.; Viveiros, A.; Nguyen, Q.; Zhong, J.C.; Turner, A.J.; Raizada, M.K.; Grant, M.B.; Oudit, G.Y. Angiotensin-converting enzyme 2: SARS-CoV-2 receptor and regulator of the renin-angiotensin system: Celebrating the 20th anniversary of the discovery of ACE2. *Circ. Res.* **2020**, *126*, 1456–1474. [CrossRef] [PubMed]
23. Tikellis, C.; Thomas, M. Angiotensin-converting enzyme 2 (ACE2) is a key modulator of the renin angiotensin system in health and disease. *Int. J. Pept.* **2012**, *2012*, 256294. [CrossRef] [PubMed]
24. Gorshkov, K.; Susumu, K.; Chen, J.; Xu, M.; Pradhan, M.; Zhu, W.; Hu, X.; Breger, J.C.; Wolak, M.; Oh, E. Quantum Dot-Conjugated SARS-CoV-2 Spike Pseudo-Virions Enable Tracking of Angiotensin Converting Enzyme 2 Binding and Endocytosis. *ACS Nano* **2020**, *14*, 12234—12247. [CrossRef] [PubMed]
25. Uversky, V.N.; Elrashdy, F.; Aljadawi, A.; Redwan, E.M. Household pets and SARS-CoV2 transmissibility in the light of the ACE2 intrinsic disorder status. *J. Biomol. Struct. Dyn.* **2020**, 1–4. [CrossRef]
26. Sang, E.R.; Tian, Y.; Gong, Y.; Miller, L.C.; Sang, Y. Integrate structural analysis, isoform diversity, and interferon-inductive propensity of ACE2 to predict SARS-CoV2 susceptibility in vertebrates. *Heliyon* **2020**, *6*, e04818. [CrossRef]
27. Zhou, H.X.; Pang, X. Electrostatic interactions in protein structure, folding, binding, and condensation. *Chem. Rev.* **2018**, *118*, 1691–1741. [CrossRef]
28. Flocco, M.M.; Mowbray, S.L. Strange bedfellows: Interactions between acidic side-chains in proteins. *J. Mol. Biol.* **1995**, *254*, 96–105. [CrossRef]
29. Rost, B. Twilight zone of protein sequence alignments. *Protein Eng.* **1999**, *12*, 85–94. [CrossRef]
30. Kinjo, A.R.; Nishikawa, K. Eigenvalue analysis of amino acid substitution matrices reveals a sharp transition of the mode of sequence conservation in proteins. *Bioinformatics* **2004**, *20*, 2504–2508. [CrossRef]
31. Krissinel, E. On the relationship between sequence and structure similarities in proteomics. *Bioinformatics* **2007**, *23*, 717–723. [CrossRef]
32. Shen, M.; Liu, C.; Xu, R.; Ruan, Z.; Zhao, S.; Zhang, H.; Wang, W.; Huang, X.; Yang, L.; Tang, Y.; et al. SARS-CoV-2 Infection of Cats and Dogs? *Preprints* **2020**, *2020*, 2020040116.
33. Shi, J.; Wen, Z.; Zhong, G.; Yang, H.; Wang, C.; Huang, B.; Liu, R.; He, X.; Shuai, L.; Sun, Z.; et al. Susceptibility of ferrets, cats, dogs, and other domesticated animals to SARS–coronavirus 2. *Science* **2020**, *368*, 1016–1020. [CrossRef] [PubMed]
34. Sit, T.H.; Brackman, C.J.; Ip, S.M.; Tam, K.W.; Law, P.Y.; To, E.M.; Yu, V.Y.; Sims, L.D.; Tsang, D.N.; Chu, D.K.; et al. Infection of dogs with SARS-CoV-2. *Nature* **2020**, *586*, 776–778. [CrossRef] [PubMed]
35. Pruitt, K.D.; Tatusova, T.; Maglott, D.R. NCBI Reference Sequence (RefSeq): A curated non-redundant sequence database of genomes, transcripts and proteins. *Nucleic Acids Res.* **2005**, *33*, D501–D504. [CrossRef] [PubMed]
36. Federhen, S. The NCBI taxonomy database. *Nucleic Acids Res.* **2012**, *40*, D136–D143. [CrossRef] [PubMed]
37. Johnson, M.; Zaretskaya, I.; Raytselis, Y.; Merezhuk, Y.; McGinnis, S.; Madden, T.L. NCBI BLAST: A better web interface. *Nucleic Acids Res.* **2008**, *36*, W5–W9. [CrossRef] [PubMed]
38. Jenuth, J.P. The NCBI. In *Bioinformatics Methods and Protocols*; Springer: Berlin/Heidelberg, Germany, 2000; pp. 301–312.
39. Likas, A.; Vlassis, N.; Verbeek, J.J. The global k-means clustering algorithm. *Pattern Recognit.* **2003**, *36*, 451–461. [CrossRef]
40. Lu, Y.; Lu, S.; Fotouhi, F.; Deng, Y.; Brown, S.J. FGKA: A fast genetic k-means clustering algorithm. In Proceedings of the 2004 ACM Symposium on Applied Computing, Nicosia, Cyprus, 14–17 March 2004; pp. 622–623.
41. Kumar, T.A. CFSSP: Chou and Fasman secondary structure prediction server. *Wide Spectr.* **2013**, *1*, 15–19.

42. Pande, A.; Patiyal, S.; Lathwal, A.; Arora, C.; Kaur, D.; Dhall, A.; Mishra, G.; Kaur, H.; Sharma, N.; Jain, S.; et al. Computing wide range of protein/peptide features from their sequence and structure. *bioRxiv* **2019**, 599126.
43. Garg, V.K.; Avashthi, H.; Tiwari, A.; Jain, P.A.; Ramkete, P.W.; Kayastha, A.M.; Singh, V.K. MFPPI–multi FASTA ProtParam interface. *Bioinformation* **2016**, *12*, 74. [CrossRef]
44. Peng, K.; Radivojac, P.; Vucetic, S.; Dunker, A.K.; Obradovic, Z. Length-dependent prediction of protein intrinsic disorder. *BMC Bioinform.* **2006**, *7*, 208. [CrossRef] [PubMed]
45. Obradovic, Z.; Peng, K.; Vucetic, S.; Radivojac, P.; Dunker, A.K. Exploiting heterogeneous sequence properties improves prediction of protein disorder. *Proteins Struct. Funct. Bioinform.* **2005**, *61*, 176–182. [CrossRef] [PubMed]
46. Peng, Z.L.; Kurgan, L. Comprehensive comparative assessment of in-silico predictors of disordered regions. *Curr. Protein Pept. Sci.* **2012**, *13*, 6–18. [CrossRef] [PubMed]
47. Meng, F.; Uversky, V.N.; Kurgan, L. Comprehensive review of methods for prediction of intrinsic disorder and its molecular functions. *Cell. Mol. Life Sci.* **2017**, *74*, 3069–3090. [CrossRef] [PubMed]
48. Pace, C.N.; Vajdos, F.; Fee, L.; Grimsley, G.; Gray, T. How to measure and predict the molar absorption coefficient of a protein. *Protein Sci.* **1995**, *4*, 2411–2423. [CrossRef] [PubMed]

Sample Availability: Data used in this manuscript is available in the NCBI public depository.

Publisher's Note: MDPI stays neutral with regard to jurisdictional claims in published maps and institutional affiliations.

© 2020 by the authors. Licensee MDPI, Basel, Switzerland. This article is an open access article distributed under the terms and conditions of the Creative Commons Attribution (CC BY) license (http://creativecommons.org/licenses/by/4.0/).

Review

Combining High-Pressure Perturbation with NMR Spectroscopy for a Structural and Dynamical Characterization of Protein Folding Pathways

Cécile Dubois, Isaline Herrada, Philippe Barthe and Christian Roumestand *

Centre de Biochimie Structurale, INSERM U1054, CNRS UMR 5048, Université de Montpellier, 34090 Montpellier, France; cecile.dubois@cbs.cnrs.fr (C.D.); isaline.herrada@yahoo.fr (I.H.); Philippe.Barthe@cbs.cnrs.fr (P.B.)

* Correspondence: christian.roumestand@cbs.cnrs.fr; Tel.: +33-4-6741-7704

Academic Editor: Marilisa Leone

Received: 15 October 2020; Accepted: 23 November 2020; Published: 26 November 2020

Abstract: High-hydrostatic pressure is an alternative perturbation method that can be used to destabilize globular proteins. Generally perfectly reversible, pressure exerts local effects on regions or domains of a protein containing internal voids, contrary to heat or chemical denaturant that destabilize protein structures uniformly. When combined with NMR spectroscopy, high pressure (HP) allows one to monitor at a residue-level resolution the structural transitions occurring upon unfolding and to determine the kinetic properties of the process. The use of HP-NMR has long been hampered by technical difficulties. Owing to the recent development of commercially available high-pressure sample cells, HP-NMR experiments can now be routinely performed. This review summarizes recent advances of HP-NMR techniques for the characterization at a quasi-atomic resolution of the protein folding energy landscape.

Keywords: protein folding; NMR; High Hydrostatic Pressure

---

1. Introduction

Since Anfinsen's early works [1], much effort has been expended in attempting to understand how amino acid sequence impacts the structure, dynamic properties, and global stability of proteins. On the other hand, much less is known about how proteins fold from single, disordered inactive polypeptide chains to unique tri-dimensional active structures. General descriptions of folding pathways cannot be predicted for arbitrary amino acid sequences but can be reached only from experimental studies, possibly coupled to molecular dynamic simulations. Thus, revealing the folding mechanism of proteins required knowledge from different fields: biology but also chemistry and physics, including computational simulations. Finally, the theoretical framework of free energy landscape theory and the funnel concept emerged [2–8], giving satisfactory models to understand this mechanism.

Folding/unfolding experiments performed in vitro have yielded much of the information concerning protein folding mechanisms. To this aim, several perturbation methods have been used, among them the addition of chemical denaturants (urea, guanidinium chloride), pH changes, or modification of the temperature of the sample are the most popular. Alternatively, pressure presents a lot of advantage to study protein unfolding. First, unfolding by pressure is reversible, essentially because high pressure, contrary to temperature, disfavors intermolecular protein interactions, preventing irreversible aggregation. This reversibility allows the measurement of thermodynamic parameters for the folding/unfolding reaction [9–11]. Second, contrary to the use of pH or chaotropic reagents, it does not change the charge or the chemical composition of the system. Pressure is actually a very straightforward perturbation: it induces unfolding because the molar volume of the folded states is larger than that of the unfolded states of protein, essentially because of the existence of solvent excluded voids in the folded states that are eliminated in the unfolded states [12,13]. Thus, following the Le Chatelier's principle, pressure shifts the folded state/unfolded state equilibrium toward the unfolded state, the one with the lower molar volume.

Pressure perturbation is generally used in combination with circular dichroism [14], fluorescence [15], or FT-IR spectroscopy [16–18]: these spectroscopic methods give global information on the state of the system, i.e., on the relative populations of the folded and unfolded states at a given pressure. Due to their extreme sensitivity to the structural environment, Nuclear Magnetic Resonance (NMR) observables constitute an attractive alternative. Global information on protein folding reaction can be obtained from simple 1D NMR spectra, as with the other spectroscopies quoted before. But more interestingly, multi-dimensional (2D) NMR provides multiple probes in the protein structure, giving access to local, residue-specific information. Thus, at the expense of some difficulties in implementation, most of which have been adequately addressed in commercial instrumentation, the combination of high-pressure perturbation with multi-dimensional NMR constitutes a powerful tool that can be used to describe the conformational landscape of proteins at a resolution which cannot be accessed by global spectroscopic observables.

2. High Pressure NMR Instrumentation

Combining high pressure with NMR spectroscopy constitutes a real challenge since several difficulties must be overcome: the system should be pressure-resistant, permeable to radiofrequency used for spin excitation, and non-magnetic in order to be safely introduced in the magnet. Of course, the conventional borosilicate or Pyrex glass tubes currently used for high-resolution liquid-state NMR do not match the "pressure-resistant" criterion, even though some commercial manufacturers offer borosilicate glass tubes, which can support moderate pressure up to 12 bar (Norell^{TM}). These tubes are essentially designed to work with gas-pressurized samples, the pressure limit being too low to allow protein denaturation, usually expected in the range 1–12 kbar at ambient temperature [19].

The first set-up really adapted for high-pressure NMR spectroscopy was the "autoclave" system developed by Benedek & Purcell in 1954 [20]. In this set-up, the sample and the radiofrequency coil are directly placed in a high-pressure non-magnetic vessel made initially of beryllium–copper alloy, later replaced by more resistant titanium alloy [21–24]. Very high pressure can be obtained with this set-up (9–10 kbar), compatible with protein denaturation. Nevertheless, it suffers from different drawbacks: among other things, the electrical coupling between the coil and the metallic chamber alters the coil efficiency, yielding low sensitivity. In addition, adding a second coil in the chamber is difficult, impeding the realization of heteronuclear experiments, standard in biomolecular NMR. An original alternative was proposed by Castro and Delsuc [25], where the metallic chamber was replaced by a composite material chamber of fiberglass and epoxy resin. Due to the insulator property of this material, the proton radiofrequency coil (excitation and detection) can be directly embedded in the chamber wall, and an additional coil for heteronuclei excitation can be glued directly on the external surface of the chamber. Nevertheless, the use of this probe was limited by its low burst pressure (1.5–2 kbar), hardly compatible with protein denaturation. Very recently, Meier et al. [26] published what can be considered as the ultimate development of the autoclave approach, at least in terms of pressure limit. They replaced the titanium chamber with a diamond anvil cell (DAC), similar to those used for HP crystallography [27]. The sample and the resonator are placed directly in the DAC, and pressures up to 0.9 Mbar can be reached with this set-up. Nevertheless, this set-up is not adapted to biomolecular RMN: the design of the radio-frequency section is not adapted to biomolecular NMR, and the sample volume (about 100 pL) is too small to yield enough sensitivity in case of biomolecules. Moreover, such very high pressure is not really useful for the study of protein denaturation. On the contrary, at above 10 kbar changes in the water structure are expected, which hamper the thermodynamic analysis.

The alternate strategy to "autoclave" systems is pressure-resistant tubes or cells that limit the pressurized region to the sample itself and can be used with commercial NMR probes.

Polyimide (Vespel) [28] and single crystal sapphire tubes [29] that can withstand pressure up to 1 kbar were proposed. If they improved comfort in the use of HP-NMR by organic chemists, their low burst pressure makes them unsuitable for the study of protein denaturation. In the mid-1970s, Yamada et al. [30,31] developed high-pressure glass (or quartz) cells allowing to work with pressure up to 3 kbar. Even if this pressure seems too low to enable denaturation for many proteins, this drawback can be circumvented by adding sub-denaturing concentration of chaotropic reagents [32] or of organic solvents to the buffer, and playing with the temperature [33], in order to tune the protein stability with the pressure range allowed by the system. The set-up consists of a long capillary enlarged on one end to form the cell itself that goes into the NMR probe, protected by a Teflon tube. The other end of the capillary is glued on a bronze–beryllium valve that allows pressure transmission from the HP-pump. The capillary length should fit with the magnet size, in order to be long enough to maintain the bronze–beryllium seal far from the magnetic center of the magnet, minimizing perturbations of the magnetic field homogeneity, thus permitting to record high-resolution spectra. It can be used with any commercial NMR probes and allows recording any through-bond or through-space homonuclear or heteronuclear (double-, triple-resonance) correlation NMR experiments. Akasaka and coworkers have used this cell to characterize the folding of numerous globular proteins [10]. Nevertheless, the manufacturing of this system remains delicate, and the sensitivity of the NMR experiments is limited by the small sample volume available (30–40 μL), requiring highly concentrated samples [34,35]. In 1996, Wand and coworkers developed a new set-up consisting of a simple two-component valve system that holds and seals a high-pressure sapphire tube [36,37]. A similar and complementary approach was proposed by the group of Kalbitzer [38], initially based on a sapphire tube and later replaced by a ceramic tube [39]. The later developments of these set-ups [40] have been integrated in the system now commercially available from Deadalus Innovation™ company. The high-pressure sample tubes are made from aluminum-toughened zirconia ceramic. They provide access to pressures up to 3 kbar and to a −15 to 115 °C temperature range. With an outer diameter of 5 mm, they are compatible with most of the commercially available probes. The inner diameter of 2.8 mm provides a working volume of about 200 μL, which allows for a sensitivity near that of 3 mm glass tubes, standard now for (ambient pressure) biomolecular NMR at high fields. The two-component valve initially proposed by Urbauer and Wand [36,37] is used to couple the ceramic tube to the high-pressure tube. Compared with Yamada's cell, this set-up provides a similar spectral quality but about a 10-fold increase in sensitivity and is incomparably easier to handle, essentially due to a wider inner diameter facilitating its filling. Thus, these ceramic tubes can be easily filled with complex viscous samples, such as those used for RDC measurement [41–43]. Pressure is generally transmitted by mineral oil, so that no physical separation is needed between the aqueous buffer containing the protein and the transmitting fluid.

3. "Global" Thermodynamic and Kinetic Parameters for Folding/Unfolding Reactions Obtained from 1D High-Pressure NMR Spectroscopy

Typically, 1D HP-NMR can be used for steady-state measurements, recorded when the ratio between folded/unfolded protein populations has reached equilibrium after a pressure jump. Such measurements give access to the global thermodynamic parameters ΔG^0 (the free-energy difference at atmospheric pressure between the folded and unfolded states of the protein) and ΔV^0 (the volume difference at atmospheric pressure between the folded and unfolded states of the protein), characteristic of the folding/unfolding reaction. In addition, kinetic measurements recorded just after the *P*-jump, while establishing equilibrium, are also possible with HP-NMR, yielding kinetic information on the folding/unfolding reaction as well as volumetric properties of protein folding transition states.

3.1. *Steady-State Measurements of Global Thermodynamic Parameters with 1D High-Pressure NMR*

The good dispersion usually observed in the ^{1}H-NMR spectrum of a folded protein is essentially due to the extreme sensibility of the proton NMR resonances (or chemical shifts) to through-space effects of neighboring groups. These effects vanished when the protein unfolds, and the ^{1}H-NMR spectrum becomes poorly resolved. For instance, the well-resolved regions characteristic of resonances of the methyl groups, or of the amide groups, in the ^{1}H spectrum of a folded protein will collapse upon unfolding (Figure 1).

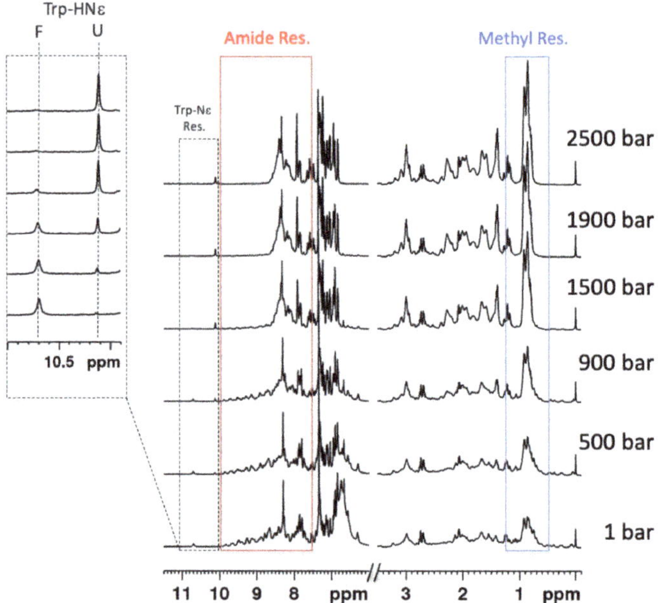

Figure 1. Evolution upon pressure of the 1D ^{1}H-NMR spectrum of Titin I27 Ig-like domain. Stacked plot of 1D spectra recorded as a function of pressure at 600 MHz and 298 K on a 1 mM sample of Titin I27 in Tris buffer pH 7.0, 1 mM DTT. A 1.7 M sub-denaturing concentration of GuHCl has been added to the sample in order to decrease the protein stability and to observe complete unfolding in the 1–2500 bar pressure range allowed by the experimental set-up (zirconium oxide ceramic tubes, Daedalus Innovation^{TM}). The solid-line frames delimit the regions corresponding to HN amide (red frame) and CH_3 methyl group resonances (blue frame). The insert corresponds to a zoom on the indole resonances region (black dashed-line frame) showing the decrease with pressure of the HN indole resonance of Trp-34 in the folded state (F) and the concomitant increase of the same resonance in the unfolded states (U).

Thus, the evolution of the 1D NMR spectrum allows us to monitor the high-pressure denaturation of a protein, as depicted in Figure 1 for the I27 Immunoglobin-like domain of the sarcomeric protein Titin [44]. As an effect of the energy barrier between the folded and unfolded protein ($\approx$2 kcal/mol in the experimental conditions reported in Figure 1), these species are in slow exchange with regard to the NMR timescale: we observe the disappearance of resonances belonging to the native state, with the concomitant appearance of new peaks that correspond to the spectrum of the unfolded states.

As it can be observed for the indole resonance of Trp-34 (Figure 2), the decrease in pressure of the peak corresponding to the folded (F) state, as well as the increase of the peak corresponding to the unfolded states (U), can be generally well-fitted by a sigmoidal curve [45,46] characteristic of a two-state equilibrium in the form of:

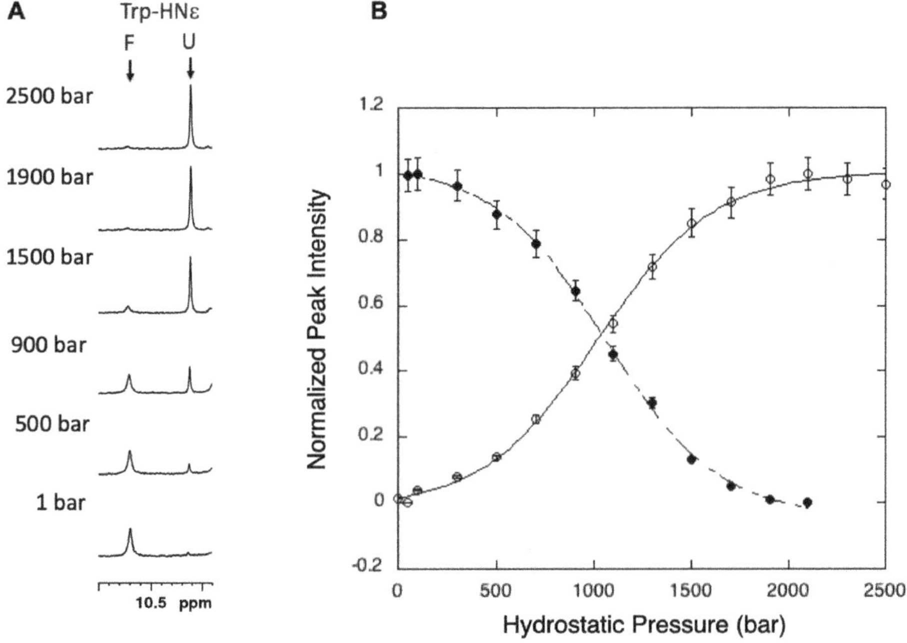

Figure 2. Monitoring the unfolding reaction of titin I27 domain with 1D HP-NMR spectroscopy (**A**): 1D HN indole region of the proton NMR spectra of Titin I27 recorded at increasing pressure. F stands for the resonance in the folded state, U for the resonance in the unfolded state. (**B**): denaturation curves obtained from the fit of the evolution with pressure of the native (open circle) and the denatured (filled circle) indole resonance of tryptophan-34 with a two-state equilibrium equation. Similar values of ΔV^0 are found for both fits.

with a characteristic F $\underset{K_f}{\overset{K_u}{\rightleftharpoons}}$ U equilibrium constant of:

$$K_{eq} = k_{f(p)}/k_{u(p)} = [U]/[F] \quad (1)$$

where $k_{f(p)}$ and $k_{u(p)}$ stands for the folding and the unfolding rate constants at a given pressure p. K_{eq} can be also expressed from the Boltzmann equation as:

$$K_{eq} = \exp(-\Delta G_{eq}/RT) \quad (2)$$

where the free energy change can be expressed as a Taylor expansion, truncated at the second order term:

$$\Delta G_{eq} = G_U - G_F = \Delta G^0 + \Delta V^0(p - p_0) - 1/2\,\Delta\beta V^0(p - p_0)^2 \quad (3)$$

Here ΔG_{eq} and ΔG^0 are the Gibbs-free energy changes from F to U at pressure p and p_0 (p_0 = 1 bar), respectively; ΔV^0 is the partial molar volume change; $\Delta\beta$ is the change in compressibility coefficient ($\Delta\beta = -(1/V^0) * \delta V/\delta p$), R is the gas constant, and T is the absolute temperature. It has been shown that for proteins the difference in compressibility between native and denatured states is negligible [47]. Thus, the expression of ΔG_{eq} simplifies to:

$$\Delta G_{eq} = \Delta G^0 + \Delta V^0(p - p_0) \quad (4)$$

Using NMR spectroscopy, the observable will be I, either the intensity (peak height) or the integral of a peak corresponding to either the folded species or of the unfolded species. In the present case (Figure 2), we chose to follow either the decrease of the peak intensity corresponding to the HN indole resonance of Trp-34 in the folded species or the increase of the corresponding resonance in the unfolded species. Alternatively, one can follow the increase of the peak at 0.86 ppm corresponding to the resonances of methyl groups in the unfolded species. Thus, the equilibrium constant can be written as:

$$K_{eq} = \frac{[U]}{[F]} = \frac{I_F - I}{I - I_U} \tag{5}$$

If we choose to follow the increase with pressure of a resonance corresponding to the unfolded species in the 1D NMR spectrum, I_F stands for the intensity of the corresponding NMR line in the folded spectrum at 1 bar ($I_F = I_{min}$), whereas I_U corresponds to the intensity of the same line at high pressure, when the protein is fully unfolded ($I_U = I_{max}$). Combining this equation with Equations (2) and (4) gives the characteristic equation for a two-state equilibrium:

$$I = \frac{I_F + I_U e^{-[\Delta G^0 + (p - p0)\Delta V^0]/RT}}{1 + e^{-[\Delta G^0 + (p - p0)\Delta V^0]/RT}} \tag{6}$$

Fitting either the sigmoidal decrease with pressure of the indole resonance in the folded state or the sigmoidal increase of the indole resonance in the unfolded state (Figure 2) yield "global" values for ΔV^0 of unfolding (−84 ± 5 mL/mol and −82.4 ± 5 mL/mol, respectively) and for ΔG^0 (2.14 ± 0.12 kcal/mol and 2.01 ± 0.13 kcal/mol, respectively), under the conditions of the study (pH = 7, 25 °C, 1.7M GuHCl). Note that the values extracted from the two different fits fall within experimental uncertainties, confirming the two-state equilibrium for the folding/unfolding reaction. Slightly different values (ΔV^0 = −78.8 ± 4 mL/mol, ΔG^0 = 2.26 ± 0.26 kcal/mol) can be measured at equilibrium when referring to the resonance corresponding to methyl groups in the unfolded states (0.96 ppm). This indicates that we are measuring "apparent" values for these thermodynamic parameters that depend of course on the global stability of the protein but that are also influenced by the local stability sensed by a given resonance in a given environment. As we will see further, this is of paramount importance for the description of protein folding pathways.

3.2. Measurements of Global Kinetic Parameters of the Folding/Unfolding Reaction with 1D High-Pressure NMR

ΔV^0 and ΔG^0 are thermodynamic parameters at atmospheric pressure, characteristics of the system at equilibrium. But a kinetic analysis of the folding/unfolding reaction is needed to obtain information on the transition state (usually described not as a unique conformer but as an ensemble of conformers, hence the term of "Transition State Ensemble" (TSE) used for proteins) of the reaction. Even though characterizing transition states in protein folding constitutes an essential step in the puzzle [48], the relations between the protein sequences, their 3D structures, and the structure (at least the hydration state) of their TSE are not yet well understood. Thus, the HP-NMR comparative study of the folding of Titin I27 module and DEN4-ED3 domain from the viral envelope of the dengue virus, two proteins with unrelated sequences but sharing a common Ig-like fold, shows similar folding intermediates but very different TSE, the transition state of Titin I27 being considerably less hydrated than the one of DEN4-ED3 [49]. Such analysis relies on the measurement of kinetic parameters after perturbation (*P*-jump) of the thermodynamic equilibrium between the folded and unfolded conformers of the protein at a given pressure, yielding the rates of folding and unfolding at atmospheric pressure. Moreover, these studies give access to the values of the activation volume between the folded or unfolded state and the TSE, related to the hydration state of the TSE.

Due to the very large volumes of activation involved in the folding/unfolding reaction, high pressure can considerably slow down the rate of folding and also possibly unfolding [13,50]. Thus, although the completion of a folding/unfolding reaction is usually a few seconds at atmospheric pressure, it can take up to a few hours at high pressure (about 12 h for Δ + PHS SNase at pressure above 1 kbar [12]). This is more than enough for the use of real-time 1D NMR to follow the folding/unfolding reaction after a *P*-jump, until steady state is achieved. Real-time 1D NMR spectroscopy consists in recording with time a series of 1D NMR spectrum at a constant repeating rate. After a *P*-jump, it is then possible to observe the exponential decrease with time of a resonance corresponding to the folded species, or the exponential growth with time of a resonance corresponding to the unfolded states, until the steady state is reached. In the case of Titin I27, we observed the exponential growth with time of the resonance at 0.96 ppm corresponding to the methyl groups of the unfolded states (Figure 3A). Alternatively, the exponential decrease with time after a positive *P*-jump of the well-resolved resonances of the shielded methyl protons can be used as a probe for such measurements, giving residue-specific information on the folding kinetics related to the local hydration of the TSE (see further) [32].

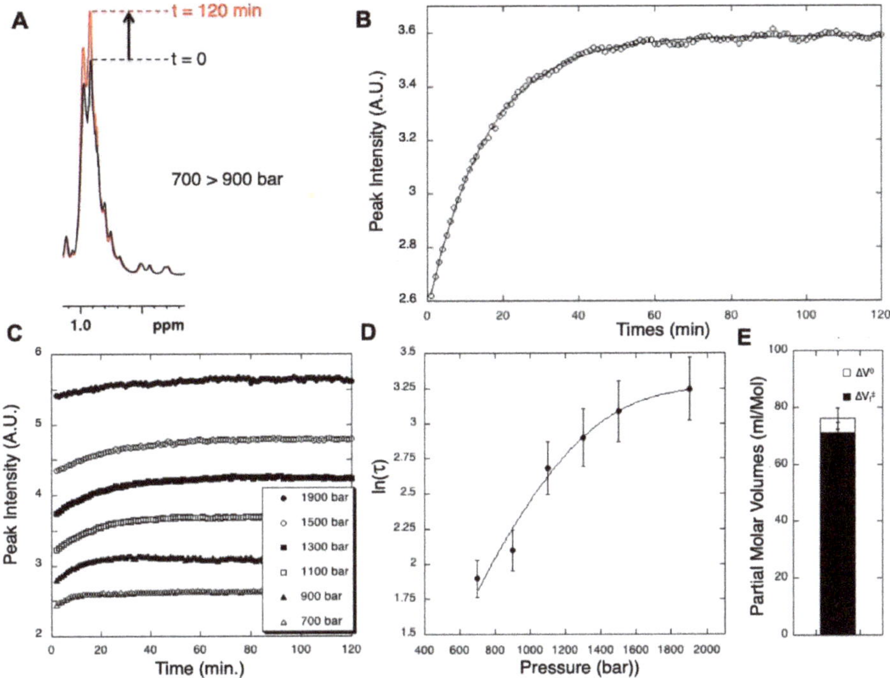

Figure 3. Measuring global kinetic parameters for the folding/unfolding reaction with real-time 1D HP-NMR spectroscopy. (**A**) Two 1D NMR proton spectra (methyl groups resonances) recorded on Titin I27 (same conditions as in Figure 1) just after a 700 to 900 bar P-Jump (black trace) and 2 h after the P-jump (red trace) 1). These two spectra represent the extreme points of a series of sixty spectra of 2 min each recorded over a period of 2 h. The arrow indicates the increase of the resonance at 0.96 ppm that corresponds to methyl groups in the unfolded species. (**B**) Measurement of the relaxation time, τ, at 900 bar, through the fit of the exponential growth of this methyl resonance. (**C**) Exponential growths of the resonance corresponding to the methyl groups in the unfolded states after successive 200 bar P-Jumps between 300 and 1900 bar, the pressure range where Titin I27 unfolds. Relaxation times $\tau_{(p)}$ can be measured from these experiments for the different pressures. (**D**) "Chevron plot" of ln(τ) measured at different pressures: the fit with Equation (8) allows to extract the folding or unfolding kinetic rate constants k_{f0} and k_{u0}, respectively, and the activation volume of folding $\Delta V^{\ddagger}_{f0}$ or of unfolding $\Delta V^{\ddagger}_{u0}$ at atmospheric pressure. (**E**) Volumetric diagram obtained for Titin I27 domain, displaying average values of $\Delta V^{\ddagger}_{f0}$ (plain bar) and ΔV^{0} (open bar). The value of the ratio $\Delta V^{\ddagger}_{f0}/\Delta V^{0}$ close to 1 deduced from this diagram indicates a dehydrated TSE.

Concerning the experimental aspects for the realization of the *P*-jump, these experiments are more demanding than those used for the steady-state analysis. Indeed, the time needed for the sample pressurization should be negligible with respect to the time needed by the folding/unfolding reaction to reach the plateau. For instance, the 200 bar *P*-jumps used for acquiring the data presented in Figure 3 needed about 10 s when performed with the Deadalus Innovation^{TM} Xtreme electric HP-pump. In this particular case, the time needed to reach the steady state after equilibrium was about 40 min (Figure 3C), so that the pressurization time can be safely neglected. For proteins with shorter folding relaxation times, Kremer et al. [51] have circumvented this limitation by pre-pressurizing a reservoir, upstream of the high-pressure cell, containing a large volume of mineral oil, much greater than the volume corresponding to the pressurization line and the high-pressure cell itself. Thus, opening an electric valve placed in between the reservoir and the cell allows an almost immediate (in the millisecond range) equilibration of the pressure between the reservoir and the sample cell. Since the volume of pressurization liquid is far greater in the reservoir than in the rest of the set-up, the final pressure reached in the sample cell when opening the valve is virtually the initial pressure in the reservoir.

The relaxation time characteristic of the kinetics after a given *P*-jump ($\tau_{(p)} = 1/(k_{u(p)} + k_{f(p)})$, where $k_{u(p)}$ and $k_{f(p)}$ are the unfolding and the folding rates at the pressure p reached at the end of the *P*-Jump, can be extracted from the fit of the exponential growth of this resonance (Figure 3B). Then, it becomes possible to extract the values at atmospheric pressure of k_{u0} and k_{f0}, as well as those of the activation volume of unfolding $\Delta V^{\ddagger}_{u0}$ (or folding, $\Delta V^{\ddagger}_{f0}$), by measuring this relaxation time after different p-jumps, between different pressures in the range where the protein unfolds (Figure 3C):

At a given pressure: $\tau_{(p)} = 1/(k_{u(p)} + k_{f(p)})$

with $k_f(p) = k_{f0}e^{-p\Delta V^{\ddagger}_{f0}/RT}$ and $k_u(p) = k_{u0}e^{-p\Delta V^{\ddagger}_{u0}/RT}$

$\tau_{(p)}$ can be rewritten as:

$$\tau_{(p)} = \left[k_{u0}e^{\left(\frac{-p\Delta V^{\ddagger}_{u0}}{RT}\right)} + k_{f0}e^{\left(\frac{-p\Delta V^{\ddagger}_{f0}}{RT}\right)}\right]^{-1} \tag{7}$$

The value of ΔV^0, the volume difference between the folded and unfolded states measured at equilibrium ΔV^0 (= $\Delta V_f - \Delta V_u$), and K_{eq} (= k_{f0}/k_{u0}) can be measured from the steady state experiments described above. One can then decrease the number of parameters for the fit:

$$\tau_{(p)} = \left[k_{u0}e^{\left(\frac{-p\Delta V^{\ddagger}_{u0}}{RT}\right)} + k_{u0}K_{eq}e^{\left(\frac{-p(\Delta V^0 + \Delta V^{\ddagger}_{u0})}{RT}\right)}\right]^{-1} \tag{8}$$

Only two variables need to be fitted: k_{u0} (or k_{f0}), the unfolding (or folding) rate at atmospheric pressure, and $\Delta V^{\ddagger}_{u0}$ (or $\Delta V^{\ddagger}_{f0}$), the activation volume for unfolding (or folding) at atmospheric pressure.

The fit is usually performed on a plot of ln(τ) as a function of pressure, displaying the characteristic "chevron plot" pattern (Figure 3D). In the case of Titin I27, the activation volume $\Delta V^{\ddagger}_{f0}$ is close to the equilibrium ΔV^0 value (Figure 3E), suggesting a dehydrated TSE where most of the native voids are present.

4. "Local" Thermodynamic and Kinetic Parameters for Folding/Unfolding Reactions Obtained from 2D High-Pressure NMR Spectroscopy

In most of the studies reported in the literature, the folding/unfolding reaction of a protein is approximated by a two-state model, excluding the existence of folding intermediates in its folding energy landscape. This is in fact a rough approximation: following the more appropriate model of foldons [52,53], most globular proteins should deviate from a two-state folding mechanism by populating folding intermediates. This apparent discrepancy comes from the very low population of the intermediate states at equilibrium, due to their low stability, which hampers their detection by the usual spectroscopic methods. In addition, spectroscopies often focus on only one observable (intrinsic fluorescence of a tryptophan residue, methyl NMR resonance, etc.), yielding values for the thermodynamic parameters that are supposed to reflect the global stability. We have seen previously that these values are also affected by the local stability of the protein: in the case of Titin I27, significantly different values were obtained for ΔV^0 and ΔG^0 from the pressure dependence of the resonances corresponding to either the unfolded state methyl groups or to the indole NH of the tryptophan side chain. Thus, a better description of the protein folding energy landscape, including the identification of folding intermediates, can be obtained by a multiple probes analysis of the folding process.

Multidimensional homo- or heteronuclear NMR spectroscopy provides an intrinsic multi-probe approach yielding residue specific information, through correlation spectroscopy involving nuclei located on the peptide backbone (^{1}H and ^{15}N of amide groups, ^{1}Hα and ^{13}Cα). Amide protons offer ideal probes to monitor the unfolding reaction: each amino acid bears an NH group, with the exception of proline, generally a minority residue in the composition of soluble proteins (< 3%), and will give rise to a specific correlation in the usually well-resolved 2D [^{1}H-^{15}N] HSQC spectrum of the native protein. In addition, proton/deuteron exchange measurement for amide protons has been extensively used to evaluate the local stability of a protein [54], bringing important information on local unfolding phenomena.

4.1. Measurement of "Local" Thermodynamic Parameters for Folding/Unfolding Reaction: Tracking Folding Intermediates in the Protein Energy Landscape with 2D High-Pressure NMR

Figure 4 displays the evolution of the correlation peaks on [^{1}H,^{15}N] HSQC experiments recorded with increasing pressures. According to the slow exchange regime between the folded and unfolded species, already visible and discussed for 1D NMR spectra, one observes the disappearance of correlation peaks belonging to the native form with the concomitant appearance of new peaks (centered at 8.5 ppm on the proton chemical-shift axis), which correspond to the spectrum of the unfolded species. As discussed above, the weak spectral dispersion of the cross-peaks corresponding to the unfolded protein is due to the loss of the through-space effects in the "random coil-like" structure of the unfolded states.

It is then possible to measure the evolution with pressure of either the intensity (peak height) or the volume of each cross-peak in the HSQC 2D spectrum: for instance, the loss of intensity of the native state resonances directly reflects the decrease in population of the folded state as detected locally by each residue. Note that, although global unfolding of a protein can obey complex models, locally the loss of the native state cross-peak intensity represents a two-state transition, that can be safely fitted with Equation (6). Thus, the fit of the local pressure unfolding curves yields residue-specific values for the apparent volume change (ΔV^0) and apparent free energy (ΔG^0) difference between the folded and unfolded states (Figure 4C,D).

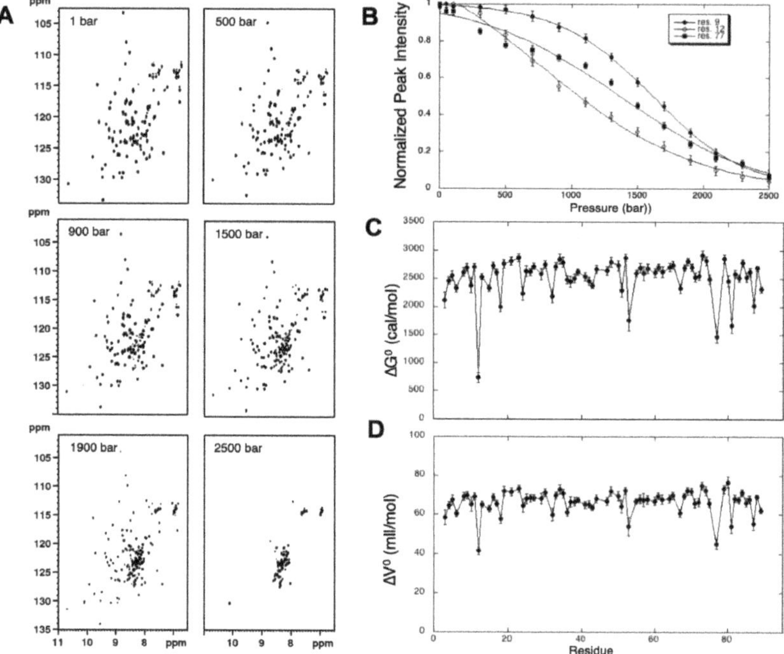

Figure 4. Monitoring unfolding of Titin I27 domain with high pressure 2D NMR. (**A**) Examples of [^{1}H-^{15}N] HSQC-NMR spectra recorded on a ^{15}N-labeled sample of Titin I27 at different pressures as indicated (same other experimental conditions as in Figure 1); (**B**) Overlay of 3 different residue-specific pressure denaturation curves obtained from the fit with Equation (6) of the cross-peak intensities measured at equilibrium from the corresponding residues. For clarity, the cross-peak intensities have been normalized. (**C**) Residue-specific ΔG^0 and (**D**) ΔV^0 measured from residue-specific pressure unfolding curves of Titin I27 domain.

Note that one usually prefers to fit the sigmoidal decay of the native resonances rather than the sigmoidal growth of the unfolded resonances, even though similar results should be obtained, as mentioned above for the indole resonance of the tryptophan residue (see Figure 2). This is because of the considerably better spectral resolution observed in the HSQC spectrum of the folded protein, which is also usually assigned, contrary to the spectrum of the unfolded states.

Large variations in the ΔV^0 and ΔG^0 values within the protein sequence sign deviation from a simple two-state unfolding transition and suggest the potential presence of folding intermediates. For instance, in the case of Titin I27, whereas a ΔV^0 for unfolding of ≈ -70 mL/mol was measured for most of the residues, ΔV^0 fell to a value < 55 mL/mol for some residues, meaning that some regions of the protein unfold earlier than others and suggesting the presence of partially folded intermediates in the protein energy landscape with some degree of stability (Figure 4). In this particular case, 2D HP-NMR clearly revealed the existence of a folding intermediate where the N-terminal β-strand is detached from the Ig-like β-sandwich. This intermediate was generally not detected in chemical denaturation studies [55] and only suspected in force spectroscopy studies [56,57] of Titin I27 multi-modules constructs. This is a clear demonstration of the potency of HP-NMR that can bring unprecedented details in the analysis of protein folding pathways.

Structural information on the folding intermediates can also be obtained from residue-specific denaturation curves [12,58,59]. To this aim, the residue specific curves must be first normalized (Figure 5). Then, at a given pressure, the value of 1 measured for a given cross-peak ($I = I_F = 1$) can be associated with a probability of 1 (100%) to find the corresponding residue "i" in the native state, whereas, at the same pressure, a residue "j" for which the corresponding cross-peak has disappeared ($I = I_U = 0$) from the HSQC spectrum has a probability equal to zero to be in a native state.

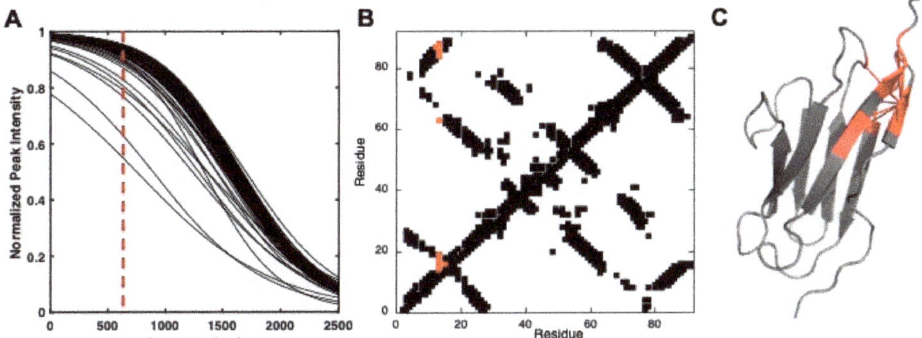

Figure 5. Pressure denaturation of Titin I27 domain. (**A**) Overlay of the normalized residue-specific denaturation curves obtained for Titin I27 domain. The vertical dashed red line at 600 bar represents the pressure used for analysis of the data presented here. (**B**) Contact map built from the best solution structure obtained for Titin I27 Ig-like domain [44]. All native contacts are displayed below the diagonal, whereas only native contacts for which a probability can be calculated from corresponding residue-specific denaturation curves are presented above the diagonal. In addition, the contacts above the diagonal have been colored in red when contact probabilities *p*(ij) lower than 0.5 are observed at 600 bar. (**C**) Ribbon representations of the solution structure of I27 where the red sticks represent contacts that are weakened ($p(\mathrm{ij}) \leq 0.5$) at 600 bar. Residues involved in these contacts are also colored in red on the ribbon.

Considering now a pressure where these two residues i and j are in an intermediate situation where the probability to be in a folded state are *p*(i) and *p*(j) ($0 < p(\mathrm{i})$ and $p(\mathrm{j}) < 1$), if these two residues are in contact in the native state (at atmospheric pressure) their probability *p*(i,j) to be in contact at this pressure is given by the geometric mean of the two individual probabilities: $p(i,j) = \sqrt{p(i) \times p(j)}$ [60]. These contact probabilities can be displayed with a color code on contact maps constructed from the 3D crystal or NMR native structure of the protein, by measuring all contacts (usually only those concerning Cα atoms of the different residues, for simplicity) between different atoms (Figure 5). When combined with molecular dynamic simulations, this approach can give a pictorial representation of the conformational ensemble. To this aim, native contact lists generated from contact maps and weighted by the probabilities of contact *p*(ij) at a given pressure are used in Go-model simulations in order to generate multiple conformers and to possibly solve the structure of folding intermediates [12].

Beside this now well-established method, the use of the pressure dependence of amide exchange rates was proposed to characterize intermediate states. Again, a residue-specific measurement of amide exchange rate constants can be obtained from the decrease in intensity of their corresponding cross-peak in the [^{1}H-^{15}N]-HSQC after dissolving a lyophilized protein sample in D_2O buffer. The use of H/D exchange measurements [54] has been proposed to identify local stabilities in globular proteins [61,62] through the values of individual amide protection factors (PF) calculated from the experimental exchange rate constants [63]. Note that the values of PF strongly depend on the physical and chemical parameters of the system: pH, temperature, and also pressure [63,64].

H/D exchange experiments combined with pressure perturbation have been used for the first time to examine the energetics of apocytochrome b562 [64]. With increasing pressure, a systematic decrease in the protection factors was observed, and changes on apparent volume for exchange (ΔV_{ex}) were estimated from the linear dependence of the free energy of exchange with pressure ($\Delta G_{ex}(\mathrm{p}) = \Delta G_{ex}{}^0 + \mathrm{p}\Delta V_{ex}$). Three regions with distinct stabilities and pressure sensitivities can be identified [64]. We have used this method for Δ+PHS SNase and several of its cavity mutants and found results in good agreement with our previous equilibrium unfolding data [65]. Nevertheless, one limitation of this method is that it applies only to solvent protected amide protons, under conditions where H/D exchange rates are still measurable (relatively low pH and low temperature).

4.2. Measurements of "Local" Kinetic Parameters of the Folding/Unfolding Reaction with 2D High-Pressure NMR

As for the steady-state parameters ΔV^0 and ΔG^0 discussed above, 2D real-time high-pressure NMR can allow a residue specific analysis for the kinetic parameters of the folding/unfolding reaction: the values of the unfolding and folding rate constants k_{u0} and k_{f0}, as well as the value of the activation volume of unfolding $\Delta V^{\ddagger}_{u0}$ (or folding, $\Delta V^{\ddagger}_{f0}$). This can be readily done by following the exponential decrease in intensity (or volume) after a *P*-Jump of cross-peaks corresponding to the native protein in a series of 2D [^{1}H,^{15}N] HSQC spectra recorded with time. These experiments can allow a residue-specific description of the TSE, with the location of internal voids already formed at this step of the folding reaction. In other words, they provide a structural description of the TSE, with the location of the "dry" folded regions ($\Delta V^{\ddagger}_{f0}/\Delta V^0$ close to 1) and of the hydrated unfolded ones ($\Delta V^{\ddagger}_{f0}/\Delta V^0$ close to 0) (Figure 6).

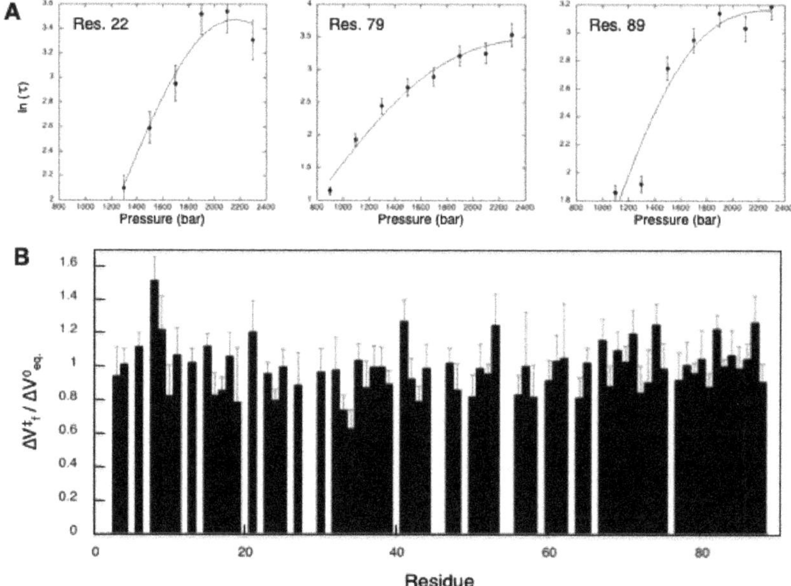

Figure 6. Residue-specific analysis of the unfolding reaction kinetics of Titin I27 domain. (**A**) Examples of residue-specific chevron plots measured for Titin I27 domain. The residue-specific relaxation times τ have been extracted from the decay with time of the intensity of cross-peaks belonging to the native protein species in a series of sixty 2D [^{1}H-^{15}N]- SOFAST-HMQC experiments (2 min measuring time each) recorded during 2 h after P-jumps of 200 bar. (**B**) Residue-specific values for the ratio $\Delta V^{\ddagger}_{f0}/\Delta V^0$ deduced from the fit of residue-specific chevron plots with Equation (6) and plotted versus the sequence of Titin I27.

Nevertheless, the time resolution of NMR spectroscopy is limited: the recording time of a 2D [^{1}H,^{15}N] HSQC ranges from 10 to 40 min, depending on the sample concentration and the digital resolution needed. In addition, such experiments can be used only for proteins with very slow relaxation times, in the range of one to several hours. For instance, this method was successfully applied to wild-type Δ+PHS SNase and a series of variants having extremely slow relaxation time (up to 12 h) [65]. This drawback has been at least partially circumvented by methodological developments during the last decade: advances have been realized in the field of "real-times" measurement of NMR multidimensional experiments [66–69], extending the application of real-time 2D NMR. Now, 2D correlation experiments can be acquired in tens of seconds, and sometimes even in less than one second, instead of tens of minutes. For example, 2D [^{1}H-^{15}N]- SOFAST-HMQC experiments [67,69] recorded in two minutes have been used to monitor the kinetics of unfolding of Titin I27 domain (Figure 6), exhibiting relaxation times of about 30 min (see Figure 3B). Similar experiments, but recorded in only 25 s, have been used for the L125A variant of Δ+PHS SNase, with relaxation times shorter than 10 min [65]. The use of "ultra-fast" 2D NMR spectroscopy [66], allowing to record a 2D spectrum in only one scan, can in principle extend the use of real-time 2D spectroscopy to proteins with shorter relaxation times, in the minute range. In addition, fast or ultra-fast experiments can be used in combination with Non Uniform Sampling (NUS) methods [70–72], which can speed up data collection. These methods allow for a decrease in the total number of points (the number of FIDs) used for sampling the indirect dimension (^{15}N dimension in the [^{1}H,^{15}N] HSQC experiments), maintaining the digital resolution at the expense of possible artifacts in the processed spectrum. Currently, a 4-fold gain in measuring time can be obtained, compared with the conventional method.

Obviously, the main limitation of this method remains the sensitivity, combined to a correct spectral resolution, of these experiments. In addition, playing with (increasing) the *P*-Jump amplitude in order to increase the sensitivity of the measurement, due to the subsequent increase in the intensity change for the cross-peaks, reaches also some limits. Indeed, the *P*-jump amplitude should remain moderate to avoid any imbalance between the folding and the unfolding reactions. Thus, an excessive positive *P*-jump will favor the unfolding reaction at the expense of the folding reaction, yielding erroneous values for the kinetic parameters. For instance, in the case of Titin I27, we have used pressure jumps of 200 bar, corresponding to about 10 percent of the pressure range needed to fully unfold the protein (2000 bar) [44].

Real-time 2D NMR spectroscopy remains inappropriate to study sub-second folding kinetics, which is the case for a lot of globular proteins. In the case of proteins with fast relaxation times (<1s), other NMR approaches are available, mainly based on 2D exchange spectroscopy techniques. The use of high-pressure ZZ-exchange experiments was introduced by Zhang et al. to obtain residue-specific folding rates for the two autonomous N-terminal and C-terminal domains of the ribosomal protein L9 [73]. This method is applicable to any proteins under experimental conditions where the folded/unfolded species exchange in a few tens to a few hundreds of milliseconds.

More recently, Charlier et al. significantly improved the pressure jump apparatus originally designed by Kremer et al. for introducing pulsed pressure perturbation in 1D and 2D NMR experiments [51], allowing for the switching of pressure on a millisecond time scale [74]. Combined with adequate 2D heteronuclear NMR experiments, this system allows measuring the rate of exchange and chemical shifts of the folded, intermediate, and unfolded states.

5. Conclusions

While pressure perturbation allows one to finely and reversibly tune the stability of a protein and to modulate the rate of a conformational exchange, NMR spectroscopy can bring the spatial and temporal resolution necessary for the description of the protein folding energy landscape. Thus, HP-NMR, combining pressure perturbation, and NMR spectroscopy can give, at a residue-level resolution, an accurate structural and dynamical description of the protein folding energy landscape, revealing the existence of intermediate states as well as the rates of the associated local rearrangements.

Beyond this fundamental interest, a better understanding of protein folding/unfolding mechanisms is mandatory in many fields. Protein misfolding is involved in most of the neurodegenerative diseases, such as Alzheimer, Parkinson, prion disease, etc. The comprehensive study of the folding mechanism of the proteins specifically involved in these diseases [75–78] might allow the rational design of more efficient drugs [79]. Giving clues on the phenomena underlying protein stability, such studies can also be meaningful for the design of industrial enzymes able to work at high pressure. Thus, understanding how a protein can accommodate mutations to gain stability, keeping its function intact, has an important economic impact. Combining NMR with high pressure is an extremely powerful approach in this particular field, providing rigorous answers to important questions.

Author Contributions: C.D. and I.H. prepared the protein samples and carried out HP-NMR experiments. P.B. made the NMR resonance assignment of Titin I27 domain and wrote MATLAB software for the analysis of HP-NMR experiments. C.R. supervises and coordinates the projects in the HP-NMR group. He conceived experiments and analysis, participated in the interpretation, and wrote the manuscript. All authors have read and agreed to the published version of the manuscript.

Funding: This work was supported by French Infrastructure for Integrated Structural Biology (FRISBI) grant No. ANR-10-INSB-05.

Conflicts of Interest: The authors declare no conflict of interest.

Abbreviations and Nomenclature

Ig-like: Immunoglobulin-like, HP-NMR: High (Hydrostatic) Pressure Nuclear Magnetic Resonance, TSE: Transition State Ensemble.

References

1. Anfinsen, C.B. Principles that govern the folding of protein chains. *Science* **1973**, *181*, 223–230. [CrossRef] [PubMed]
2. Onuchic, J.N.; Nymeyer, H.; Garcia, A.E.; Chahine, J.; Socci, N.D. The energy landscape theory of protein folding: Insights into folding mechanisms and scenarios. *Adv. Protein Chem.* **2000**, *53*, 87–152. [PubMed]
3. Bryngelson, J.D.; Wolynes, P.G. Spin glasses and the statistical mechanics of protein folding. *Proc. Natl. Acad. Sci. USA* **1987**, *84*, 7524–7528. [CrossRef] [PubMed]
4. Shea, J.E.; Brooks, C.L., III. From folding theories to folding proteins: A review and assessment of simulation studies of protein folding and unfolding. *Annu. Rev. Phys. Chem.* **2001**, *52*, 499–535. [CrossRef] [PubMed]
5. Leopold, P.E.; Montal, M.; Onuchic, J.N. Protein folding funnels: A kinetic approach to the sequence-structure relationship. *Proc. Natl. Acad. Sci. USA* **1992**, *89*, 8721–8725. [CrossRef] [PubMed]
6. Daggett, V.; Fersht, A.R. Is there a unifying mechanism for protein folding? *TIBS* **2003**, *28*, 18–25. [CrossRef]
7. Cheung, M.S.; Chavez, L.L.; Onuchic, J.N. The energy landscape for protein folding and possible connections to function. *Polymer* **2004**, *45*, 547–555. [CrossRef]
8. Onuchic, J.N.; Wolynes, P.G. Theory of protein Folding. *Curr. Opin. Struct. Biol.* **2004**, *14*, 70–75. [CrossRef]
9. Kamatari, Y.O.; Kitahara, R.; Yamada, H.; Yokoyama, S.; Akasaka, K. High-pressure NMR spectroscopy for characterizing folding intermediates and denatured states of proteins. *Methods* **2004**, *34*, 133–143. [CrossRef]
10. Akasaka, K. Probing conformational fluctuations of proteins by pressure perturbation. *Chem. Rev.* **2006**, *106*, 1814–1835. [CrossRef]
11. Akasaka, K.; Kitahara, R.; Kamatari, Y.O. Exploring the folding energy landscape with pressure. *Arch. Biochem. Biophys.* **2013**, *531*, 110–115. [CrossRef] [PubMed]
12. Roche, J.; Caro, J.A.; Norberto, D.R.; Barthe, P.; Roumestand, C.; Schlessman, J.L.; Garcia, A.E.; Garcia-Moreno, B.; Royer, C.A. Cavities determine the pressure unfolding of proteins. *Proc. Natl. Acad. Sci. USA* **2012**, *109*, 6945–6950. [CrossRef] [PubMed]
13. Rouget, J.B.; Aksel, T.; Roche, J.; Saldana, J.L.; Garcia, A.E.; Barrick, D.; Royer, C.A. Size and sequence and the volume change of protein folding. *J. Am. Chem. Soc.* **2011**, *133*, 6020–6027. [CrossRef] [PubMed]
14. Lerch, M.T.; Horwitz, J.; McCoy, J.; Hubbell, W.L. Circular dichroism and site-directed spin labeling reveal structural and dynamical features of high-pressure states of myoglobin. *Proc. Natl. Acad. Sci. USA* **2013**, *110*, 4714–4722. [CrossRef] [PubMed]
15. Dellarole, M.; Royer, C.A. High-pressure fluorescence applications. *Methods Mol. Biol.* **2014**, *1076*, 53–74.
16. Dwyer, C.L. High Pressure NMR and IR Spectroscopy in Organometallic Chemistry. In *Compréhensive Organometallic Chemistry III—From Fundamentals to Applications*, 3rd ed.; Michael, D., Robert, P., Crabtree, H., Eds.; Elsevier Ltd.: Amsterdam, The Netherlands, 2007; Volume 1, pp. 483–507.
17. Torrent, J.; Rubens, P.; Ribó, M.; Heremans, K.; Vilanova, M. Pressure versus temperature unfolding of ribonuclease A: An FTIR spectroscopic characterization of 10 variants at the carboxy-terminal site. *Protein Sci.* **2001**, *10*, 725–734. [CrossRef] [PubMed]

18. Nicolini, C.; Ravindra, R.; Ludolph, B.; Winter, R. Characterization of the temperature- and pressure-induced inverse and reentrant transition of the minimum elastin-like polypeptide GVG(VPGVG) by DSC, PPC, CD, and FT-IR spectroscopy. *Biophys. J.* **2004**, *86*, 1385–1392. [CrossRef]
19. Smeller, L. Pressure-temperature phase diagrams of biomolecules. *Biochim. Biophys. Acta* **2002**, *1595*, 11–29. [CrossRef]
20. Benedek, G.B.; Purcell, E.M. Nuclear magnetic resonance in liquids under high pressure. *J. Chem. Phys.* **1954**, *22*, 2003–2012. [CrossRef]
21. Ballard, L.; Reiner, C.; Jonas, J. High-resolution NMR probe for experiments at high-pressures. *J. Magn. Res.* **1996**, *123*, 81–86. [CrossRef]
22. Ballard, L.; Yu, A.; Reiner, C.; Jonas, J. A high-pressure, high-resolution NMR probe for experiments at 500 MHz. *J. Magn. Res.* **1998**, *133*, 190–193. [CrossRef] [PubMed]
23. Jonas, J. High-resolution nuclear magnetic resonance studies of proteins. *Biochem. Biophys. Acta* **2002**, *1595*, 145–159. [CrossRef]
24. Jiri, J. High-Pressure Studies Using NMR Spectroscopy. In *Encyclopedia of Spectroscopy and Spectrometry*, 2nd ed.; Lindon, J.C., Tranter, G.E., Koppenaal, D., Eds.; Academic Press (Elsevier Ltd.): Cambridge, MA, USA, 2010; pp. 854–864.
25. Castro, P.; Delsuc, M.A. An NMR probe with a high-pressure chamber made from composite materials. *Magn. Reson. Chem.* **1998**, *36*, 833–838. [CrossRef]
26. Meier, T.; Khandarkhaeva, S.; Petitgirard, S.; Körber, T.; Lauerer, A.; Rössler, E.; Dubrovinsky, L. NMR at pressures up to 90 GPa. *J. Magn Reson.* **2018**, *292*, 44–47. [CrossRef]
27. Fourme, R.; Girard, E.; Akasaka, K. High-pressure macromolecular crystallography and NMR: Status, achievements and prospects. *Curr. Opin. Struct. Biol.* **2012**, *22*, 636–642. [CrossRef] [PubMed]
28. Vanni, H.; Earl, W.L.; Merbach, A.E. Two approaches to high-resolution high-pressure nuclear magnetic resonance. *J. Magn. Reson.* **1978**, *29*, 11–19. [CrossRef]
29. Roe, D.J. Sapphire NMR tube for high-resolution studies at elevated pressure. *J. Magn. Reson.* **1985**, *63*, 388–391.
30. Yamada, H. Pressure-resisting glass cell for high-pressure, high resolution NMR measurements. *Rev. Sci. Instrum.* **1974**, *45*, 640–642. [CrossRef]
31. Yamada, H.; Nishikawa, K.; Honda, M.; Shimura, T.; Akasaka, K.; Tabayashi, K. Pressure-resisting cell for high-pressure, high-resolution nuclear magnetic resonance measurements at very high magnetic fields. *Rev. Sci. Instrum.* **2001**, *72*, 1463–1471. [CrossRef]
32. Kitahara, R.; Royer, C.; Yamada, H.; Boyer, M.; Saldana, J.L.; Akasaka, K.; Roumestand, C. Equilibrium and pressure-jump relaxation studies of the conformational transitions of P13MTCP1. *J. Mol. Biol.* **2002**, *320*, 609–628. [CrossRef]
33. Vajpai, N.; Nisius, L.; Wiktor, M.; Grzesiek, S. High-pressure NMR reveals close similarity between cold and alcohol protein denaturation in ubiquitin. *Proc. Natl. Acad. Sci. USA* **2013**, *110*, 368–376. [CrossRef] [PubMed]
34. Lassalle, M.W.; Akasaka, K. The use of high-pressure nuclear magnetic resonance to study protein folding. *Methods Mol. Biol.* **2007**, *350*, 21–38. [PubMed]
35. Akasaka, K. High pressure NMR spectroscopy. *Subcell. Biochem.* **2015**, *72*, 707–721. [PubMed]
36. Urbauer, J.L.; Ehrhardt, M.R.; Bieber, R.J.; Flynn, P.F.; Wand, A.J. High-resolution triple-resonance NMR spectroscopy of a novel calmodulin-peptide complex at kilobar pressures. *J. Am. Chem. Soc.* **1996**, *118*, 11329–11330. [CrossRef]
37. Wand, A.J.; Earhardt, M.R.; Urbauer, J.L. Apparatus and Method for High Pressure NMR Spectroscopy. U.S. Patent 6,362,624, 26 March 2002.
38. Arnold, M.R.; Kalbitzer, H.R.; Kremer, W. High-sensitivity sapphire cells for high pressure NMR spectroscopy on proteins. *J. Magn. Reson.* **2003**, *161*, 127–131.
39. Erlach, M.B.; Munte, C.E.; Kremer, W.; Hartl, R.; Rochelt, D.; Niesner, D.; Kalbitzer, H.R. Ceramic cells for high pressure NMR spectroscopy of proteins. *J. Magn. Reson.* **2010**, *204*, 196–199. [CrossRef]
40. Peterson, R.W.; Wand, J.A. Self-contained high-pressure cell, apparatus, and procedure for the preparation of encapsulated proteins dissolved in low viscosity fluids for nuclear magnetic resonance spectroscopy. *Rev. Sci. Instrum.* **2005**, *76*, 1–7. [CrossRef]

41. Brunner, E.; Arnold, M.R.; Kremer, W.; Kalbitzer, H.R. Pressure stability of phospholipid bicelles: Measurement of residual dipolar couplings under extreme conditions. *J. Biomol. NMR* **2001**, *21*, 173–176. [CrossRef]
42. Fu, Y.; Wand, A.J. Partial alignment and measurement of residual dipolar couplings of proteins under high hydrostatic pressure. *J. Biomol. NMR* **2013**, *56*, 353–357. [CrossRef]
43. Sibille, N.; Dellarole, M.; Royer, C.; Roumestand, C. Measuring residual dipolar couplings at high hydrostatic pressure: Robustness of alignment media to high pressure. *J. Biomol. NMR* **2014**, *58*, 9–16. [CrossRef]
44. Herrada, I.; Barthe, P.; Vanheusden, M.; DeGuillen, K.; Mammri, L.; Delbecq, S.; Rico, F.; Roumestand, C. Monitoring Unfolding of Titin I27 Single and Bi Domain with High-Pressure NMR Spectroscopy. *Biophys. J.* **2018**, *11*, 341–352. [CrossRef] [PubMed]
45. Lassalle, M.W.; Yamada, H.; Akasaka, K. The pressure-temperature free energy-landscape of staphylococcal nuclease monitored by (1)H-NMR. *J. Mol. Biol.* **2000**, *298*, 293–302. [CrossRef] [PubMed]
46. Hata, H.; Kono, R.; Fujidawa, M.; Kitahara, R.; Kamatari, Y.O.; Akasaka, K.; Xu, Y. High pressure NMR study of dihydrofolate reductase from deep-sea bacterium Moritella profunda. *Cell Mol. Biol.* **2004**, *50*, 311–316. [PubMed]
47. Ravindra, R.; Winter, R. On the temperature-pressure free-energy landscape of proteins. *ChemPhysChem* **2003**, *4*, 359–365. [CrossRef]
48. Fersht, A.R. Characterizing transition states in protein folding: An essential step in the puzzle. *Curr. Opin. Struct. Biol.* **1995**, *5*, 79–84. [CrossRef]
49. Saotome, T.; Doret, M.; Kulkarni, M.; Yang, Y.S.; Barthe, P.; Kuroda, Y.; Roumestand, C. Folding of the Ig-Like Domain of the Dengue Virus Envelope Protein Analyzed by High-Hydrostatic-Pressure NMR at a Residue-Level Resolution. *Biomolecules* **2019**, *9*, 309. [CrossRef]
50. Vidugiris, G.J.A.; Markley, J.L.; Royer, C.A. Evidence for a molten globule-like transition state in protein folding from determination of activation volumes. *Biochemistry* **1995**, *34*, 4909–4912. [CrossRef]
51. Kremer, W.; Arnold, M.; Munte, C.E.; Hartl, R.; Erlach, M.B.; Koehler, J.; Meier, A.; Kalbitzer, H.R. Pulsed pressure perturbations, an extra dimension in NMR spectroscopy of proteins. *J. Am. Chem. Soc.* **2011**, *133*, 13646–13651. [CrossRef]
52. Englander, S.W.; Mayne, L.; Krishna, M.M.G. Protein folding and misfolding: Mechanism and principles. *Q. Rev. Biophys.* **2007**, *40*, 287–326. [CrossRef]
53. Lindberg, M.O.; Oliveberg, M. Malleability of protein folding pathways: A simple reason for complex behavior. *Curr. Opin. Struct. Biol.* **2007**, *17*, 21–29. [CrossRef]
54. Englander, S.W.; Kallenbach, N.R. Hydrogen exchange and structural dynamics of proteins and nucleic acids. *Q. Rev. Biophys.* **1983**, *4*, 521–655. [CrossRef] [PubMed]
55. Fowler, S.B.; Clarke, J. 2001. Mapping the folding pathway of an immunoglobulin domain: Structural detail from Phi value analysis and movement of the transition state. *Structure* **2001**, *9*, 355–366. [CrossRef]
56. Marszalek, P.E.; Lu, H.; Li, H.; Carrion-Vazquez, M.; Oberhauser, A.F.; Schulten, K.; Fernandez, J.M. Mechanical unfolding intermediates in titin modules. *Nature* **1999**, *402*, 100–103. [CrossRef] [PubMed]
57. Rico, F.; Gonzalez, L.; Casuso, I.; Puig-Vidal, M.; Scheuring, S. High-speed force spectroscopy unfolds titin at the velocity of molecular dynamics simulations. *Science* **2013**, *342*, 741–743. [CrossRef]
58. Roche, J.; Dellarole, M.; Caro, J.A.; Guca, E.; Norberto, D.R.; Yang, Y.S.; Garcia, A.E.; Roumestand, C.; García-Moreno, B.; Royer, C.A. Remodeling of the folding free-energy landscape of staphylococcal nuclease by cavity-creating mutations. *Biochemistry* **2012**, *51*, 9535–9546. [CrossRef]
59. de Oliveira, G.A.; Silva, J.L. A hypothesis to reconcile the physical and chemical unfolding of proteins. *Proc. Natl. Acad. Sci. USA* **2015**, *112*, E2775–E2784. [CrossRef]
60. Fossat, M.J.; Dao, T.P.; Jenkins, K.; Dellarole, M.; Yang, Y.S.; McCallum, S.A.; Garcia, A.E.; Barrick, D.; Roumestand, C.; Royer, C.A. High-Resolution Mapping of a Repeat Protein Folding Free Energy Landscape. *Biophys. J.* **2016**, *111*, 2368–2376. [CrossRef]
61. Roder, H.; Elove, G.A.; Englander, S.W. Structural characterization of folding intermediates in cytochrome-c by H-exchange labeling and proton NMR. *Nature* **1988**, *335*, 700–704. [CrossRef] [PubMed]
62. Bai, Y.; Sosnick, T.R.; Mayne, L.; Englander, S.W. Protein folding intermediates-native state hydrogen exchange. *Science* **1995**, *269*, 192–197. [CrossRef] [PubMed]

63. Bai, Y.; Milne, J.S.; Mayne, L.; Englander, S.W. Primary structure effects on peptide group hydrogen exchange. *Proteins* **1993**, *17*, 75–86. [CrossRef] [PubMed]
64. Fuentes, E.J.; Wand, A.W. Local stability and dynamics of apocytochrome b562 examined by the dependence of hydrogen exchange on hydrostatic pressure. *Biochemistry* **1998**, *37*, 9877–9883. [CrossRef] [PubMed]
65. Roche, J.; Dellarole, M.; Caro, J.A.; Norberto, D.R.; Garcia, A.E.; Garcia-Moreno, B.; Roumestand, C.; Royer, C.A. Effect of internal cavities on folding rates and routes revealed by real-time pressure-jump NMR spectroscopy. *J. Am. Chem. Soc.* **2013**, *135*, 14610–14618. [CrossRef] [PubMed]
66. Frydman, L.; Scherf, T.; Lupulescu, A. The acquisition of multidimensional NMR spectra within a single scan. *Proc. Natl. Acad. Sci. USA* **2002**, *99*, 15858–15862. [CrossRef] [PubMed]
67. Schanda, P.; Brutscher, B. Very fast two-dimensional NMR spectroscopy for real-time investigation of dynamic events in proteins on the time scale of seconds. *J. Am. Chem Soc.* **2005**, *127*, 8014–8015. [CrossRef]
68. Gal, M.; Schanda, P.; Brutscher, B.; Frydman, L. UltraSOFAST HMQC-NMR and repetitive acquisition of 2D protein spectra at Hz rates. *J. Am. Chem. Soc.* **2007**, *129*, 1372–1377. [CrossRef]
69. Schanda, P.; Forge, V.; Brutscher, B. Protein folding and unfolding studied at atomic resolution by fast two-dimensional NMR spectroscopy. *Proc. Natl. Acad. Sci. USA* **2007**, *104*, 11257–11262. [CrossRef]
70. Hyberts, S.G.; Arthanari, H.; Wagner, G. Applications of Non-Uniform Sampling and Processing. In *Novel Sampling Approaches in Higher Dimensional NMR (Topics in Current Chemistry, vol. 316)*; Billeter, M., Orekhov, V., Eds.; Springer: Berlin, Germany, 2012; pp. 125–128.
71. Hyberts, S.G.; Arthanari, H.; Robson, S.A.; Wagner, G. Perspectives in magnetic resonance: NMR in the post-FFT era. *J. Magn. Reson.* **2014**, *241*, 60–73. [CrossRef]
72. Palmer, M.R.; Suiter, C.L.; Henry, G.E.; Rovnyak, J.; Hoch, J.C.; Polenova, T.; Rovnyak, D. Sensitivity of Nonuniform Sampling NMR. *J. Phys. Chem. B* **2015**, *119*, 6502–6515. [CrossRef] [PubMed]
73. Zhang, Z.; Kitazawa, S.; Peran, I.; Stenzoski, N.; McCallum, S.A.; Raleigh, D.P.; Royer, C.A. High pressure ZZ-exchange NMR reveals key features of protein folding transition states. *J. Am. Chem. Soc.* **2016**, *138*, 15260–15266. [CrossRef]
74. Charlier, C.; Alderson, T.R.; Courtney, J.M.; Ying, J.; Anfinrud, P.; Bax, A. Study of protein folding under native conditions by rapidly switching the hydrostatic pressure inside an NMR cell. *Proc. Natl. Acad. Sci. USA* **2018**, *115*, E4169–E4178. [CrossRef]
75. Kuwata, K.; Li, H.; Yamada, H.; Legname, G.; Prusiner, S.B.; Akasaka, K.; James, T.L. Locally disordered conformer of the hamster prion protein: A crucial intermediate to PrPSc? *Biochemistry* **2002**, *41*, 12277–12283. [CrossRef] [PubMed]
76. Niraula, T.N.; Konno, T.; Li, H.; Yamada, H.; Akasaka, K.; Tachibana, H. Pressure-dissociable reversible assembly of intrinsically denatured lysozyme is a precursor for amyloid fibrils. *Proc. Natl. Acad. Sci. USA* **2004**, *101*, 4089–4093. [CrossRef] [PubMed]
77. Kamatari, Y.O.; Yokoyama, S.; Tachibana, H.; Akasaka, K. Pressure-jump NMR study of dissociation and association of amyloid protofibrils. *J. Mol. Biol.* **2005**, *349*, 916–921. [CrossRef] [PubMed]
78. Munte, C.E.; Beck-Erlach, M.; Kremer, W.; Koehler, J.; Kalbitzer, H.R. Distinct conformational states of the alzheimer b-amyloid peptide can be detected by high-pressure NMR spectroscopy. *Angew. Chem. Int. Ed. Engl.* **2013**, *52*, 8943–8947. [CrossRef]
79. Louis, J.M.; Roche. Evolution under drug pressure remodels the folding free-energy landscape of mature HIV-1 protease. *J. Mol. Biol* **2016**, *428*, 2780–2792. [CrossRef]

Publisher's Note: MDPI stays neutral with regard to jurisdictional claims in published maps and institutional affiliations.

© 2020 by the authors. Licensee MDPI, Basel, Switzerland. This article is an open access article distributed under the terms and conditions of the Creative Commons Attribution (CC BY) license (http://creativecommons.org/licenses/by/4.0/).

Review

Structure-Function Insights of Jaburetox and Soyuretox: Novel Intrinsically Disordered Polypeptides Derived from Plant Ureases

Matheus V. Coste Grahl [1,†], Fernanda Cortez Lopes [2,†], Anne H. Souza Martinelli [3,†], Celia R. Carlini [1,4,*] and Leonardo L. Fruttero [5,6,*]

1 Graduate Program in Medicine and Health Sciences, Brain Institute of Rio Grande do Sul (InsCer), Pontifícia Universidade Católica do Rio Grande do Sul (PUCRS), Porto Alegre CEP 90610-000, Brazil; matheusgrahl@hotmail.com

2 Graduate Program in Cellular and Molecular Biology, Center of Biotechnology, Universidade Federal do Rio Grande do Sul (UFRGS), Av. Bento Gonçalves 9500, Building 43431, Porto Alegre CEP 91501-970, RS, Brazil; fernandacortezlopes@gmail.com

3 Department of Biophysics & Deparment of Molecular Biology and Biotechnology-Biosciences Institute (IB), Universidade Federal do Rio Grande do Sul, UFRGS, Porto Alegre 91501-970, RS, Brazil; ahsmartinelli@yahoo.com.br

4 Brain Institute and School of Medicine, Pontifícia Universidade Católica do Rio Grande do Sul (PUCRS), Porto Alegre 90610-000, RS, Brazil

5 Departamento de Bioquímica Clínica, Facultad de Ciencias Químicas, Universidad Nacional de Córdoba, Córdoba CP 5000, Argentina

6 Centro de Investigaciones en Bioquímica Clínica e Inmunología (CIBICI), Consejo Nacional de Investigaciones Científicas y Técnicas (CONICET), Córdoba CP 5000, Argentina

* Correspondence: celia.carlini@pucrs.br (C.R.C.); lfruttero@fcq.unc.edu.ar (L.L.F.); Tel.: +55-51-3320-3485 (C.R.C.); +54-351-535-3850 (L.L.F.)

† These authors have contributed equally to the paper.

Academic Editor: Marilisa Leone

Received: 6 October 2020; Accepted: 6 November 2020; Published: 16 November 2020

Abstract: Intrinsically disordered proteins (IDPs) and intrinsically disordered regions (IDRs) do not have a stable 3D structure but still have important biological activities. Jaburetox is a recombinant peptide derived from the jack bean (Canavalia ensiformis) urease and presents entomotoxic and antimicrobial actions. The structure of Jaburetox was elucidated using nuclear magnetic resonance which reveals it is an IDP with small amounts of secondary structure. Different approaches have demonstrated that Jaburetox acquires certain folding upon interaction with lipid membranes, a characteristic commonly found in other IDPs and usually important for their biological functions. Soyuretox, a recombinant peptide derived from the soybean (*Glycine max*) ubiquitous urease and homologous to Jaburetox, was also characterized for its biological activities and structural properties. Soyuretox is also an IDP, presenting more secondary structure in comparison with Jaburetox and similar entomotoxic and fungitoxic effects. Moreover, Soyuretox was found to be nontoxic to zebra fish, while Jaburetox was innocuous to mice and rats. This profile of toxicity affecting detrimental species without damaging mammals or the environment qualified them to be used in biotechnological applications. Both peptides were employed to develop transgenic crops and these plants were active against insects and nematodes, unveiling their immense potentiality for field applications.

Keywords: biopesticides; antifungal activity; insecticidal activity; mechanism of action; transgenic crops

1. Introduction

Intrinsically disordered proteins (IDPs) and intrinsically disordered regions (IDRs) fail to form a stable tridimensional structure, challenging the century-old paradigm that a biological function is a specific property of a unique structure. In spite of the lack of an ordered structure, these proteins exhibit vital biological activities and can be found in all organisms, especially in eukaryotes [1,2]. IDPs and IDRs differentiate from structured proteins and domains, and they are characterized by notable conformational flexibility and structural plasticity. One of the differences of IDPs/IDRs and structured proteins include the amino acid composition since IDPs/IDRs are rich in disorder-promoting residues, such as Arg, Pro, Gln, Gly, Glu, Ser, Ala, and Lys. Also, IDPs/IDRs have low sequence complexity, high net charge, low mean hydropathy and are highly dynamic [2,3].

Many IDPs gain secondary structure when binding onto surfaces, for example, to a cell membrane [2]. Since IDPs/IDRs cannot fold spontaneously and some of them require partners to acquire a more ordered structure, these proteins do not have a code that defines the capacity of foldable proteins to fold spontaneously into a biologically active structure [4]. A typical IDP/IDR has a multitude of elements for potentially foldable, partially foldable, differently foldable, or unfoldable protein segments [5,6]. Their folding can be acquired after the interaction with proteins, nucleic acids, membranes, or small molecules. These conformation modifications can be driven also by changes in the IDPs environment as well as post-translational modifications. These IDPs can remain substantially disordered or become tightly folded after interaction [7–9].

IDPs are difficult proteins to study, due to their dynamic conformational landscapes changing between different structures on a variety of time scales [10,11]. Biophysical studies are crucial to clarify the relationship of the IDPs biological functions and their structures. Recent advances in heteronuclear multidimensional nuclear magnetic resonance (NMR) have allowed the complete assignment of resonances for several IDPs. NMR can also provide data about mobility of the unstructured regions [4,11]. In this way, NMR is possibly the most powerful technique for structural studies of these disordered proteins [12]. Furthermore, computational studies assumed an increasingly importance in interpreting these challenging experimental data [11].

2. Jaburetox and Soyuretox: Historical Aspects and Potential as Biopesticides

2.1. Transgenic Plants Expressing Biocide Polypeptides and Plant Defense

Hunger continues to afflict mainly the poorer countries around the world. In 2016, 10.7% of the world population were chronically undernourished (www.worldhunger.org). According to data provided by the World Bank (www.databank.worldbank.org), over the last 15 years, the word population has increased with an annual growth rate of ~1% (1.075 in 2019), from 6.59 (2006) to 7.67 (2019) billions inhabitants. With the increase in life expectancy, from 69.2 (2006) to 72.5 (2018) years, particularly in the richer countries, even considering a decrease of growth rate, estimates are that the world population could reach 9.7 billion (www.population.un.org) in 2050. By then, food demand will be 60% higher (www.webforum.org). As agricultural land is finite (it increased only from 47.18 in 2006 to 48.43 in 2016 millions square kilometers according to the World Bank), and is expected to shrink due to urbanization, climate change and soil degradation, increases in food production will require an even more efficient agriculture. The output of agriculture is hampered, however, by losses in the field or after harvesting, by a variety of insect pests, nematodes, fungi and diseases induced by bacteria or viruses [13,14]. Herbivores alone feeding on foliage, sap and root can decrease more than 20% of net plant productivity and food losses to insects are expected to even grow in a scenario of global warming [15].

To efficiently control insect pests in agriculture, combining different strategies is frequently required, including the use of resistant crop varieties. When there is no natural plant genotypes genetically resistant to insect pests, development of genetically modified (GM) resistant plants is an option. A milestone in the development of insect-resistant crops was established in the late 1980's, by genetically engineering a tobacco plant to express an entomotoxic protein from the bacterium *Bacillus thuringiensis* (Bt) [16,17]. In 2018, the list of genetically modified plants that were commercialized had 26 species, either tolerant to herbicides or with increased resistance to insects (most expressing *Bt* toxins), grown or imported in 75 countries (www.gmoanswers.com). *Bt* toxins (or Cry proteins) have some restrictions, such as the low toxicity against sap-sucking insects [18,19]. Moreover, an increasing level of resistance of insects against Cry insecticidal proteins has been reported [20]. Fortunately, there is a number of plant entomotoxic proteins that can be used instead or in synergy with the *Bt* technology to control insect pests in new generations of transgenic plants yet to be developed [21].

2.2. Plant Proteins and Peptides with Insecticidal and Fungitoxic Properties

Insect pests and phytopathogenic fungi are detrimental to several crops and cause significant economic losses in agriculture worldwide. To cope with herbivory and fungal diseases, plants have evolved sophisticated defense mechanisms. Plant tissues accumulate, constitutively or after induction, various classes of defensive compounds that confer resistance against herbivores and infection by fungi, bacteria, viruses, as well as nematodes. The most known plant proteins involved in defense mechanisms against insect pests include lectins [22], ribosome-inactivating proteins of types 1 and 2 [23], inhibitors of proteolytic enzymes and glycohydrolases [24], modified forms of storage proteins [25,26], among others. Several plant peptides display antifungal properties such as defensins [27], lipid transport proteins [28], chitinases [29], lectins [30], thionins [31], cyclopeptide alkaloids [32] and other less common types. For a general review on these topics please refer to Dang and Van Damme [33] and Grossi-De-Sá et al. [21].

Ureases represent another group of plant proteins with insecticidal and antifungal properties which widen the proposed physiological roles of these enzymes [34–37]. Ureases (urea amidohydrolase; EC 3.5.1.5) are well conserved and nickel-dependent enzymes that catalyze urea hydrolysis into ammonia and carbon dioxide, synthesized by plants, fungi and bacteria [38–41]. Canatoxin is a less abundant urease isoform isolated from *Canavalia ensiformis* (jack bean) seeds [42,43]. Structurally similar to the seed's major urease, both proteins display insecticidal and antifungal properties independent of their ureolytic activity [35,38,39]. Soybean (*Glycine max*) and pigeon pea (*Cajanus cajan*) ureases were also shown to display insecticidal [44–46] and antifungal [44,47,48] properties. Noteworthy, ureases are insecticidal against hemipteran pests (such as the stink bug *Nezara viridula* and the cotton stainer bug *Dysdercus peruvianus*), which were not susceptible to the entomotoxic activity of Cry toxins from *B. thuringiensis* [18,19]. Since these proteins are abundant in many edible vegetables, particularly in legumes, they can be generally regarded as biosafe [49].

2.3. Ureases and Derived Peptides as Sources of Insecticidal and Fungitoxic (Poly)Peptides

The insecticidal [50] and fungitoxic [51] effects of canatoxin were described before its characterization as an isoform of jack bean urease (JBU) [43]. In the first study of canatoxin's insecticidal effect, it became clear that only insects relying on cathepsin-like digestive enzymes (such as the cowpea weevil *Callosobruchus maculatus* and the kissing bug *Rhodnius prolixus*) are sensitive to the toxin, while insects with digestion based on trypsin-like enzymes (such as the tobacco hornworn *Manduca sexta* or the fruitfly *Drosophila melanogaster*) show no susceptibility. The hypothesis of a proteolytic activation of the toxin was then proposed [50]. The hydrolysis of canatoxin with *C. maculatus* digestive enzymes yielded a 10 kDa entomotoxic peptide named pepcanatox [52]. Our group has demonstrated through inhibition of cathepsin-like enzymes, that the enzymatic activity of cathepsin B (cysteine proteinase) and cathepsin D (aspartic proteinases) is necessary for the release of toxic fragments of canatoxin [50,52]. Cathepsin B is a cysteine proteinase that can act as an exopeptidase or endopeptidase at acidic pH [53]. Cleavage by cathepsin B has a preference for basic and hydrophobic amino acids [54]. Meanwhile, cathepsin D cleavage occurs at acidic pH and has a preference for hydrophobic residues [55]. Subsequently, the major proteolytic activities of midgut homogenates of *D. peruvianus* nymphs were shown in vitro to catalyze the release of pepcanatox from JBU [56]. Cysteine, aspartic and metalloproteinases are present in both homogenates. Fluorogenic substrates containing JBU partial sequences flanking the N-terminal or the C-terminal portion of the entomotoxic peptide were efficiently cleaved by the *D. peruvianus* nymph midgut homogenates. Different classes of enzymes in the homogenates cleaved both substrates suggesting that in vivo the release of the entomotoxic peptide results from the concerted action of at least two different proteinases [56].

Jaburetox-2Ec, a recombinant peptide with 93 amino acids (~11 kDa) equivalent to pepcanatox, was produced heterologously in *Escherichia coli* from the corresponding sequence of the JBU isoform JBURE-II [57]. Here the term "peptide" is used solely to emphasize the fact that it is a fragment of a much larger protein regardless of its molecular mass. Later on, the peptide called simply Jaburetox was developed, with the same urease-derived sequence and the 6 His tail found in Jaburetox-2Ec, but lacking the V5 epitope present in the latter [58]. Both peptides, Jaburetox-2Ec and Jaburetox, display equivalent insecticidal activity, evidencing that the epitope V5 is not implied in their entomotoxicity [58].

Since one of the most well studied mechanisms of defense against insect pest is digestive enzyme inhibition, this possibility was explored by our group for ureases and derived peptides. As described by Carlini et al. [50] and Ferreira da-Silva et al. [52], canatoxin showed no inhibitory effect on the proteolytic (cathepsin B or D-like) or α-amylase activities. Moreover, the peptides derived from Canatoxin's digestion with cathepsin-like enzymes, including pepcanatox, did not display either proteolytic or amylase inhibitory properties [59]. Although Jaburetox itself was not tested, taking into account its virtually identical sequence when compared to pepcanatox, it is safe to assume that Jaburetox has no inhibitory effects upon digestive enzymes.

Jaburetox is lethal to several insects susceptible to canatoxin (the cotton stainer bug *D. peruvianus*, the kissing bugs *R. prolixus* and *Triatoma infestans*) and also kills insects that are resistant to intact ureases, such as lepidopterans (fall armyworm *Spodoptora frugiperda*, cotton bollworm *Helicoverpa armigera*) and dipterans (*Aedes aegypti*) [57,60], because the hydrolysis of the protein to release the peptide is no longer required.

Concerning the antifungal property of ureases and derived peptides, the most abundant jack bean isoform, JBU, was shown to display antifungal properties against a panel of 16 phytopathogenic filamentous fungi species of 11 genera, blocking spore germination and/or mycelial growth, and inhibiting multiplication of yeasts at submicromolar concentrations [47,61]. Jaburetox also displayed antifungal properties against filamentous fungi and yeasts [61].

Antifungal effects were observed in vitro also for isoforms of soybean urease (SBU) [47,62]. The participation of ureases in plant defense against fungal diseases was demonstrated in urease-null soybean plants obtained by gene silencing [63]. Later, the peptide called Soyuretox, homologous to Jaburetox, but derived from the ubiquitous isoform of the soybean urease, was heterologously expressed in *E. coli*, characterized structurally and its entomotoxic and antifungal effects were demonstrated [64].

Jaburetox showed no acute toxicity to mice and rats [57] and was found not toxic in a risk assessment study [65] while Soyuretox was not toxic to zebrafish embryos [64]. These data suggest that these peptides may be safe alternatives to attain resistance to insect herbivory and/or fungal disease in transgenic plants. In the following sections, the structural aspects (Figure 1) and biological profile of Jaburetox and Soyuretox are reviewed.

3. Structural Aspects of Jaburetox and Soyuretox and Its Interaction with Membranes

Since the discovery of the insecticidal effect of Jaburetox, our group carried out different approaches aimed to identify the active region of the molecule and characterize the structure involved in this toxicity. In this direction, a first modeling for Jaburetox was performed employing an ab initio approach. Using this strategy, 10 different models of possible conformations of the peptide were proposed. In all of them, a β-hairpin motif appeared in the C-terminal region of the molecule, similar to those found in pore-forming peptides. Based on this, it was hypothesized that the β-hairpin motif could be involved in the entomotoxic activity of Jaburetox [57].

Two years later, molecular aspects of Jaburetox and its capability to interact with model membranes were demonstrated for the first time. In this study, Jaburetox displayed the ability to interact with and disrupt large unilamellar vesicles (LUVs) composed mainly by acidic lipids. Moreover, the interaction with LUVs were found to depend on the aggregation state of the polypeptide [66]. Dynamic light scattering and small angle X-ray scattering techniques were used to demonstrate the Jaburetox-lipid interaction with platelet-like multilamellar liposomes (PML) [67]. A tridimensional model indicated that Jaburetox could anchor at a polar/non polar lipid interface, and it was suggested that the β-hairpin motif could be the responsible for this membrane-disturbing effect [66]. The presence of the Jaburetox's β-hairpin motif in the structure of JBU was confirmed by crystallography by the Ponnuraj group in India, who elucidated the 3D structure of JBU [68].

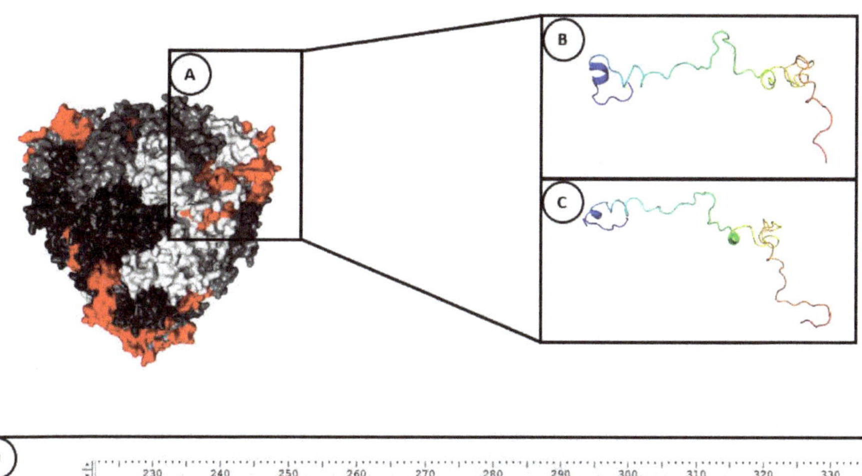

Figure 1. Structural representation of entomotoxic peptides and Jack Bean Urease. (**A**) Graphical representation of the location of the peptides (red) in the protein structure of Jack Bean Urease (pdb: 3LA4). Each monomer of the JBU hexamer is represented with a different shade of grey. (**B**) Jaburetox, (**C**) Soyuretox and (**D**) comparison of the primary sequences.

In order to identify the regions of the molecule involved in the biological activities of Jaburetox, truncated versions of the peptide were obtained by site-directed mutagenesis [58]. Three mutated versions were constructed and overexpressed in *E. coli*: (1) a peptide lacking the β-hairpin motif (residues 61–74 was deleted, the resulting peptide was called jbtx Δβ-hairpin); (2) a peptide corresponding to the N-terminal portion (residues 1–44, called jbtx N-ter); and (3) a peptide corresponding to the C-terminal region (residue 45–93, called jbtx C-ter). All these peptides were tested for their insecticidal activity and the ability to interact with LUVs. The peptide jbtx Δβ-hairpin displayed the same insecticidal effect of the whole peptide, demonstrating that β-hairpin was not involved in this toxicity. The peptide jbtx N-ter, corresponding to the N-terminal half of Jaburetox, showed insecticidal effect comparable to that of the native peptide, while jbtx C-ter, representing the C-terminal half, was largely inactive on two different insect models [43]. On the other hand, both jbtx N-ter and jbtx C-ter were able to interact with LUVs, the C-terminal half being more effective. As the amphiphilic β-hairpin is present in the C-terminal domain, this motif could be in part involved in membrane interaction. In contrast, for the entomotoxic activity, the half peptide jbtx N-ter contains the most important domain, as observed in the experiments. Three dimensional models for Jaburetox were proposed using bioinformatics and the polypeptide and its mutants were submitted to a molecular dynamic simulation in aqueous system for 500 ns. The results suggested that the whole peptide becomes more unstructured, particularly at its N-terminal portion, and contained a few secondary structure elements with a major part of molecule in random coil conformation at the end of the 500 ns simulation. Interestingly, the β-hairpin structure was conserved in the C-terminal half. When the mutants were submitted to simulation, the jbtx N-ter peptide became totally unfolded while the jbtx C-ter showed a stabilization with β-sheet structures after 500 ns molecular dynamics [58].

Jaburetox and its truncated peptides were also studied using an electrophysiological approach to test their ability to form channels in planar lipids bilayers (PLBs). Two different membrane compositions, to produce neutral net charged and negative net charged lipid interfaces, were tested. All peptides, Jaburetox, jbtx N-ter, jbtx C-ter and jbtx Δβ-hairpin, were able to form channels in both types of bilayers, observed within 30 min after addition of 5–15 μg/mL of each peptide. All channels showed similar biophysical properties, being highly selective to cations, and displayed two conducting states: 7 pS–18 pS and 32 pS–79 pS (smaller channel and main channels, respectively). Similar to Jaburetox, jbtx N-ter was more active at negative voltages while the others did not show voltage dependence. Multiple levels of currents were observed during the experiments using high doses of peptides, suggesting the presence of several identical channels or simultaneous activity of oligomers, corroborating previously reported data on the tendency of Jaburetox and its truncated peptides to form aggregates and the fact that the peptides' oligomerization state influences their biological effects [58,66].

A structural characterization of Jaburetox in solution was carried out using light scattering, circular dichroism (CD) spectroscopy and NMR, and demonstrated the intrinsically disordered nature of the polypeptide [69]. Light scattering studies of the hydrodynamics properties of Jaburetox showed that the peptide in a neutral solution is found in a single oligomeric form with a molar mass of 11.03 kDa. It exhibited a large hydrodynamic radius for a peptide of this molecular mass, a feature suggestive of a disordered polypeptide. CD spectroscopy revealed a typical random coil conformation of Jaburetox and no strong negative signals above 205 nm, thus indicating that Jaburetox presents small amounts of secondary structure. Computational analysis predicted a propensity to disorder mainly at the Jaburetox's N-terminal domain (Figure 1). As Jaburetox tends to aggregate, a thermal scanning fluorimetric assay was performed to identify the best conditions to stabilize the peptide. In the presence of the reducing agent tris(2-carboxyethyl)phosphine (TCEP), the tendency of Jaburetox to aggregate was reduced allowing the NMR assays. The heteronuclear single quantum coherence (HSQC) spectrum unveiled low signal dispersion in the proton dimension which is an indicative of a disordered state of the polypeptide. The chemical shifts were used to predict the residue-specific propensity to form a secondary structure (SSP). SSP analysis predicted that Jaburetox is widely disordered with a small tendency to form α- structures, in addition, with slightly smaller SSP values in the N-terminal (larger predicted disorder) compared to the C-terminal portion (smaller predicted disorder), corroborating the disorder predictors (Figure 2). Data acquired from the nuclear Overhauser enhancement (NOE), which reflect the global fold state of the protein, did not demonstrate the presence of a stable tertiary structure. Still, some elements of secondary structure were detected in parts of the molecule, as a small α-helical motif in the N-terminal region (residues A12-V16), and two turn-like structures, one located in the middle of the polypeptide (residues R48-G56) and the other in its C-terminal region (residues I63-E74). However, the β-hairpin, evidenced in the JBU crystal and in the bioinformatics studies, was not observed in the 3D structure. An in-cell NMR spectroscopy was employed to investigate the molecular folding of the peptide in a physiological condition. A HSQC spectra recorded on *E. coli* cells overexpressing Jaburetox confirmed its disordered folding inside the cells [69].

In order to elucidate if Jaburetox could acquire some conformation when in contact with membranes, a study using artificial and biological membranes was performed [62]. When incubated with sodium dodecyl sulfate (SDS) micelles, a change in the secondary structure of Jaburetox was detected in its CD spectra. Moreover, NMR HSQC confirmed changes in its tertiary structure. Some conformational changes were observed in the molecule, mainly in its N-terminal region, when in contact with LUVs and micelles, prepared with different net charges and molar ratios of phospholipids. This effect was more visible with negatively charged LUVs and micelles, despite without major acquisition of tertiary structure, as determined by NMR. Fluorescence microscopy was employed to show the interaction between Jaburetox labelled with fluorescein isothiocyanate (Jaburetox-FITC) and cockroach's nervous cord (NC) as a biological membrane, revealing a great intensity of Jaburetox-FITC attached to the ganglia. When Jaburetox-FITC was pre-incubated with LUVs and bicelles, before addition to insect NC membrane, the fluorescence decreased about 50% and 70%, respectively, compared to the initial values, suggesting that the vesicles competed with the insect membranes as a target for the binding of Jaburetox. These data provided evidence that the interaction between Jaburetox and phospholipids did not induce a complete transition from unfolded to folded state but it could have facilitated the peptide's anchorage in cell membranes for posterior acquisition of a folded state [62].

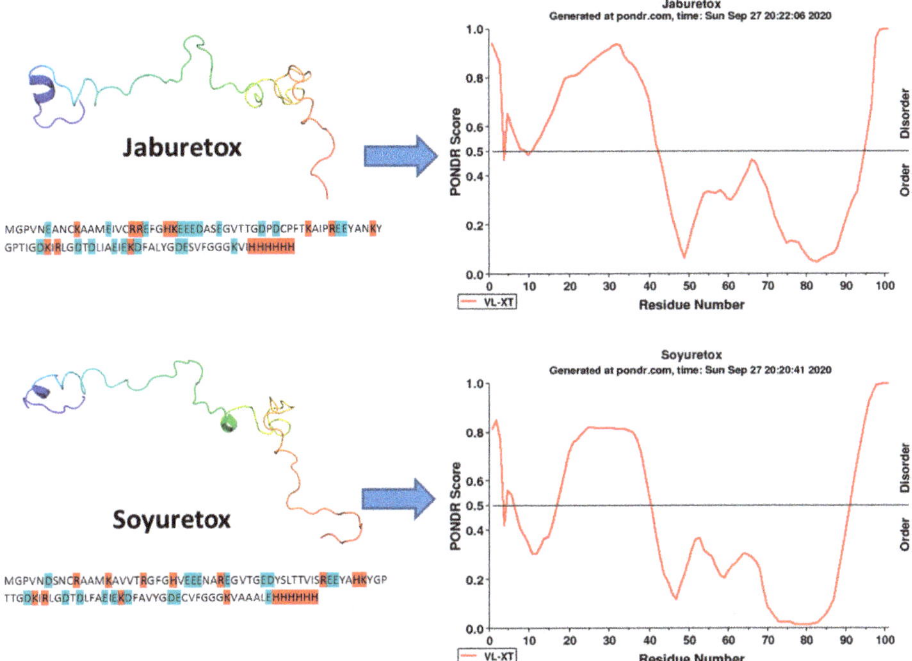

Figure 2. Disorder profile of Jaburetox and Soyuretox. Algorithm of disorder prediction VL-XT PONDR® was applied to compare Jaburetox and Soyuretox amino acid sequences. The amino acid sequences (including the 6-His tags) were submitted to http://www.pondr.com/ to generate the graphics. PONDR score above 0.5 indicates disorder. Jaburetox is slightly more disordered than Soyuretox, especially in its N-terminal region. Jaburetox structure is based on conformer 1 (PDB 2MM8). The modeled structure of Soyuretox was obtained with Modeller 9.19 using Jaburetox as a model. The primary sequences of the peptides appear below each 3D-structure, negatively charged amino acids are highlighted in blue and positively charged in red.

The secondary and tertiary structures of Soyuretox were determined using bioinformatics tools, CD spectra and NMR experiments. Jaburetox was used in the same study for comparison. Both peptides share 68% of identity in their amino acid sequences (Figure 1). The secondary structure of Soyuretox and Jaburetox was analyzed by CD spectroscopy at pH 6.5 and 8.0. At pH 6.5 solution, both peptides maintained a disordered behavior. Nonetheless, while Jaburetox kept its disordered state at pH 8.0, Soyuretox acquired some secondary structure in the alkaline medium. A 3D model of Soyuretox was obtained using the structure of JBU as template and a molecular dynamics (MD) was carried out. After 500 ns of MD simulation, Soyuretox became more globular in solution and showed changes in its secondary structure, with loss of helices and beta strands [64]. In spite of the fact that Jaburetox and Soyuretox behave similarly, Soyuretox demonstrated a tendency to aggregate at pH 6.5 and 8.0 (at the protein concentrations necessary for NMR the experiment), preventing the assignment of the NMR signals [64,69]. The HSQC NMR spectrum obtained for Soyuretox showed a low signal dispersion in the proton dimension, typical of intrinsically disordered states. The ability of Soyuretox to interact with SDS micelles was studied by CD, revealing that in presence of 10 mM SDS (above critical micellar concentration) there was an increase in the peptide's content of secondary structure. A more ordered structure of Soyuretox in the presence of SDS micelles (10 mM) was also confirmed in the peptide's HSQC NMR spectrum, exhibiting a widening of signal dispersion. Under these conditions, Soyuretox kept its intrinsically disordered state [64]. A compilation of the main data obtained in the structural studies of Jaburetox and Soyuretox is present in Table 1.

Table 1. Compiled data from structural studies of Jaburetox and Soyuretox.

Approach	Peptide(s)	Data Obtained	Reference
Ab initio modeling	Jaburetox	A β-hairpin motif was observed in the C-terminal region of the molecule. It was hypothesized that the β-hairpin motif could be involved in the entomotoxic activity.	[57,66]
Molecular dynamics simulation (in aqueous system for 500 ns)	Jaburetox, its N- and C-termini peptides and Soyuretox	Jaburetox became more unstructured in its N-terminal portion, containing a few secondary structural elements and the major part of molecule in random coil. The β-hairpin structure was conserved in the C-terminal domain. The N-terminal peptide became totally unfolded and C-terminal showed a stabilization with β-sheet structures. Soyuretox became more globular in solution and showed changes in its secondary structure, with loss of helices and beta strands.	[58,64]
Dynamic Light Scattering and Small Angle X-ray Scattering	Jaburetox	Demonstrated the ability of Jaburetox to interact with lipids using platelet-like multilamellar liposomes (PML).	[67]
Dynamic Light Scattering	Jaburetox	Jaburetox in a neutral solution is found in a single oligomeric form, exhibiting a large hydrodynamic radius, suggestive of a disordered polypeptide.	[69]
Circular dichroism (CD) spectroscopy	Jaburetox and Soyuretox	Jaburetox showed a typical random coil conformation and small amount of secondary structure under native state. Jaburetox increased its secondary structure content when in contact with SDS-micelles and large unilamellar vesicles (LUVs) composed by phospholipids of different net charges. Jaburetox and Soyuretox showed disordered behavior at pH 6.5. Soyuretox acquired some secondary structure at pH 8.	[69] [62] [64]
Nuclear Magnetic Resonance (NMR) spectroscopy	Jaburetox	The heteronuclear single quantum coherence (HSQC) spectrum unveiled low signal dispersion in the proton dimension; the SSP analysis of chemical shifts predicted that Jaburetox is widely disordered with a small tendency to form α-structures. The 3D structure obtained from nuclear Overhauser enhancement (NOE) do not demonstrated the presence of a stable tertiary structure.	[69]
	Soyuretox	The HSQC NMR spectrum obtained for Soyuretox showed a low signal dispersion in the proton dimension. A more ordered structure of Soyuretox in the presence of SDS micelles (10 mM) was also confirmed in the peptide's HSQC NMR spectrum, demonstrating a widening of signal dispersion.	[64]

4. Biological Studies

The basis of the apparent selectivity of Jaburetox and Soyuretox towards certain organisms without affecting others is not well understood. As mentioned before, studies of acute toxicity injecting or feeding Jaburetox in murine models resulted in no effect [57]. Standardized in vitro assays indicated that Jaburetox did not cause cito- or genotoxicity in human and other mammalian cell lines [70]. Regarding Soyuretox, embryotoxicity assays in the zebrafish model demonstrated the lack of effect of the peptide on several developmental and behavioral parameters [64]. On the other hand, both peptides presented potent insecticidal and antimicrobial effects (discussed in the following subsections). A possible explanation for this selectivity could be related to the fact that these peptides have more affinity for certain types of membrane lipids than others. Thus, the presence of these and possibly other molecules in susceptible cells could be necessary for the peptide to act upon. The biosecurity profile of Jaburetox seems promising since, in addition to the lack of toxic effects observed so far in vertebrates, a risk assessment study could not identify potential adverse reactions associated to the peptide [65]. Finally, the amino acid sequences of Jaburetox, Soyuretox and their homologs are present in ureases and these enzymes are abundant in several edible plants, including some that are eaten raw [71,72].

4.1. Entomotoxic Effects

The indiscriminate use of pesticides has caused a negative impact in the environment and in the health of consumers. Besides, it has facilitated the emergence of resistance in more than 600 insect and mite species [73,74]. Those phenomena have driven the study of new insecticidal molecules that, ideally, are environment-friendly and affect the target pest species without harming beneficial insects, humans and other animals [75]. According to the Insecticide Resistance Action Committee (IRAC), there are more than thirty types of modes of action currently described, including sodium channel modulators, juvenile hormone inhibitors and miscellaneous non-specific (multi-site) inhibitors, among others [73]. In the cases of Jaburetox and Soyuretox, their mechanism of action is still not completely understood. As discussed above, what we do know is that Jaburetox interacts with lipid membranes, especially those of acidic nature [66], and present cell membrane-disturbing activities [76]. We have reported that Jaburetox act upon several insect organs and at different levels, i.e., by altering the activity of various enzymes and/or the protein and the gene expression of several proteins [77,78]. The property of this peptide to target different organs, cell types, and proteins probably reflects its intrinsically disordered nature that would allow to accommodate the interaction of Jaburetox with different binding partners. Concerning Soyuretox, we know less about its sites of action, but as far as entomotoxicity goes, current evidence suggests that Soyuretox has properties similar to those of Jaburetox [64]. As commented before, Jaburetox and Soyuretox share 68% of their sequence whereas the N-terminal regions of the peptides are the most divergent. However, they can conserve functions and other features without necessarily presenting a conserved sequence as it has been described for different IDPs [79]. In an attempt to systematize and integrate the data obtained by our laboratory and collaborators regarding the entomotoxic effects of Jaburetox and Soyuretox, these findings were grouped hoping to throw light to some general trends, and trying to lay working hypothesis and new avenues for future research. The effects of Jaburetox and Soyuretox on different insect species are summarized in Table 2.

Table 2. Effects of Jaburetox and Soyuretox in different species of insects.

Species	Stage(s)	Assay	Toxic Peptide(s)	Effect(s)	Reference
Dysdercus peruvianus	Nymphs	Feeding	Jaburetox	Lethality	[57,80]
	Nymphs	Feeding and injection	Soyuretox	Lethality	[64]
Oncopeltus fasciatus	Nymphs	Injection	Jaburetox	Lethality	[58]
	Nymphs	Injection, feeding	Jaburetox	Lethality	[58,80]
Rhodnius prolixus	Nymphs and adults	Injection, feeding, in vitro	Jaburetox and Soyuretox	Effects on diuresis, enzymatic activities, expression of genes, cell activation and immune response, interaction of Jaburetox with the central nervous system and the salivary glands, among others (see Figure 3)	[58,64,77,78,81–83], unpublished results
	Nymphs and adults	Injection	Jaburetox	Lethality	[80]

Table 2. *Cont.*

Species	Stage(s)	Assay	Toxic Peptide(s)	Effect(s)	Reference
Triatoma infestans	Adults	Injection	Jaburetox	Lethality, behavioral alterations, neurotoxicity, localization of the peptide in the central nervous system, interaction of the peptide with UDP-*N*-acetylglucosamine pyrophosphorylase (UAP), inhibition of nitric oxide synthase (NOS) activity	[84]
Phoetalia pallida	Adults	Injection	Jaburetox	Blockade of evoked contractions of coxal muscle	[58]
		In vitro	Jaburetox	Interaction of the peptide with the central nervous system	[62]
Nauphoeta cinerea	Adults	Injection	Jaburetox	Alteration of locomotor behavior, leg and antennae grooming, neuromuscular blockade, cardiotoxicity and alterations in nerve and muscle electrophysiological profiles	[85]
	Adults	Feeding, injection, in vitro	Jaburetox	Absence of lethality, modulation of NOS, UAP and acetylcholinesterase activity in the central nervous system	[86]
Blatella germanica	Nymphs	Feeding	Jaburetox	Lethality	[57]
Spodoptera frugiperda	Larvae	Feeding	Jaburetox	Lethality and weight reduction	[57]
	Larvae	Feeding on transgenic corn plants	Jaburetox	Weight reduction, reduced feed consumption, sterility of females and lethality	[87]
	Larvae	Feeding	Jaburetox	Lethality and delay in larval development	[87]
Helicoverpa armigera	Larvae	Feeding on transgenic tobacco plants	Jaburetox	Lethality and reduced feed consumption	[88]
Aedes aegypti	Larvae	Feeding	Jaburetox	Lethality	[60]

4.1.1. Lethality

Like other insecticidal proteins, including the widely used Cry proteins of *B. thuringiensis* [89], the "parent" proteins of Jaburetox, canatoxin and JBU, or SBU, in the case of Soyuretox, need a step of proteolytic activation to act upon insects of different orders. Even though these urease isoforms present entomotoxic effects *per se*, they are not lethal when fed to insects with trypsin-based digestion [35]. This caveat can be surpassed by employing Jaburetox as first demonstrated by Mulinari et al. [57]. Since that finding, several species were tested for lethality and almost all of them were susceptible to Jaburetox. This peptide was effective via injection and oral administration against juveniles and adults. Moreover, the doses employed were very low when compared to other entomotoxic proteins derived from plants [35]. There are less information available for Soyuretox, although the effect of a dose comparable to those employed with Jaburetox also resulted lethal in *D. peruvianus* [64]. The only exception to this trend so far has been the cockroach *Nauphoeta cinerea*, since feeding or injecting Jaburetox did not result in mortality [86]. As we will discuss later, the effect on the activities of the central nervous system enzymes was different in cockroaches of this species when compared to susceptible insects such as the kissing bug *R. prolixus*. This difference could explain, at least in part, the resistant phenotype of *N. cinerea*.

4.1.2. Effects on the Central Nervous and Neuromuscular Systems

Behavior alterations consistent with a neurotoxic activity of Jaburetox in *T. infestans* [84] led our group to further investigate the effects on the insect's central nervous system. Immunohistochemical techniques evidenced the labelled somata of the antennal lobe and of the suboesophageal ganglion 3 h after the injection of the peptide into the hemocoel. Co-immunoprecipitation assays followed by tandem mass spectrometry identified the enzyme UDP-*N*-acetylglucosamine pyrophosphorylase as a Jaburetox-interacting protein in the bug's brain and associated ganglia [84]. This enzyme provides the activated precursor for the synthesis of chitin and for glycosylation pathways of glycoproteins and other derived products [90].

The central nervous system was also a target for Jaburetox in the triatominae *R. prolixus*, and the interaction between the fluorescently-labeled peptide and the organ could be observed in vitro. Furthermore, the activities of different enzymes in the central nervous system were altered after feeding the peptide to fifth instar nymphs [78].

The interaction of Jaburetox with the nervous cord tissue of *N. cinerea* was observed in vitro [62]. Injection of the peptide into the cockroaches *Phoetalia pallida* and *N. cinerea* led to the blockade of evoked contractions of coxal muscle [58,85]. In *N. cinerea*, this effect was amplified by chloral hydrate, (a drug known to reinforce effects mediated by GABA receptors), suggesting that Jaburetox could be activating the gabaergic neurotransmission [85]. As reported in the kissing bugs, the enzymatic activities of several enzymes were modulated in adult cockroaches upon injection [86].

The effect of Jaburetox was tested on *Xenopus laevis* oocytes overexpressing the Nav 1.1 channels from the cockroach *Blatella germanica* [85]. Voltage-clamp analysis showed a 50% increase in the sodium currents upon Jaburetox treatment while no alteration in the kinetics of the Nav 1.1 channel activation was noticed. Muscle and nerve action potentials recorded in the isolated leg of the locust *Locusta migratoria* decreased transiently about 20% in Jaburetox-treated preparations, returning to basal values after 20 min although by then the contraction of the tarsus has stopped. The absence of a fast decrease in the resting membrane potential during the voltage clamp studies, especially considering the membrane-disturbing effects of Jaburetox, suggested that the main component of its neurotoxicity could involve alteration of the gating properties of sodium channels [85]. These aspects were recently reviewed by Barreto et al. [91].

4.1.3. Effects on Behavior

Behavioral alterations were first reported in adult individuals of the triatomine *T. infestans* by Galvani et al. [84]. The injection of a lethal dose of Jaburetox that would eventually kill the insects within 18 h, led to early transient symptoms that included paralysis of the legs, proboscis extension and abnormal movements of the antennae. In fact, those findings pointed out to neurotoxicity, a phenomenon which was later on confirmed by diverse approaches and ascribed, at least in part, to alterations of the nitrinergic system in the central nervous system of this species [84]. In the case of the cockroach *N. cinerea* [85], the locomotor behavior was also altered after Jaburetox injection, and adult insects exhibited a significant decrease in the travelled distance accompanied by a corresponding increase in the stopping time. The leg and antenna grooming activities were also modified, with significant increments upon injection. These and other toxic effects were attributed to an initial activation of voltage-gated sodium channels [85].

4.1.4. Effects on Enzymatic Pathways

Since the finding that the injection of Jaburetox diminished the enzyme activity of Nitric Oxide Synthase (NOS) in the central nervous system of *T. infestans* [84], a series of approaches were undertaken in order to try to understand the basis of this alteration. Besides its function in nitrinergic signaling in the central nervous system, NO participates in the immune response of insects due to its capacity of inducing oxidation of heme groups and nitrosylation of amino acid residues in proteins of pathogens [92]. The diminution of NOS activity upon Jaburetox injection was not related to the protein levels, since no differences in band intensities were seen in Western blots of brain homogenates of vehicle-injected and Jaburetox-injected insects [84]. The NOS activity also decreased when the homogenates were incubated with the peptide in vitro, suggesting a direct effect of the Jaburetox on the enzyme [84]. Similar results were obtained in the central nervous system of the related triatome *R. prolixus*, with both in vivo and in vitro Jaburetox treatments leading to a decrease in NOS enzyme activity without affecting its gene expression. Nevertheless, the effect of Jaburetox was different on the activity of NOS in the salivary glands and hemocytes, where the expression of its gene was increased, indicating an organ-specific effect [78]. Jaburetox-induced alterations of NOS is not restricted to triatomines, since the peptide also induced a decrease in NOS enzyme activity in the central nervous system of the cockroach *N. cinerea*, upon in vivo or in vitro treatments, without affecting the protein expression [86]. The regulation of NO production mediated by Jaburetox is complex and could involve more than one level, for example, affecting directly the enzyme as the in vitro assays pointed out and/or, indirectly, through modifications on the expression of its gene, or even altering the membrane properties of target cells [77].

Considering that Jaburetox interacted with UAP in the central nervous system of *T. infestans*, the effect of the peptide on this enzyme was also tested. It was observed that the UAP enzyme was affected either after in vivo or in vitro treatments with Jaburetox, in this case causing a significant increase in activity without modifying the expression of its corresponding gene. Again, an organ-specific effect was demonstrated, with different responses of UAP in the central nervous system as compared to the salivary glands and hemocytes [77]. When a recombinant version of the *R. prolixus'* UAP was incubated in vitro with Jaburetox, no modification of the enzyme activity was observed, suggesting that other factor(s) present in the tissue homogenates are probably required for the peptide to exert its regulatory effect [78].

Taking into account the various functions of the main product of UAP, UDP-*N*-acetylglucosamine, several physiological processes can be influenced by the toxin. One of such process is chitin synthesis, catalyzed by chitin synthase and serving UDP-*N*-acetylglucosamine as substrate. The effect of Jaburetox on the expression of chitin synthase gene was then explored. It was found that Jaburetox treatment led to a diminution of chitin synthase expression in the central nervous system, salivary glands, anterior midgut, Malpighian tubules and fat body, but not in the hemocytes or the posterior midgut of *R. prolixus* [77,78]. The profile of UAP modulation by the Jaburetox in *N. cinerea* was different, since its activity was only affected 18 h after injection. This distinct response when compared to the triatomines could be related to the fact that this cockroach is so far the only species found to be resistant to the acute lethal effect of Jaburetox [86]. As the activity of acetylcholinesterase, an enzyme involved in the resistance of insects to organic pesticides, increased upon treatment of the cockroach with Jaburetox, the lack of lethality could be a reflex of an analogous mechanism(s) disabling the toxic effects of the peptide [86]. However, Jaburetox is far from being innocuous to *N. cinerea*, as the paralyzing effect and alterations in behavior caused by the peptide could lead to death as well, due to inability to hide and avoid danger, or to find food.

4.1.5. Effects on Diuresis

The evaluation of the effects of Jaburetox on diuresis was one of the first investigations carried out to understand the mechanism of action of the toxin. In 2009, Stanisçuaski et al. [81] conducted studies on *R. prolixus'* Malpighian tubules to explore, in vitro, the effects of Jaburetox on serotonin-induced diuresis. The authors demonstrated that Jaburetox and also JBU are capable of interacting with membrane factors that end up inhibiting diuresis by triggering different signaling cascades. While JBU effect is mediated by the activation of the eicosanoid cascade and is dependent on Ca^{++} ions, the antidiuretic effect of Jaburetox is mediated by an increase in cyclic guanosine monophosphate (cGMP) levels. This increment leads to the interruption of ion transport by blockage of the apical V-ATPase and disruption of the transepithelial potential across the tubule's membrane through an unknown pathway, leading (directly or indirectly) to the inhibition of water secretion and consequent impairment of diuresis [81].

4.1.6. Effects on the Immune System

As opposed to vertebrates, insects do not have a developed acquired immunity. Instead, they have a robust innate immunity that can be subclassified into cellular and humoral responses [88]. Cellular immunity is characterized by the action of defense cells (hemocytes) in aggregation, phagocytosis and encapsulation processes. The humoral immune response comprises the activation of antimicrobial peptides, of reactive oxygen species (ROS), and of enzyme complexes that regulate melanization and the coagulation cascade, among them the phenoloxidase (PO) [93,94].

In this context, previous data from our group indicated that JBU and the ubiquitous Soybean Urease (uSBU) are able to induce activation of the insect immune response in *R. prolixus* [48,95]. Defferrari and coworkers demonstrated that JBU is capable of activating both, the cellular and humoral immune responses. The activation of cellular aggregation induced by JBU is mediated by the cyclooxygenase (COX) pathway and required extracellular Ca^{++} ions. JBU also elicited the darkening of the hemolymph, an immune response associated with the melanization reaction triggered by the PO. At cellular level, immunolocalization assays demonstrated that the toxin is capable of inducing cytoskeleton damage and nuclear condensation in hemocytes [95]. Additionally, Martinelli and collaborators reported that uSBU in vivo and in vitro is also capable of inducing hemocyte aggregation in *R. prolixus* [48].

Based on these studies, cellular and biochemical approaches were carried out in order to evaluate the effects of Jaburetox and Soyuretox on the immune response of *R. prolixus*. Like JBU and uSBU, Jaburetox and Soyuretox induced Ca^{++}-dependent aggregation of hemocytes in vivo and in vitro, mediated by the COX pathway [64,82]. Despite the aggregation, Fruttero et al. and Moyetta et al. demonstrated that the phagocytic capacity of hemocytes is not altered by the toxin [77,82]. In addition, Jaburetox also generated chromatin condensation, cytoskeleton disorganization and caspase 3 activation in the hemocytes, indicating the induction of apoptosis by the toxin [82]. The interaction of Jaburetox with the hemolymphatic cells was also seen upon in vivo and in vitro treatments, and the peptide was found in different subcellular locations [77].

Besides affecting the cellular immune response, Jaburetox also modulates the humoral immunity. In *R. prolixus*, the toxin induced an increment in the PO activity in vivo, without altering the activity of other effectors, such as the antibacterial cecropins and lysozymes [82]. Jaburetox triggered in hemocytes the increment of NOS gene expression. NO produced by the enzyme is known to induce the formation of free radicals that aid in immune defenses. However, these changes in gene expression were not accompanied by the corresponding modifications in protein levels in hemocytes or in enzymatic activity of NOS assayed in vitro, after the exposure to the toxin [77,82]. Through fluorescence assays with specific probes, it was observed that cells aggregated in the presence of Jaburetox had a greater local production of NO [77,82]. In 2020, Grahl et al. demonstrated in cultured hemocytes that a high dose of Jaburetox (6 μM) induced a significant increase of ROS production without altering cell viability [83].

When the Jaburetox-treated insects were injected with the pathogenic bacterium *Staphylococcus aureus*, the bacterial clearance was significantly reduced, indicating an immunosuppressive effect. Thus, the cellular and humoral immune activations triggered by Jaburetox do not protect the insect against posterior bacterial challenges [82]. These responses are similar to those elicited by bacterial and protozoan pathogens, raising the possibility that Jaburetox is recognized by the innate insect immunity as a pathogen-associated molecular pattern.

Another important immune response is the release of extracellular nucleic acid traps [96]. This immune mechanism of vertebrates and invertebrates is characterized by ROS-dependent release of chromatin into the cytoplasm, promoting the association of the nuclear material with antimicrobial proteins. Thereafter, this complex is released to the extracellular medium to withstand infections [96,97]. In this context, considering the changes in gene expression and nuclear condensation induced by Jaburetox in *R. prolixus*, experiments were designed to evaluate the impact of Jaburetox on the interactions of nucleic acids (DNA and RNA) extracted from the same insect species and used to mimic extracellular nucleic acid traps. It was observed that injection of the toxin together with RNA caused an increase in hemocyte aggregation, however when the toxin is injected together with DNA, no aggregation was seen. Concerning humoral responses, Jaburetox plus RNA yielded an increased PO activity only 6 h after injection, while Jaburetox plus DNA sustained an augmented humoral response both at 6 and 18 h after injection [83].

The effect of extracellular nucleic acids on the Jaburetox-induced immunosuppressive effect against pathogenic bacteria was also studied. Immunocompetence assays injecting Jaburetox alone, or Jaburetox plus DNA or RNA before the injection of bacteria, demonstrated that both RNA and DNA counteracted Jaburetox effects, and restored the bacterial-clearance capacity of the insects [83].

Finally, to better understand the immunological modulation caused by nucleic acids, the toxin ability to induce the release of extracellular nucleic acid traps was evaluated. It was seen that Jaburetox was not able to induce the release of RNA or DNA, either upon in vitro or in vivo treatments. The incapacity of the insects to release extracellular nucleic acids after Jaburetox treatment could partly explain the immunosuppressive effect of the peptide and the weakened response of the treated insects against a bacterial challenge [83]. Since *R. prolixus* has been an instrumental model to understand the effects of the urease-derived peptides, we have summarized all our findings in the Figure 3.

4.2. Antifungal and Antibacterial Activity

So far there are not many examples in the literature of IDPs with antifungal activity, despite the two decades elapsed since the definition of IDPs, around the 2000s [98–101].

As mentioned above, Jaburetox presents antifungal activity against filamentous fungi and yeasts [61]. Postal and co-authors observed the toxicity of Jaburetox against the phytopathogenic filamentous fungi *Mucor* sp. (at 10 μM) and *Penicillium herguei* (at 20 μM). *Rhizoctonia solani* was not susceptible to Jaburetox in the tested doses. Regarding yeasts, Jaburetox at 9 μM inhibited the multiplication of *Saccharomyces cerevisiae*, *Candida parapsilosis* and *Pichia membranifaciens* and at 18 μM, the peptide inhibited *Candida tropicalis*, *C. albicans* and *Kluyveromyces marxiannus*. Fluorescence microscopy of *S. cerevisiae* evidenced an increase in membrane permeability in Jaburetox-treated cells, using the SYTOX Green stain. In *C. tropicalis*, exposition to Jaburetox also induced the formation of pseudohyphae. These microscopy experiments were conducted at lower doses of Jaburetox (0.36–0.72 μM). In another work of our group, Broll and co-workers [62] showed that FITC-labeled Jaburetox interacted with *S. cerevisiae* cells, and remained bound to membrane cell debris even after yeast lysis. These results suggested that the target of Jaburetox is present on the yeast external membrane [62].

The antimicrobial activity of Jaburetox against some bacteria such as *Bacillus cereus*, *Escherichia coli*, *Pseudomonas aeruginosa* and *Staphylococcus aureus* was observed in preliminary assays in the dose range of 0.25 μM to 13.5 μM [60]. As described earlier, Jaburetox was shown to permeabilize model membranes, as LUVs and PLBs, composed by different phospholipids and net charges; phosphatidylglycerol (PG) and phosphatidic acid (PA) with negative charges and the neutral phosphatidylethanolamine (PE), phosphatidylcholine (PC) and cholesterol (Ch) [66,76]. Many microorganisms contain negatively charged lipids in their membrane compositions, as PG and cardiolipin (CL) [102,103]. The main phospholipids found in the bacterium *S. aureus* membrane are PG, CL, and lysophosphatidylglycerol (LPG) [104,105]. In yeasts, a study using eight *C. albicans* azole-resistant and azole-sensitive strains demonstrated that the major phospholipids compositions in the plasma membrane of all the isolates were PC, PE, phosphatidylinositol (PI) and phosphatidylserine (PS). The percentage of phospholipids varied individually [106]. Interestingly, both microorganisms are susceptible to Jaburetox [60,61].

Soyuretox was also investigated regarding its antifungal activity. It was found to be active against *C. albicans*, *C. parapsilosis* and *S. cerevisiae* at 9 μM and 18 μM concentrations, similar to the fungitoxic doses reported for Jaburetox. For *C. albicans*, at the minimal inhibitory of 5 μM, production of superoxide anions was detected as part of the fungitoxic mode of action of Soyuretox. Binding of Soyuretox to *C. albicans* cells was observed by immunofluorescence [64].

The mechanism of antifungal activity of Jaburetox and Soyuretox remains elusive. It is known that the peptides permeabilize the fungal membrane, and cause change of fungal morphology, inducing formation of pseudohyphae, structures considered a stress and defense response mechanism of yeasts [48]. Moreover, the peptides induced intracellular production of superoxide anions in yeasts, causing oxidative stress. Our data suggest that these peptides probably interact with lipids in the fungal membrane (Figure 4). Although it still lacks experimental demonstration, it is plausible that the IDP nature of these peptides could be relevant for their antimicrobial activities, as changes in order-disorder states upon ligand binding could possibly modulate their fungitoxic action.

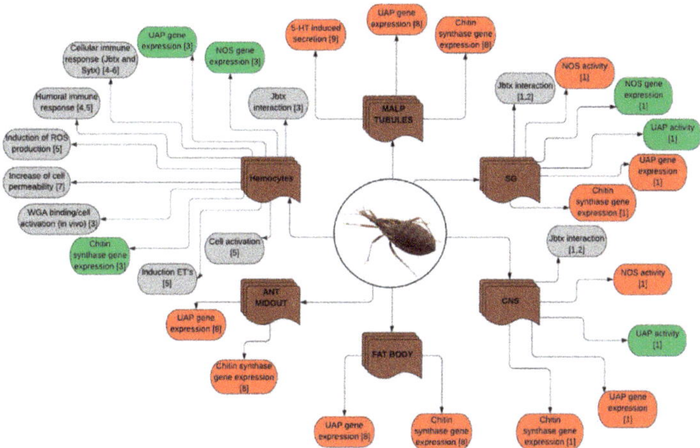

Figure 3. Tissue-specific effects of urease-derived peptides in *Rhodnius prolixus*. In the flowchart, the brown boxes are the organs or the cells affected by the urease-derived peptides while the grey boxes indicate no change. The green boxes represent an increase in the assessed effect while the red boxes indicate a decrease. Acronyms stand for: 5-HT, Serotonin; ANT MIDGUT, Anterior midgut; CNS, Central nervous system; ET'S, Extracellular traps; JBTX, Jaburetox; MALP TUBULES, Malpighian tubules; NOS, Nitric oxide synthase; ROS, Reactive oxygen species; SG, Salivary glands; SYTX, Soyuretox; UAP, UDP-*N*-acetylglucosamine pyrophosphorylase; WGA, Wheat germ agglutinin. The numbers between brackets indicate the corresponding references: (1) Fruttero et al., [78]; (2) unpublished results; (3) Moyetta et al., [77]; (4) Fruttero et al., [82]; (5) Coste Grahl et al., [83]; (6) Kappaun et al., [64]; (7) and (8) unpublished results; (9) Staniscuaski et al., [81].

Jaburetox and Soyuretox are peptides prone to aggregation. In the studies aiming to characterize their 3D structures, TCEP, a potent reducing agent, was used to avoid aggregation during NMR experiments. It was demonstrated that Jaburetox and its truncated peptides (jbtx N-ter and jbtx C-ter) tend to form aggregates in solution and that their oligomerization state interfered in biological activities and membrane interactions [58,66,76]. Aggregation is known as an important factor in mode of action of antimicrobial peptides [102].

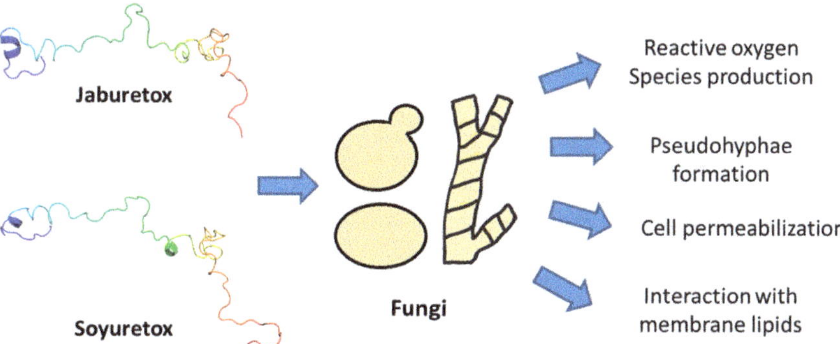

Figure 4. Schematic representation of the antifungal effects of Jaburetox and Soyuretox against filamentous fungi and yeasts.

5. Structural Aspects of Other Intrinsically Disordered Bioactive Polypeptides

Even though approximately 30% of eukaryote proteins have disordered regions composed of fifty or more amino acid residues [99], there are relatively few reports of antimicrobial and insecticidal IDPs in the literature. Without aiming to be exhaustive, we briefly discuss here some examples and establish comparisons to Jaburetox and Soyuretox when relevant.

Plants are an inexhaustible source of bioactive molecules, including those that are part of their highly evolved defense mechanisms [34]. One important example are the cyclotides, naturally occurring macrocyclic peptides found in several families of plants. They present a unique head-to-tail cyclized backbone, stabilized by three disulfide bonds forming a cystine knot. This arrangement makes cyclotides exceptionally stable against chemical, thermal and biological degradation. These macromolecules are able to cross cellular membranes and control intracellular protein-protein interactions, enabling them to act upon different targets [107]. Cyclotides present diverse host-defense roles including insecticidal activity and it is believed that this property is derived from their ability to bind to membranes and form pores [108]. Kalata B1 is the most studied cyclotide, derived from the African plant *Oldelandia affinis* [109,110]. Daly et al. [111] reported that the N-terminal pro-domain of the kalata B1 precursor is intrinsically unstructured. This terminal region induces the self-association of the precursor to form a dimeric structure, which can, in turn, be determinant for the role of the N-terminal as a vacuolar-targeting signal. According to the authors, the disorder in the terminal region could be linked either to the fact that it is a functional segment with higher mobility or because it partially folds upon binding to a target, as could be also the case for Jaburetox and Soyuretox [64,69] Thus, the pore-forming capacity seems to be part of the toxic mechanism of both, urease-derived peptides and cyclotides. However, the intrinsically disordered nature of the cyclotides does not seem to be related to their insecticidal effect but rather to a role in signaling. The relevance of IDPs in signal transduction in plants is better documented and, in this case, it is believed that the intrinsically disordered nature is necessary to confer the low affinity and high specificity needed to perform the required interactions [112].

Concerning antifungal IDPs, histatins are a family of small, histidine-rich, cationic proteins present in mammalian saliva that constitute the first line of defense against oral candidiasis caused by *C. albicans* and to other pathogenic fungi. Histatin 5, an intrinsically disordered model protein, is the major histatin component of the unstimulated parotid secretion and the most potent antifungal protein of all the histatin family [113,114]. Histatin-5 has antifungal activity against *C. albicans* at 15 μM [115], a similar fungitoxic concentration for Jaburetox and Soyuretox [61,64]. The physiological concentration of histatin-5 in human saliva is 15 to 50 μM, while the concentration of protein required to kill half of maximum number of cells (ED_{50}) is 1.4 μM. There is an extensive debate regarding the mode of action of this protein, with evidences pointing against pore formation or membrane lysis. The targets of histatin-5 appear to be intracellular and, once taken up by cells, it affects mitochondrial functions causing oxidative stress and ultimately killing the cells by ion imbalance and volume dysregulation induced by osmotic stress [116]. In addition, this peptide is related to depletion of intracellular ATP content and also oxidative damage due to ROS formation in intracellular organelles [117]. The production of oxidative molecules by histatin-5 is a common aspect with the mode of actions of Jaburetox and Soyuretox, which induced ROS generation both in insect hemocytes [83] and in *C. albicans* cells [64]. Since the fungitoxic mode of action of histatin-5 is not completely understood, it is not clear how the intrinsically disordered nature of the protein participates in the process. Nevertheless, histatin-5 mechanism of action against *C. albicans* is similar to what is known so far for Jaburetox and Soyuretox, including membrane interaction and permeabilization, and ROS formation. There is also evidence that Jaburetox is taken up by hemocytes [77], thus suggesting intracellular targets.

Hornerin is an IDP of 254 kDa that belongs to the S100-fused-type family. This protein is believed to be one of the main reasons why healthy human skin is remarkably resistant towards the infection by *Pseudomonas aeruginosa*, an environmental opportunistic pathogen widespread in water and soil [118]. Recently, fragments of hornerin were characterized as potent microbicidal agents and that this feature is maintained, independent of the amino acid sequence, provided they are linear cationic peptides containing a high percentage of disorder-promoting amino acids and a low percentage of order-promoting ones. The authors reported that the antimicrobial capacity of these cationic intrinsically disordered antimicrobial peptides (CIDAMPs) depends on their chain length, net charge, lipidation and environmental conditions [119]. The CIDAMPs have an intracellular mode of action, as hornerin transverses bacterial membranes by an energy-dependent mechanism and accumulates in the cytoplasm. The molecular targets of CIDAMPs seems to be different sites of the protein synthesis machinery [120]. The described features of CIDAMPS and the other IDP active peptides are summarized in Table 3.

Table 3. Other intrinsically disordered proteins with biological activities.

IDP	Source	Biological Activity	Disorder Region	Reference
Kalata B1	African plant *Oldelandia affinis*	Signaling, ability to bind and to form pores in membranes	N-terminal pro-domain	[111]
Histatin 5	Mammalian saliva	Antifungal activity	No defined structure in solution	[116,117]
Hornerin	Human skin	Antibacterial activity	Almost all the protein is unstructured, except the N-terminus. Cationic peptides generated from hornerin present antimicrobial activity	[119]

In future works we intend to evaluate Jaburetox-derived peptides, its N- and C-termini portions as generated by Martinelli and co-authors [58] against fungi, in order to identify the fungitoxic region of the molecule. There are differences in secondary structure of the two terminal regions, the N-terminus being more disordered than the C-terminus [69] and this difference could be important to the antifungal activity. In addition, the C-terminal domain of Jaburetox interacts more effectively with lipid membranes [58]. More studies are required to answer these questions.

6. Biotechnological Applications and Perspectives

The use of GM crops resistant to pests such as fungi, nematodes and insects is an appealing strategy considering the current need of efficiently increasing the yield of the agricultural production with less impact in the environment and health [121]. Since their discovery in the 80′s, the use of transgenic crops has been dominated by the Bt technology. Nevertheless, some insect species are not susceptible to them and its intensive application has led to the development of resistance [19,122]. Considering the fact that IDPs do not need to fold in a proper way to be biologically active, this feature can potentially be an advantage regarding their expression in transgenic plants. In the case of Jaburetox, its disordered structure gives the peptide the capacity to withstand a vast range of temperatures and pH without losing its biological activity [69], a desirable feature for a biotechnological tool. Moreover, the conformational flexibility of Jaburetox and Soyuretox allows them to interact with several binding partners with different subcellular distributions, leading ultimately to diverse targets. This feature gives them the ability of avoiding or at least delaying the generation of resistance.

In this context, three types of transgenic crops expressing urease-derived peptides have been developed with promising results [87,88,123]. Soybean plants overexpressing Soyuretox were challenged with the root-knot nematode *Meloidogyne javanica*, a major agricultural pest in several countries [123]. As a result, the average reproductive factor of the nematode was significantly reduced. On the other hand, Didoné [87] reported that maize (*Zea mays*) expressing Jaburetox fed to the important polyphagous pest *S. frugiperda* led not only to a 39% lethality of larvae, but also to other sub-lethal statistically significant effects, such as body weight reduction, decreased ingestion and remarkably, fertility decline. In addition, Ceccon [88] demonstrated that Jaburetox-expressing tobacco plants also produced high mortality and a pronounced reduction in the leaf consumption by the lepidopteran *H. armigera*. These authors started the development of gene stacking strategy, with plants expressing simultaneously Jaburetox and a double-stranded RNA complementary to the *rieske* gene, which has the advantage of diminishing the possibility of resistance development events [87]. The use of plants overexpressing Soyuretox or Jaburetox would, in principle, be a way to avoid harming beneficial or innocuous insects, since only those species that fed on the plants would be affected. Nevertheless, since off-target effects could be an issue related to the broad insecticidal activity of Jaburetox/Soyuretox, transgenic crops can be improved using tissue-specific or damage-induced promoters [124,125].

These latest studies [87,88,123] are a proof of concept that IDPs in general, and the urease-derived peptides in particular, are very attractive pesticides that can be engineered for use as an effective and environmental-friendly strategy, alone or in combination with other IDPs or toxic molecules. The multidisciplinary research approaches employed by our group and collaborators improved significantly the understanding of the structural and biological aspects of these IDPs and encourage us to pursue a full comprehension of their mechanism of action that would, ultimately, facilitate their application in the field.

Author Contributions: All authors wrote the original draft. M.V.C.G., F.C.L., A.H.S.M. and L.L.F. designed the tables and figures. C.R.C. and L.L.F. revised the manuscript. C.R.C. acquired funding. All authors have read and agreed to the published version of the manuscript.

Funding: Work in C.R.C. laboratory is supported by the Brazilian agencies: Conselho Nacional de Desenvolvimento Científico e Tecnológico (CNPq) [proj. 44.6052/2014-1 and Edital Jovem Talento–BJT 40.0189/2014-3, Science Without Borders]; Coordenação de Aperfeiçoamento de Pessoal de Nível Superior (CAPES) [Finance Code 001, Edital Toxinologia 63/2010 proj. 1205/2011 and Portal de Periódicos]; Fundação de Amparo à Pesquisa do Estado do Rio Grande do Sul (FAPERGS, Edital PRONEX 008/2009, proj. 10/0014-2). M.V.C.G. received a CAPES Ph. D. fellowship. F.C.L. received a Ph. D. fellowship from CAPES and is a post-doctoral fellow from CAPES (PNPD). A.H.S.M. received Ph. D. fellowship from CNPq and a post-doctoral fellowship from CAPES. C.R.C. is a Research Productivity Fellow from CNPq. L.L.F. is a member of the National Research Council (CONICET, Argentina).

Acknowledgments: The authors thank Rodrigo Ligabue-Braun (Departamento de Farmacociências, UFCSPA, Porto Alegre, RS, Brazil) for assistance with bioinformatics and conceptual discussions.

Conflicts of Interest: The authors declare no conflict of interest.

References

1. Oldfield, C.J.; Dunker, A.K. Intrinsically disordered proteins and intrinsically disordered protein regions. *Annu. Rev. Biochem.* **2014**, *83*, 553–584. [CrossRef] [PubMed]
2. Uversky, V.N. Intrinsically disordered proteins and their environment: Effects of strong denaturants, temperature, pH, Counter ions, membranes, binding partners, osmolytes, and macromolecular crowding. *Protein J.* **2009**, *28*, 305–325. [CrossRef] [PubMed]
3. Uversky, V.N. Intrinsically disordered proteins and their "Mysterious" (meta)physics. *Front. Phys.* **2019**, *7*, 8–23. [CrossRef]
4. Uversky, V.N. Intrinsically Disordered Proteins: Targets for the Future? *Struct. Biol. Drug Discov. Methods Tech. Pract.* **2020**, 587–612. [CrossRef]
5. Uversky, V.N. Unusual biophysics of intrinsically disordered proteins. *Biochim. Biophys. Acta* **2013**, *1834*, 932–951. [CrossRef] [PubMed]
6. Uversky, V.N. A decade and a half of protein intrinsic disorder: Biology still waits for physics. *Protein Sci.* **2013**, *22*, 693–724. [CrossRef]
7. Uversky, V.N. What does it mean to be natively unfolded? *Eur. J. Biochem.* **2002**, *269*, 2–12. [CrossRef]
8. Tompa, P.; Fuxreiter, M. Fuzzy complexes: Polymorphism and structural disorder in protein–protein interactions. *Trends Biochem. Sci.* **2008**, *33*, 2–8. [CrossRef]
9. Uversky, V.N. Multitude of binding modes attainable by intrinsically disordered proteins: A portrait gallery of disorder-based complexes. *Chem. Soc. Rev.* **2011**, *40*, 1623–1634. [CrossRef]

10. Dunker, A.K.; Babu, M.M.; Barbar, E.; Blackledge, M.; Bondos, S.E.; Dosztányi, Z.; Dyson, H.J.; Forman-Kay, J.; Fuxreiter, M.; Gsponer, J.; et al. What's in a name? Why these proteins are intrinsically disordered. *Intrinsically Disord. Proteins* **2013**, *1*, e24157. [CrossRef]
11. Eliezer, D. Biophysical characterization of intrinsically disordered proteins. *Curr. Opin. Struct. Biol.* **2009**, *19*, 23–30. [CrossRef] [PubMed]
12. Chen, J.; Liu, X.; Chen, J. Targeting Intrinsically Disordered Proteins through Dynamic Interactions. *Biomolecules* **2020**, *10*, 743. [CrossRef]
13. Mitchell, C.; Brennan, R.M.; Graham, J.; Karley, A.J. Plant defense against herbivorous pests: Exploiting resistance and tolerance traits for sustainable crop protection. *Front. Plant Sci.* **2016**, *7*, 1132. [CrossRef]
14. Oerke, E.C. Crop losses to pests. *J. Agric. Sci.* **2006**, *144*, 31–43. [CrossRef]
15. Gregory, P.J.; Johnson, S.N.; Newton, A.C.; Ingram, J.S.I. Integrating pests and pathogens into the climate change/food security debate. *J. Exp. Bot.* **2009**, *60*, 2827–2838. [CrossRef] [PubMed]
16. Ahmad, P.; Ashraf, M.; Younis, M.; Hu, X.; Kumar, A.; Akram, N.A.; Al-Qurainy, F. Role of transgenic plants in agriculture and biopharming. *Biotechnol. Adv.* **2012**, *30*, 524–540. [CrossRef]
17. Palma, L.; Muñoz, D.; Berry, C.; Murillo, J.; Caballero, P.; Caballero, P. *Bacillus thuringiensis* toxins: An overview of their biocidal activity. *Toxins* **2014**, *6*, 3296–3325. [CrossRef]
18. Chougule, N.P.; Bonning, B.C. Toxins for transgenic resistance to hemipteran pests. *Toxins* **2012**, *4*, 405–429. [CrossRef]
19. Heckel, D.G. How do toxins from *Bacillus thuringiensis* kill insects? An evolutionary perspective. *Arch. Insect Biochem. Physiol.* **2020**, *104*, e21673. [CrossRef]
20. Wei, J.; Zhang, Y.; An, S. The progress in insect cross-resistance among *Bacillus thuringiensis* toxins. *Arch. Insect Biochem. Physiol.* **2019**, *102*, e21547. [CrossRef]
21. Grossi-de-Sá, M.F.; Pelegrini, P.B.; Vasconcelos, I.M.; Carlini, C.R.; Silva, M.S. Entomotoxic Plant Proteins: Potential Molecules to Develop Genetically Modified Plants Resistant to Insect-Pests. In *Plant Toxins*; Carlini, C.R., Ligabue-Braun, R., Eds.; Springer Netherlands: Dordrecht, The Netherlands, 2017; pp. 415–447. ISBN 978-94-007-6464-4. [CrossRef]
22. Tsaneva, M.; Van Damme, E.J.M. 130 years of Plant Lectin Research. *Glycoconj. J.* **2020**, *37*, 533–551. [CrossRef] [PubMed]
23. Bolognesi, A.; Bortolotti, M.; Maiello, S.; Battelli, M.G.; Polito, L. Ribosome-inactivating proteins from plants: A historical overview. *Molecules* **2016**, *21*, 1627. [CrossRef]
24. Zhu-Salzman, K.; Zeng, R. Insect Response to Plant Defensive Protease Inhibitors. *Annu. Rev. Entomol.* **2015**, *60*, 233–252. [CrossRef] [PubMed]
25. Souza Cândido, E.; Pinto, M.F.S.; Pelegrini, P.B.; Lima, T.B.; Silva, O.N.; Pogue, R.; Grossi-de-Sá, M.F.; Franco, O.L. Plant storage proteins with antimicrobial activity: Novel insights into plant defense mechanisms. *FASEB J.* **2011**, *25*, 3290–3305. [CrossRef] [PubMed]
26. Sales, M.P.; Gerhardt, I.R.; Grossi-de-Sa, M.F.; Xavier-Filho, J. Do legume storage proteins play a role in defending seeds against Bruchids? *Plant Physiol.* **2000**, *124*, 515–522. [CrossRef] [PubMed]
27. Vriens, K.; Cammue, B.P.A.; Thevissen, K. Antifungal plant defensins: Mechanisms of action and production. *Molecules* **2014**, *19*, 12280–12303. [CrossRef]
28. Carvalho, A.d.O.; Gomes, V.M. Role of plant lipid transfer proteins in plant cell physiology—A concise review. *Peptides* **2007**, *28*, 1144–1153. [CrossRef]
29. Grover, A. Plant Chitinases: Genetic Diversity and Physiological Roles. *CRC. Crit. Rev. Plant Sci.* **2012**, *31*, 57–73. [CrossRef]
30. Breitenbach Barroso Coelho, L.C.; Marcelino dos Santos Silva, P.; Felix de Oliveira, W.; de Moura, M.C.; Viana Pontual, E.; Soares Gomes, F.; Guedes Paiva, P.M.; Napoleão, T.H.; dos Santos Correia, M.T. Lectins as antimicrobial agents. *J. Appl. Microbiol.* **2018**, *125*, 1238–1252. [CrossRef]
31. Stec, B. Plant thionins—The structural perspective. *Cell. Mol. Life Sci.* **2006**, *63*, 1370–1385. [CrossRef]
32. Khan, H.; Mubarak, M.S.; Amin, S. Antifungal Potential of Alkaloids as an Emerging Therapeutic Target. *Curr. Drug Targets* **2016**, *18*. [CrossRef] [PubMed]
33. Dang, L.; Van Damme, E.J.M. Toxic proteins in plants. *Phytochemistry* **2015**, *117*, 51–64. [CrossRef] [PubMed]
34. Carlini, C.R.; Grossi-De-Sá, M.F. Plant toxic proteins with insecticidal properties. A review on their potentialities as bioinsecticides. *Toxicon* **2002**, *40*, 1515–1539. [CrossRef]

35. Carlini, C.R.; Ligabue-Braun, R. Ureases as multifunctional toxic proteins: A review. *Toxicon* **2016**, *110*, 90–109. [CrossRef]
36. Carlini, C.R.; Polacco, J.C. Toxic properties of urease. *Crop Sci.* **2008**, *48*, 1665–1672. [CrossRef]
37. Stanisçuaski, F.; Carlini, C.R. Plant ureases and related peptides: Understanding their entomotoxic properties. *Toxins* **2012**, *4*, 55–67. [CrossRef]
38. Kappaun, K.; Piovesan, A.R.; Carlini, C.R.; Ligabue-Braun, R. Ureases: Historical aspects, catalytic, and non-catalytic properties—A review. *J. Adv. Res.* **2018**, *13*, 3–17. [CrossRef]
39. Ligabue-Braun, R.; Carlini, C.R. Moonlighting Toxins: Ureases and Beyond. In *Plant Toxins*; Springer Netherlands: Dordrecht, The Netherlands, 2015; pp. 1–21. [CrossRef]
40. Mazzei, L.; Musiani, F.; Ciurli, S. Urease. In *RSC Metallobiology series "The Biological Chemistry of Nickel"*; The Royal Society of Chemistry: London, UK, 2017; ISBN 1555812139. [CrossRef]
41. Mobley, H.L.T.; Island, M.D.; Hausinger, R.P. Molecular biology of microbial ureases. *Microbiol. Rev.* **1995**, *59*, 451–480. [CrossRef]
42. Carlini, C.R.; Guimarães, J.A. Isolation and characterization of a toxic protein from *Canavalia ensiformis* (jack bean) seeds, distinct from concanavalin A. *Toxicon* **1981**, *19*, 667–675. [CrossRef]
43. Follmer, C.; Barcellos, G.B.S.; Zingali, R.B.; Machado, O.L.T.; Alves, E.W.; Barja-Fidalgo, C.; Guimarães, J.A.; Carlini, C.R. Canatoxin, a toxic protein from jack beans (*Canavalia ensiformis*), is a variant form of urease (EC 3.5.1.5): Biological effects of urease independent of its ureolytic activity. *Biochem. J.* **2001**, *360*, 217–224. [CrossRef]
44. Balasubramanian, A.; Durairajpandian, V.; Elumalai, S.; Mathivanan, N.; Munirajan, A.K.; Ponnuraj, K. Structural and functional studies on urease from pigeon pea (*Cajanus cajan*). *Int. J. Biol. Macromol.* **2013**, *58*, 301–309. [CrossRef] [PubMed]
45. Follmer, C.; Real-Guerra, R.; Wasserman, G.E.; Olivera-Severo, D.; Carlini, C.R. Jackbean, soybean and *Bacillus pasteurii* ureases: Biological effects unrelated to ureolytic activity. *Eur. J. Biochem.* **2004**, *271*, 1357–1363. [CrossRef] [PubMed]
46. Real-Guerra, R.; Staniscuaski, F.; Carlini, C.R. Soybean Urease: Over a Hundred Years of Knowledge. In *A Comprehensive Survey of International Soybean Research—Genetics, Physiology, Agronomy and Nitrogen Relationships*; Board, J.E., Ed.; InTech: London, UK, 2013. [CrossRef]
47. Becker-Ritt, A.B.; Martinelli, A.H.S.; Mitidieri, S.; Feder, V.; Wassermann, G.E.; Santi, L.; Vainstein, M.H.; Oliveira, J.T.A.; Fiuza, L.M.; Pasquali, G.; et al. Antifungal activity of plant and bacterial ureases. *Toxicon* **2007**, *50*, 971–983. [CrossRef] [PubMed]
48. Martinelli, A.H.S.; Lopes, F.C.; Broll, V.; Defferrari, M.S.; Ligabue-Braun, R.; Kappaun, K.; Tichota, D.M.; Fruttero, L.L.; Moyetta, N.R.; Demartini, D.R.; et al. Soybean ubiquitous urease with purification facilitator: An addition to the moonlighting studies toolbox. *Process Biochem.* **2017**, *53*, 245–258. [CrossRef]
49. Follmer, C. Insights into the role and structure of plant ureases. *Phytochemistry* **2008**, *69*, 18–28. [CrossRef] [PubMed]
50. Carlini, C.R.; Oliveira, A.E.; Azambuja, P.; Xavier-Filho, J.; Wells, M.A. Biological effects of canatoxin in different insect models: Evidence for a proteolytic activation of the toxin by insect cathepsinlike enzymes. *J. Econ. Entomol.* **1997**, *90*, 340–348. [CrossRef]
51. Oliveira, A.E.A.; Gomes, V.M.; Sales, M.P.; Fernandes, K.V.S.; Carlini, C.R.; Xavier-Filho, J. The toxicity of jack bean [*Canavalia ensiformis* (L.) DC.] canatoxin to plant pathogenic fungi. *Rev. Bras. Biol.* **1999**, *59*, 59–62. [CrossRef]
52. Ferreira-DaSilva, C.T.; Gombarovits, M.E.C.; Masuda, H.; Oliveira, C.M.; Carlini, C.R. Proteolytic activation of canatoxin, a plant toxic protein, by insect cathepsin-like enzymes. *Arch. Insect Biochem. Physiol.* **2000**, *44*, 162–171. [CrossRef]
53. Polgár, L.; Csoma, C. Dissociation of ionizing groups in the binding cleft inversely controls the endo- and exopeptidase activities of cathepsin B. *J. Biol. Chem.* **1987**, *262*, 14448–14453.
54. Turk, D.; Gunčar, G.; Podobnik, M.; Turk, B. Revised definition of substrate binding sites of papain-like cysteine proteases. *Biol. Chem.* **1998**, *379*, 137–147. [CrossRef]
55. Sun, H.; Lou, X.; Shan, Q.; Zhang, J.; Zhu, X.; Zhang, J.; Wang, Y.; Xie, Y.; Xu, N.; Liu, S. Proteolytic Characteristics of Cathepsin D Related to the Recognition and Cleavage of Its Target Proteins. *PLoS ONE* **2013**, *8*. [CrossRef] [PubMed]

56. Piovesan, A.R.; Stanisçuaski, F.; Marco-Salvadori, J.; Real-Guerra, R.; Defferrari, M.S.; Carlini, C.R. Stage-specific gut proteinases of the cotton stainer bug *Dysdercus peruvianus*: Role in the release of entomotoxic peptides from *Canavalia ensiformis* urease. *Insect Biochem. Mol. Biol.* **2008**, *38*, 1023–1032. [CrossRef] [PubMed]
57. Mulinari, F.; Stanisçuaski, F.; Bertholdo-Vargas, L.R.; Postal, M.; Oliveira-Neto, O.B.; Rigden, D.J.; Grossi-de-Sá, M.F.; Carlini, C.R. Jaburetox-2Ec: An insecticidal peptide derived from an isoform of urease from the plant *Canavalia ensiformis*. *Peptides* **2007**, *28*, 2042–2050. [CrossRef] [PubMed]
58. Martinelli, A.H.S.; Kappaun, K.; Ligabue-Braun, R.; Defferrari, M.S.; Piovesan, A.R.; Stanisçuaski, F.; Demartini, D.R.; Dal Belo, C.A.; Almeida, C.G.M.; Follmer, C.; et al. Structure-function studies on jaburetox, a recombinant insecticidal peptide derived from jack bean (*Canavalia ensiformis*) urease. *Biochim. Biophys. Acta Gen. Subj.* **2014**, *1840*, 935–944. [CrossRef]
59. Gombarovits, M. Peptídeos entomotóxicos gerados a partir da CNTX: Obtenção, isolamento, propriedades biológicas e caracterização físico-química. Ph.D. Thesis, Universidade Federal do Rio de Janeiro, Rio de Janeiro, Brazil, 1999.
60. Becker-Ritt, A.B.; Portugal, C.S.; Carlini, C.R. Jaburetox: Update on a urease-derived peptide. *J. Venom. Anim. Toxins Incl. Trop. Dis.* **2017**, *23*, 1–8. [CrossRef] [PubMed]
61. Postal, M.; Martinelli, A.H.S.; Becker-Ritt, A.B.; Ligabue-Braun, R.; Demartini, D.R.; Ribeiro, S.F.F.; Pasquali, G.; Gomes, V.M.; Carlini, C.R. Antifungal properties of *Canavalia ensiformis* urease and derived peptides. *Peptides* **2012**, *38*, 22–32. [CrossRef]
62. Broll, V.; Martinelli, A.H.S.; Lopes, F.C.; Fruttero, L.L.; Zambelli, B.; Salladini, E.; Dobrovolska, O.; Ciurli, S.; Carlini, C.R. Structural analysis of the interaction between Jaburetox, an intrinsically disordered protein, and membrane models. *Colloids Surfaces B Biointerfaces* **2017**, *159*, 849–860. [CrossRef] [PubMed]
63. Wiebke-Strohm, B.; Pasquali, G.; Margis-Pinheiro, M.; Bencke, M.; Bücker-Neto, L.; Becker-Ritt, A.B.; Martinelli, A.H.S.; Rechenmacher, C.; Polacco, J.C.; Stolf, R.; et al. Ubiquitous urease affects soybean susceptibility to fungi. *Plant Mol. Biol.* **2012**, *79*, 75–87. [CrossRef] [PubMed]
64. Kappaun, K.; Martinelli, A.H.S.; Broll, V.; Zambelli, B.; Lopes, F.C.; Ligabue-Braun, R.; Fruttero, L.L.; Moyetta, N.R.; Bonan, C.D.; Carlini, C.R.; et al. Soyuretox, an intrinsically disordered polypeptide derived from soybean (Glycine max) ubiquitous urease with potential use as a biopesticide. *Int. J. Mol. Sci.* **2019**, *20*, 5401. [CrossRef]
65. Sá, C.A.; Vieira, L.R.; Pereira Almeida Filho, L.C.; Real-Guerra, R.; Lopes, F.C.; Souza, T.M.; Vasconcelos, I.M.; Staniscuaski, F.; Carlini, C.R.; Urano Carvalho, A.F.; et al. Risk assessment of the antifungal and insecticidal peptide Jaburetox and its parental protein the Jack bean (*Canavalia ensiformis*) urease. *Food Chem. Toxicol.* **2020**, *136*, 110977. [CrossRef]
66. Barros, P.R.; Stassen, H.; Freitas, M.S.; Carlini, C.R.; Nascimento, M.A.C.; Follmer, C. Membrane-disruptive properties of the bioinsecticide Jaburetox-2Ec: Implications to the mechanism of the action of insecticidal peptides derived from ureases. *Biochim. Biophys. Acta Proteins Proteomics* **2009**, *1794*, 1848–1854. [CrossRef]
67. Micheletto, Y.M.S.; Moro, C.F.; Lopes, F.C.; Ligabue-Braun, R.; Martinelli, A.H.S.; Marques, C.M.; Schroder, A.P.; Carlini, C.R.; da Silveira, N.P. Interaction of jack bean (*Canavalia ensiformis*) urease and a derived peptide with lipid vesicles. *Colloids Surfaces B Biointerfaces* **2016**, *145*, 576–585. [CrossRef]
68. Balasubramanian, A.; Ponnuraj, K. Crystal Structure of the First Plant Urease from Jack Bean: 83 Years of Journey from Its First Crystal to Molecular Structure. *J. Mol. Biol.* **2010**, *400*, 274–283. [CrossRef]
69. Lopes, F.C.; Dobrovolska, O.; Real-Guerra, R.; Broll, V.; Zambelli, B.; Musiani, F.; Uversky, V.N.; Carlini, C.R.; Ciurli, S. Pliable natural biocide: Jaburetox is an intrinsically disordered insecticidal and fungicidal polypeptide derived from jack bean urease. *FEBS J.* **2015**, *282*, 1043–1064. [CrossRef]
70. Portugal, C.S. Avaliação dos efeitos do peptídeo recombinante Jaburetox em linhagens celulares e *Drosophila melanogaster*. Ph.D. Thesis, Universidade Luterana do Brasil, Canoas, Brazil, 2017.
71. Sirko, A.; Brodzik, R. Plant ureases: Roles and regulation. *Acta Biochim. Pol.* **2000**, *47*, 1189–1195. [CrossRef]
72. Kumar, V.; Wagenet, R.J. Urease activity and kinetics of urea transformation in soils. *Soil Sci.* **1984**, *137*, 263–269. [CrossRef]
73. Sparks, T.C.; Crossthwaite, A.J.; Nauen, R.; Banba, S.; Cordova, D.; Earley, F.; Ebbinghaus-Kintscher, U.; Fujioka, S.; Hirao, A.; Karmon, D.; et al. Insecticides, biologics and nematicides: Updates to IRAC's mode of action classification—A tool for resistance management. *Pestic. Biochem. Physiol.* **2020**, *167*, 104587. [CrossRef]

74. Mall, D.; Larsen, A.E.; Martin, E.A. Investigating the (Mis)match between natural pest control knowledge and the intensity of pesticide use. *Insects* **2018**, *9*, 2. [CrossRef]
75. Hardy, M.C. Resistance is not futile: It shapes insecticide discovery. *Insects* **2014**, *5*, 227–242. [CrossRef]
76. Piovesan, A.R.; Martinelli, A.H.S.; Ligabue-Braun, R.; Schwartz, J.L.; Carlini, C.R. *Canavalia ensiformis* urease, Jaburetox and derived peptides form ion channels in planar lipid bilayers. *Arch. Biochem. Biophys.* **2014**, *547*, 6–17. [CrossRef]
77. Moyetta, N.R.; Broll, V.; Perin, A.P.A.; Uberti, A.F.; Coste Grahl, M.V.; Staniscuaski, F.; Carlini, C.R.; Fruttero, L.L. Jaburetox-induced toxic effects on the hemocytes of *Rhodnius prolixus* (Hemiptera: *Reduviidae*). *Comp. Biochem. Physiol. Part C Toxicol. Pharmacol.* **2017**, *200*, 17–26. [CrossRef]
78. Fruttero, L.L.; Moyetta, N.R.; Krug, M.S.; Broll, V.; Grahl, M.V.C.; Real-Guerra, R.; Stanisçuaski, F.; Carlini, C.R. Jaburetox affects gene expression and enzyme activities in *Rhodnius prolixus*, a Chagas' disease vector. *Acta Trop.* **2017**, *168*, 54–63. [CrossRef]
79. Wallmann, A.; Kesten, C. Common functions of disordered proteins across evolutionary distant organisms. *Int. J. Mol. Sci.* **2020**, *21*, 2105. [CrossRef]
80. Tomazetto, G.; Mulinari, F.; Stanisçuaski, F.; Settembrini, B.; Carlini, C.R.; Záchia Ayub, M.A. Expression kinetics and plasmid stability of recombinant *E. coli* encoding urease-derived peptide with bioinsecticide activity. *Enzyme Microb. Technol.* **2007**, *41*, 821–827. [CrossRef]
81. Stanisçuaski, F.; Te Brugge, V.; Carlini, C.R.; Orchard, I. In vitro effect of *Canavalia ensiformis* urease and the derived peptide Jaburetox-2Ec on *Rhodnius prolixus* Malpighian tubules. *J. Insect Physiol.* **2009**, *55*, 255–263. [CrossRef] [PubMed]
82. Fruttero, L.L.; Moyetta, N.R.; Uberti, A.F.; Grahl, M.V.C.; Lopes, F.C.; Broll, V.; Feder, D.; Carlini, C.R. Humoral and cellular immune responses induced by the urease-derived peptide Jaburetox in the model organism *Rhodnius prolixus*. *Parasit. Vectors* **2016**, *9*, 1–14. [CrossRef]
83. Coste Grahl, M.V.; Perin, A.P.A.; Lopes, F.C.; Porto, B.N.; Uberti, A.F.; Canavoso, L.E.; Stanisçuaski, F.; Fruttero, L.L. The role of extracellular nucleic acids in the immune system modulation of *Rhodnius prolixus* (Hemiptera: *Reduviidae*). *Pestic. Biochem. Physiol.* **2020**, *167*, 104591. [CrossRef]
84. Galvani, G.L.; Fruttero, L.L.; Coronel, M.F.; Nowicki, S.; Demartini, D.R.; Defferrari, M.S.; Postal, M.; Canavoso, L.E.; Carlini, C.R.; Settembrini, B.P. Effect of the urease-derived peptide Jaburetox on the central nervous system of *Triatoma infestans* (Insecta: Heteroptera). *Biochim. Biophys. Acta Gen. Subj.* **2015**, *1850*, 255–262. [CrossRef]
85. Dos Santos, D.S.; Zanatta, A.P.; Martinelli, A.H.S.; Rosa, M.E.; de Oliveira, R.S.; Pinto, P.M.; Peigneur, S.; Tytgat, J.; Orchard, I.; Lange, A.B.; et al. Jaburetox, a natural insecticide derived from Jack Bean Urease, activates voltage-gated sodium channels to modulate insect behavior. *Pestic. Biochem. Physiol.* **2019**, *153*, 67–76. [CrossRef]
86. Perin, A.P.A.; Noronha, M.S.; Moyetta, N.R.; Coste Grahl, M.V.; Fruttero, L.L.; Staniscuaski, F. Jaburetox, a urease-derived peptide: Effects on enzymatic pathways of the cockroach *Nauphoeta cinerea*. *Arch. Insect Biochem. Physiol.* **2020**, *105*, e21731. [CrossRef]
87. Didoné, D.A. Development of Jaburetox-expressing maize plants for resistance to pest lepidopterans. Ph.D. Thesis, Universidade de Passo Fundo, Passo Fundo, Brazil, 2018.
88. Ceccon, C.C. Transgenic tobacco plants expressing hairpin and jaburetox RNA as strategies for *Helicoverpa armigera* control. Ph.D. Thesis, Universidade de Passo Fundo, Passo Fundo, Brazil, 2019.
89. Chattopadhyay, P.; Banerjee, G. Recent advancement on chemical arsenal of Bt toxin and its application in pest management system in agricultural field. *3 Biotech.* **2018**, *8*, 1–12. [CrossRef] [PubMed]
90. Mio, T.; Yabe, T.; Arisawa, M.; Yamada-Okabe, H. The Eukaryotic UDP-*N*-Acetylglucosamine Pyrophosphorylases: Gene Cloning, Protein Expression, and Catalytic Mechanism. *J. Biol. Chem.* **1998**, *273*, 14392–14397. [CrossRef] [PubMed]
91. Barreto, Y.C.; Rosa, M.E.; Zanatta, A.P.; Borges, B.T.; Hyslop, S.; Vinadé, L.H.; Dal Belo, C.A. Entomotoxicity of jaburetox: Revisiting the neurotoxic mechanisms in insects. *J. Venom Res.* **2020**, *10*, 1–15.
92. Radi, R. Nitric oxide, oxidants, and protein tyrosine nitration. *Proc. Natl. Acad. Sci. USA* **2004**, *101*, 4003–4008. [CrossRef]
93. Azambuja, P.; Garcia, E.S.; Waniek, P.J.; Vieira, C.S.; Figueiredo, M.B.; Gonzalez, M.S.; Mello, C.B.; Castro, D.P.; Ratcliffe, N.A. *Rhodnius prolixus*: From physiology by Wigglesworth to recent studies of immune system modulation by *Trypanosoma cruzi* and *Trypanosoma rangeli*. *J. Insect Physiol.* **2017**, *97*, 45–65. [CrossRef]

94. Hillyer, J.F. Insect immunology and hematopoiesis. *Dev. Comp. Immunol.* **2016**, *58*, 102–118. [CrossRef]
95. Defferrari, M.S.; Da Silva, R.; Orchard, I.; Carlini, C.R. Jack bean (*Canavalia ensiformis*) urease induces eicosanoid-modulated hemocyte aggregation in the Chagas' disease vector *Rhodnius prolixus*. *Toxicon* **2014**, *82*, 18–25. [CrossRef]
96. Nascimento, M.T.C.C.; Silva, K.P.; Garcia, M.C.F.F.; Medeiros, M.N.; Machado, E.A.; Nascimento, S.B.; Saraiva, E.M. DNA extracellular traps are part of the immune repertoire of Periplaneta americana. *Dev. Comp. Immunol.* **2018**, *84*, 62–70. [CrossRef]
97. Brinkmann, V.; Reichard, U.; Goosmann, C.; Fauler, B.; Uhlemann, Y.; Weiss, D.S.; Weinrauch, Y.; Zychlinsky, A. Neutrophil Extracellular Traps Kill Bacteria. *Science* **2004**, *303*, 1532–1535. [CrossRef]
98. Uversky, V.N.; Gillespie, J.R.; Fink, A.L. Why are "natively unfolded" proteins unstructured under physiologic conditions? *Proteins Struct. Funct. Bioinform.* **2000**, *41*, 415–427. [CrossRef]
99. Dunker, A.K.; Lawson, J.D.; Brown, C.J.; Williams, R.M.; Romero, P.; Oh, J.S.; Oldfield, C.J.; Campen, A.M.; Ratliff, C.M.; Hipps, K.W.; et al. Intrinsically disordered protein. *J. Mol. Graph. Model.* **2001**, *19*, 26–59. [CrossRef]
100. Tompa, P. Intrinsically unstructured proteins. *Trends Biochem. Sci.* **2002**, *27*, 527–533. [CrossRef]
101. Wright, P.E.; Dyson, H.J. Intrinsically unstructured proteins: Re-assessing the protein structure-function paradigm. *J. Mol. Biol.* **1999**, *293*, 321–331. [CrossRef] [PubMed]
102. Bahar, A.A.; Ren, D. Antimicrobial peptides. *Pharmaceuticals* **2013**, *6*, 1543–1575. [CrossRef]
103. Lei, J.; Sun, L.C.; Huang, S.; Zhu, C.; Li, P.; He, J.; Mackey, V.; Coy, D.H.; He, Q.Y. The antimicrobial peptides and their potential clinical applications. *Am. J. Transl. Res.* **2019**, *11*, 3919–3931.
104. Sohlenkamp, C.; Geiger, O. Bacterial membrane lipids: Diversity in structures and pathways. *FEMS Microbiol. Rev.* **2015**, *40*, 133–159. [CrossRef]
105. White, D.C.; Frerman, F.E. Extraction, characterization, and cellular localization of the lipids of *Staphylococcus aureus*. *J. Bacteriol.* **1967**, *94*, 1854–1867. [CrossRef]
106. Löffler, J.; Einsele, H.; Hebart, H.; Schumacher, U.; Hrastnik, C.; Daum, G. Phospholipid and sterol analysis of plasma membranes of azole-resistant *Candida albicans* strains. *FEMS Microbiol. Lett.* **2000**, *185*, 59–63. [CrossRef]
107. Narayani, M.; Babu, R.; Chadha, A.; Srivastava, S. Production of bioactive cyclotides: A comprehensive overview. *Phytochem. Rev.* **2020**, *19*, 787–825. [CrossRef]
108. Huang, Y.H.; Colgrave, M.L.; Daly, N.L.; Keleshian, A.; Martinac, B.; Craik, D.J. The biological activity of the prototypic cyclotide Kalata B1 is modulated by the formation of multimeric pores. *J. Biol. Chem.* **2009**, *284*, 20699–20707. [CrossRef]
109. Saether, O.; Craik, D.J.; Campbell, I.D.; Sletten, K.; Juul, J.; Norman, D.G. Elucidation of the Primary and Three-Dimensional Structure of the Uterotonic Polypeptide Kalata B1. *Biochemistry* **1995**, *34*, 4147–4158. [CrossRef] [PubMed]
110. Plan, M.R.R.; Göransson, U.; Clark, R.J.; Daly, N.L.; Colgrave, M.L.; Craik, D.J. The cyclotide fingerprint in *Oldenlandia affinis*: Elucidation of chemically modified, linear and novel macrocyclic peptides. *ChemBioChem* **2007**, *8*, 1001–1011. [CrossRef]
111. Daly, N.L.; Gunasekera, S.; Clark, R.J.; Lin, F.; Wade, J.D.; Anderson, M.A.; Craik, D.J. The N-terminal pro-domain of the kalata B1 cyclotide precursor is intrinsically unstructured. *Biopolymers* **2016**, *106*, 825–833. [CrossRef] [PubMed]
112. Marín, M.; Ott, T. Intrinsic disorder in plant proteins and phytopathogenic bacterial effectors. *Chem. Rev.* **2014**, *114*, 6912–6932. [CrossRef]
113. Oppenheim, F.G.; Xu, T.; McMillian, F.M.; Levitz, S.M.; Diamond, R.D.; Offner, G.D.; Troxler, R.F. Histatins, a novel family of histidine-rich proteins in human parotid secretion. Isolation, characterization, primary structure, and fungistatic effects on *Candida albicans*. *J. Biol. Chem.* **1988**, *263*, 7472–7477.
114. Jephthah, S.; Staby, L.; Kragelund, B.B.; Skepö, M. Temperature Dependence of Intrinsically Disordered Proteins in Simulations: What are We Missing? *J. Chem. Theory Comput.* **2019**, *15*, 2672–2683. [CrossRef]
115. Tsai, H.; Raj, P.A.; Bobek, L.A. Candidacidal activity of recombinant human salivary histatin-5 and variants. *Infect. Immun.* **1996**, *64*, 5000–5007. [CrossRef]
116. Puri, S.; Edgerton, M. How does it kill?: Understanding the candidacidal mechanism of salivary histatin 5. *Eukaryot. Cell* **2014**, *13*, 958–964. [CrossRef]

117. Melino, S.; Santone, C.; Di Nardo, P.; Sarkar, B. Histatins: Salivary peptides with copper (II)-and zinc (II)-binding motifs: Perspectives for biomedical applications. *FEBS J.* **2014**, *281*, 657–672. [CrossRef] [PubMed]
118. Wu, Z.; Meyer-Hoffert, U.; Reithmayer, K.; Paus, R.; Hansmann, B.; He, Y.; Bartels, J.; Gläser, R.; Harder, J.; Schröder, J.M. Highly complex peptide aggregates of the S100 fused-type protein hornerin are present in human skin. *J. Investig. Dermatol.* **2009**, *129*, 1446–1458. [CrossRef]
119. Latendorf, T.; Gerstel, U.; Wu, Z.; Bartels, J.; Becker, A.; Tholey, A.; Schröder, J.M. Cationic Intrinsically Disordered Antimicrobial Peptides (CIDAMPs) Represent a New Paradigm of Innate Defense with a Potential for Novel Anti-Infectives. *Sci. Rep.* **2019**, *9*, 1–15. [CrossRef] [PubMed]
120. Gerstel, U.; Latendorf, T.; Bartels, J.; Becker, A.; Tholey, A.; Schröder, J.M. Hornerin contains a Linked Series of Ribosome-Targeting Peptide Antibiotics. *Sci. Rep.* **2018**, *8*, 1–15. [CrossRef] [PubMed]
121. Tyczewska, A.; Woźniak, E.; Gracz, J.; Kuczyński, J.; Twardowski, T. Towards Food Security: Current State and Future Prospects of Agrobiotechnology. *Trends Biotechnol.* **2018**, *36*, 1219–1229. [CrossRef] [PubMed]
122. Tabashnik, B.E.; Brévault, T.; Carrière, Y. Insect resistance to Bt crops: Lessons from the first billion acres. *Nat. Biotechnol.* **2013**, *31*, 510–521. [CrossRef] [PubMed]
123. Rechenmacher, C.; Wiebke-Strohm, B.; de Oliveira-Busatto, L.A.; Weber, R.L.M.; Corso, M.C.M.; Lopes-Caitar, V.S.; Silva, S.M.H.; Dias, W.P.; Marcelino-Guimarães, F.C.; Carlini, C.R.; et al. Endogenous soybean peptide overexpression: An alternative to protect plants against root-knot nematodes. *Biotechnol. Res. Innov.* **2019**, *3*, 10–18. [CrossRef]
124. Biłas, R.; Szafran, K.; Hnatuszko-Konka, K.; Kononowicz, A.K. Cis-regulatory elements used to control gene expression in plants. *Plant Cell. Tissue Organ Cult.* **2016**, *127*, 269–287. [CrossRef]
125. Muthusamy, S.K.; Sivalingam, P.N.; Sridhar, J.; Singh, D.; Haldhar, S.M. Biotic stress inducible promoters in crop plants-a review. *J. Agric. Ecol.* **2017**, *4*, 14–24.

Publisher's Note: MDPI stays neutral with regard to jurisdictional claims in published maps and institutional affiliations.

© 2020 by the authors. Licensee MDPI, Basel, Switzerland. This article is an open access article distributed under the terms and conditions of the Creative Commons Attribution (CC BY) license (http://creativecommons.org/licenses/by/4.0/).

Review

ATP Analogues for Structural Investigations: Case Studies of a DnaB Helicase and an ABC Transporter

Denis Lacabanne [1,2,†], Thomas Wiegand [1,*,†], Nino Wili [1], Maria I. Kozlova [3], Riccardo Cadalbert [1], Daniel Klose [1], Armen Y. Mulkidjanian [3,4], Beat H. Meier [1,*] and Anja Böckmann [5,*]

1. Laboratory of Physical Chemistry, ETH Zurich, 8093 Zurich, Switzerland; denis.lacabanne@mrc-mbu.cam.ac.uk (D.L.); nino.wili@phys.chem.ethz.ch (N.W.); Riccardo.Cadalbert@nmr.phys.chem.ethz.ch (R.C.); daniel.klose@phys.chem.ethz.ch (D.K.)
2. Medical Research Council Mitochondrial Biology Unit University of Cambridge, Cambridge Biomedical Campus, Keith Peters Building, Hills Road, Cambridge CB2 0XY, UK
3. Department of Physics, Osnabrueck University, 49069 Osnabrueck, Germany; makozlova@uni-osnabrueck.de (M.I.K.); armen.mulkidjanian@uni-osnabrueck.de (A.Y.M.)
4. School of Bioengineering and Bioinformatics and Belozersky Institute of Physico-Chemical Biology, Lomonosov Moscow State University, 119234 Moscow, Russia
5. Molecular Microbiology and Structural Biochemistry UMR 5086 CNRS/Université de Lyon, Labex Ecofect, 69367 Lyon, France

* Correspondence: thomas.wiegand@phys.chem.ethz.ch (T.W.); beme@ethz.ch (B.H.M.); bockmann@ibcp.fr (A.B.)

† These authors contributed equally to this work.

Academic Editor: Marilisa Leone

Received: 17 October 2020; Accepted: 9 November 2020; Published: 12 November 2020

Abstract: Nucleoside triphosphates (NTPs) are used as chemical energy source in a variety of cell systems. Structural snapshots along the NTP hydrolysis reaction coordinate are typically obtained by adding stable, nonhydrolyzable adenosine triphosphate (ATP) -analogues to the proteins, with the goal to arrest a state that mimics as closely as possible a physiologically relevant state, e.g., the pre-hydrolytic, transition and post-hydrolytic states. We here present the lessons learned on two distinct ATPases on the best use and unexpected pitfalls observed for different analogues. The proteins investigated are the bacterial DnaB helicase from *Helicobacter pylori* and the multidrug ATP binding cassette (ABC) transporter BmrA from *Bacillus subtilis*, both belonging to the same division of P-loop fold NTPases. We review the magnetic-resonance strategies which can be of use to probe the binding of the ATP-mimics, and present carbon-13, phosphorus-31, and vanadium-51 solid-state nuclear magnetic resonance (NMR) spectra of the proteins or the bound molecules to unravel conformational and dynamic changes upon binding of the ATP-mimics. Electron paramagnetic resonance (EPR), and in particular W-band electron-electron double resonance (ELDOR)-detected NMR, is of complementary use to assess binding of vanadate. We discuss which analogues best mimic the different hydrolysis states for the DnaB helicase and the ABC transporter BmrA. These might be relevant also to structural and functional studies of other NTPases.

Keywords: solid-state NMR; ELDOR-detected NMR; ATP hydrolysis; ATP analogues; DnaB helicase; ABC transporter

---

1. Introduction

Nucleosides triphosphates (NTPs), such as ATP (adenosine triphosphate) and GTP (guanosine triphosphate), are used as energy source or as allosteric effector by a number of proteins, involved for instance, in metabolism, active transport, cell division or DNA/RNA synthesis. However, the mechanism of NTP hydrolysis in proteins are still poorly understood, especially its coupling to functional events, such as movement of proteins along nucleic acids. Indeed, detailed mechanistic insight is lacking for a number of systems, including even intensively studied systems such as dyneins [1], ABC importers [2]/exporters [3] or DNA helicases [4]. Experimentally, catching the events that occur during the NTP hydrolysis is highly challenging. Structural techniques such as X-ray crystallography, cryo-electron microscopy (cryo-EM) and nuclear magnetic resonance (NMR) mainly provide static snapshots of protein states of typically highly complex reaction coordinates of biomolecular reactions. These can then be combined with molecular dynamics simulations (MD) to obtain further information about the dynamics of such processes and to establish the chronological sequence [3,5–8]. In this context, it is highly desirable to better investigate how ATP/GTP analogues, usually with a modified or replaced γ-phosphate group, can mimic the intermediate catalytic states in order to obtain relevant snapshots of the reactions involving NTP hydrolysis. Indeed, it is well known that mimics can never fully represent naturally occurring states, as the modifications of NTPs change their conformation as well as their chemical properties other than their tendency to hydrolyse. While it is important to use NTP mimics described to be strongly hydrolysis-resistant, the true hydrolysis state must often be confirmed experimentally.

The choice of the adequate analogue is thus of importance in structural studies, but guidelines are sparse and can be highly protein-dependent. We herein focus on analogues often used to access three important states of ATP-hydrolysis: the pre-hydrolytic state, the transition state and the post-hydrolytic state (see Scheme 1A for the artificial ATP hydrolysis scheme highlighting analogues used to mimic the different states). We describe how NMR and EPR can be used to gain detailed information on the analogue used and the conformational and dynamic state it induces in the protein. We investigate this for two proteins, the bacterial DnaB helicase from *Helicobacter pylori* involved in DNA replication, and an ABC transporter implicated in multidrug resistance, BmrA (*Bacillus subtilis* multidrug resistance ATP binding cassette transporter), which share high similarities in their ATP binding sites [9]. Solid-state NMR and EPR are highly suitable to study large, noncrystalline protein assemblies, which are represented by DnaB and BmrA. The proteins are, in their multimeric states and, for BmrA, embedded in a *Bacillus subtilis* lipid membrane, sedimented directly into the solid-state NMR rotor in an external ultracentrifuge [10], a sample preparation approach that allows for the study of the investigated analogues. The protein samples prepared by this approach are highly concentrated in the NMR rotor (protein concentration of around 400 mg/mL), and have been shown to be stable over several years [11]. A description of the NMR techniques developed to investigate such molecular machines is given in detail in reference [12].

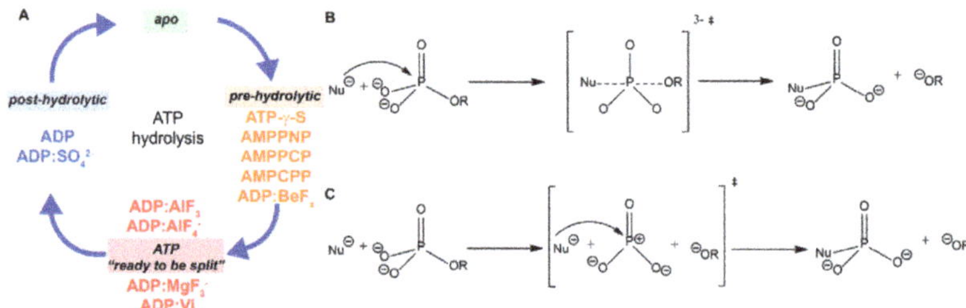

Scheme 1. Artificial ATP hydrolysis scheme and associative and dissociative mechanism of ATP hydrolysis. (**A**) Artificial ATP hydrolysis scheme showing ATP analogues used to mimic the pre-hydrolytic, the "ATP-is-ready-to-be-split" and the post-hydrolytic state. Schematic representation of the associative (**B**) and dissociative (**C**) mechanism of ATP hydrolysis. Nu^- represents a nucleotide, e.g., an OH^-.

Figure 1 and Table 1 summarize the most important NMR experiments and the nomenclature used herein and gives the information content of NMR spectra and the underlying NMR observables. The standard experiment to establish a chemical-shift fingerprint of the protein is the ^{13}C-^{13}C DARR, a two-dimensional correlation experiment using the dipolar assisted rotational recoupling (DARR) scheme [13,14]. Besides delivering a first sample quality check (Figure 1A), isolated peaks in such spectra, often found in the alanine or threonine regions, can serve to follow the conformational changes along the reaction coordinate. Differences in the cross-peak positions (encoding the chemical shift) in such spectra characterize the different protein conformations, produced by incubating the protein with ligands. Such changes are denoted as chemical-shift perturbations (CSPs) (Figure 1B). Additionally, appearing or disappearing resonances might be observed in the spectra, pointing to dynamic changes of the protein (Figure 1C). ^{31}P NMR experiments allow for direct detection of nucleotides, such as ATP mimics or DNA/RNA [12] (Figure 1D). The ^{31}P chemical-shift values react very sensitively to small conformational changes, e.g., in the phosphate backbone of ATP mimics. ^{31}P direct-pulsed experiments (recorded with short repetition times, Figure 1E) are used to detect unbound nucleotides present in the water phase in contact with the protein (interacting water) or the supernatant of the NMR rotor [15]. ^{31}P cross-polarization (CP) based experiments are employed to detect immobilized nucleotides, particularly those bound to the protein (Figure 1E). ELDOR-detected NMR (EDNMR) is a pulsed EPR-technique and allows for the measurement of hyperfine couplings of paramagnetic spin centers to nearby spin-active nuclei [16–19]. We herein use this technique to detect NMR-active nuclei in the vicinity of the ATP-cofactor (for this the diamagnetic Mg^{2+} has to be substituted by paramagnetic Mn^{2+}), particularly focusing on ^{31}P and ^{51}V nuclear spins [20]. If a ^{51}V nucleus is in spatial proximity to the Mn^{2+} ions, the ^{51}V resonance should be detected in the EDNMR spectra (Figure 1F). Note that the hyperfine coupling to the ^{51}V is often not resolved, in contrast to the ^{31}P couplings.

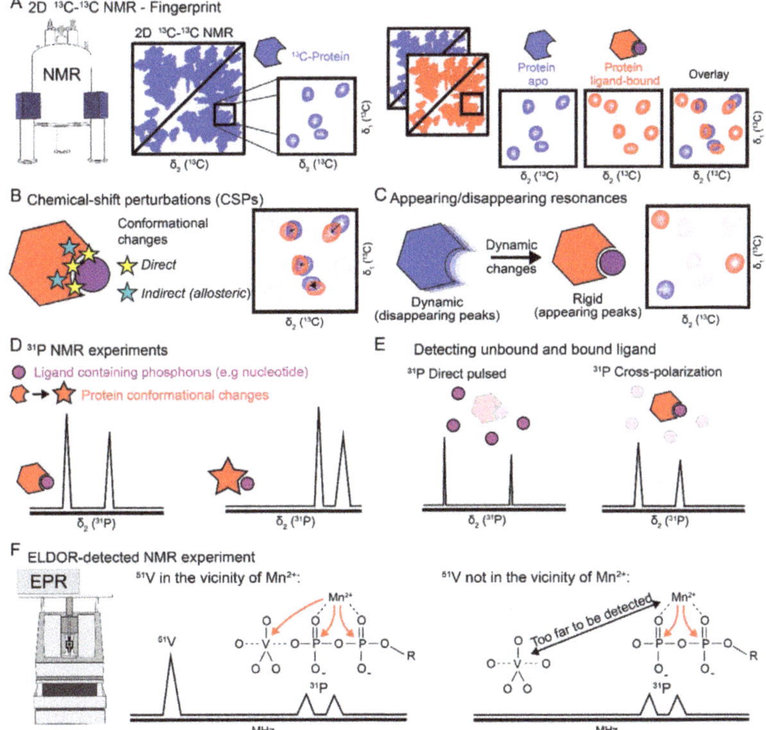

Figure 1. Magnetic-resonance approaches used to investigate the helicase DnaB and the ABC transporter BmrA in presence of ATP mimics. The employed techniques comprise 2D ^{13}C-^{13}C NMR spectral fingerprints (**A**), ^{13}C chemical-shifts perturbations (**B**), appearing/disappearing resonances due to dynamic changes (**C**), ^{31}P NMR experiments (**D**) to detect bound and unbound ligands (**E**) and ELDOR-detected NMR experiments (**F**), Details are given in the text.

Table 1. Overview of magnetic-resonance techniques applied in this work and information extracted. A cross indicates that this information is contained in the experimental outcome, a blank indicates that the type of extracted information is not accessible by the experiment.

Magnetic-Resonance Technique	Protein Conformational Changes	Protein Dynamic Changes	Nucleotide Binding	Nucleotide Conformation	Vanadate Binding	Metal Co-Factor Binding
NMR $^{13}C/^{15}N$ fingerprints	x	x	x			
CSPs	x		x			
Disappearing/appearing resonances		x	x			
^{31}P CP-MAS			x			
^{31}P chemical-shift values			x	x	x	
^{51}V NMR					x	
EDNMR					x	x

We here make use of these magnetic-resonance approaches to study the states of DnaB and BmrA induced by phosphate-modified NTP analogues widely used to mimic the three major states of the NTP hydrolysis reaction and report on the efficiency of the analogues to actually mimic the desired states.

2. The Different Hydrolysis States and the ATP-Mimics Used to Induce Them

The pre-hydrolytic state, where ATP is bound to the protein, is often already associated with protein conformational changes [21–24]. In this state, the γ-phosphate adopts a tetrahedral geometry (note that a similar discussion also holds for GTP). This geometry can change to a trigonal-bipyramidal geometry generating a pentavalent terminal phosphate group [25,26]. Most of the analogues are mimicking the pre-hydrolytic state because their γ-phosphate (in case of nonhydrolyzable analogues) or the γ-phosphate-mimicking group adopts a tetrahedral geometry [27]. The most commonly used nonhydrolyzable analogues are: AMPPNP (adenylyl imidodiphosphate) [28], AMPPCP (adenylyl methylenediphosphate) [29], AMPCPP (alpha,beta-methylene-triphosphonate) and ATP-γ-S (adenosine 5′-(gamma-thiotriphosphate)) [30]. In addition, the pre-hydrolytic state appears to be mimicked by ADP-BeF_x [27] (Figures 2 and 3A). BeF_x forms a strictly tetrahedral complex (specific to the pre-hydrolytic state); a penta-coordinated bipyramidal geometry (describing the transition state, see below) is excluded in this case [26,27]. The nonhydrolyzable ATP analogues are not completely resistant to hydrolysis. While the rate of hydrolysis of these analogues is indeed significantly lower, several of them can still be hydrolysed by many ATPases [31–38]. This behaviour can differ from protein to protein, and an analogue can fail to mimic a pre-hydrolytic state or may mimic a different/uncomplete pre-hydrolytic state in certain cases [26,39–41]. This difference can be observed between distinct protein families, or even within the same family [42], as shown in this work for the two model systems discussed.

The transition state (the "ATP-is-ready-to-be-split" state) can be accessed by an associative and dissociative mechanism, which represent the two extreme cases discussed in the literature [43–48]. In the case of an associative mechanism, the phosphorus possesses a pentavalent geometry. The nucleophilic attack of a water molecule at the γ-phosphate, forming a H_2O-P bond, in this scenario occurs before the leaving group departs and before the P-O bond breaks (similar to a S_{N2} nucleophilic substitution, see Scheme 1B,C). In contrast, in the case of a dissociative mechanism, the nucleophilic attack of a water molecule at the γ-phosphate occurs after the leaving group was released, generating a metaphosphate intermediate before it collapses onto the acceptor nucleophile (similar to a S_{N1} reaction). The transition state can be simulated by employing three prominent mimic groups in combination with ATP or ADP: aluminium fluoride (ADP:AlF_x) [27,49,50], magnesium fluoride (ADP:MgF_x) [51], and vanadate (ADP:Vi) [52]. In some enzymes, ATP hydrolysis is required prior to the binding of the transition-state mimic [53,54]. In structural studies, aluminium fluoride is most frequently used as a mimic of the γ-phosphate in the transition state, as evidenced by analysing the number of deposited structures in the PDB database (Figure 2A). When the analogue in the presence of ADP is complexed with the protein, ADP:AlF_x is believed to mimic the transition state of an ATP molecule. Two configurations of this analogue have been observed: ADP:AlF_3 and ADP:AlF_4^-. In the ADP:AlF_4^- mimic (two-thirds of in the

PDB deposited AlF_x-containing structures) the AlF_4^- group is in a squared-planar geometry and forms an octahedral complex with two oxygen ligands in the apical positions. While one ligand is provided by the β-phosphate, the other ligand comes from the hydrolytic water molecule in the attack position next to the phosphorus atom. It is believed that such a structure mimics the interaction of the catalytic water molecule with the γ-phosphate in the anionic transition state for phosphoryl transfer [49,55]. AlF_3 (one-third of in the PDB deposited AlF_x-containing structures) is in a trigonal-planar geometry forming a bipyramidal complex resembling the geometry of the transition state [49,55].

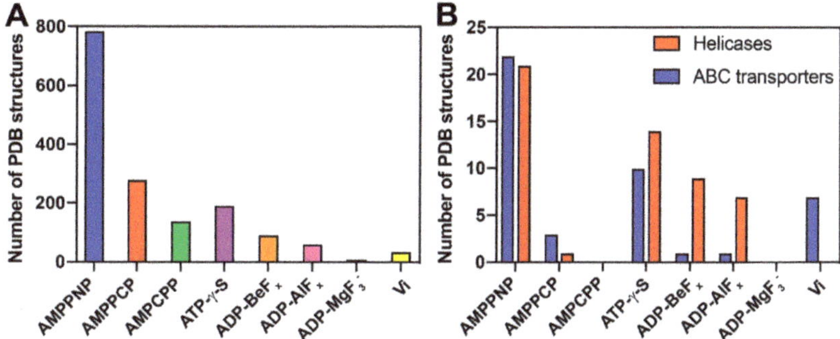

Figure 2. Distribution of the different NTP analogues based on the number of structures in the Protein Data Bank in December 2019. Number of Protein Data Bank structures for each analogue (**A**). Number of ABC transporters (blue, search query "ABC transporter + protein data bank accession codes of the ATP-analogue") and helicases (red, search query "DNA helicase + PDB ID of the ATP-analogue") Protein Data Bank structures for each analogue (**B**).

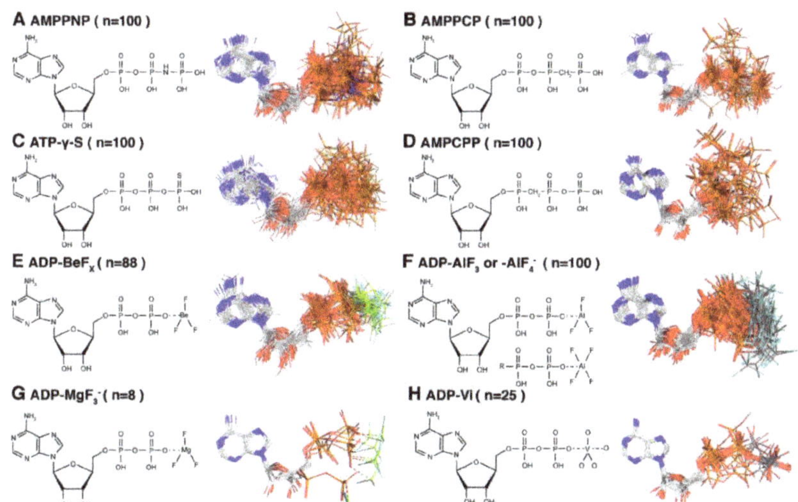

Figure 3. ATP analogues mainly used for structure determination. Left panel is the chemical structure and the right panel shows the protein-bound-structure of AMPPNP (**A**), AMPPCP (**B**), ATP-γ-S (**C**), AMPCPP (**D**), ADP:BeF_x (**E**), ADP:AlF_x (**F**), ADP:MgF_3^- (**G**), ADP:Vi (**H**) from the Protein Data Bank (https://www.rcsb.org). The protein-bound-structures of ATP analogues were generated after alignment of their adenosine moieties. In order to compare the molecules among each other, only the first 100 structures with the smallest RMSD were selected for AMPPNP, AMPPCP, ATP-γ-S and AMPCPP.

MgF_3^- shows nearly the same geometry as AlF_3 but carries a negative charge similarly to the anionic γ-phosphate in the transition state. AlF_3 and MgF_3^- are structurally similar and have similar scattering factors for X-rays; therefore, it has been suggested that MgF_3^- is present in some crystal structures, which are indicated as containing NDP:AlF_3 [56]. Indeed, Mg^{2+} ions are usually present in the samples as cofactors of NTP hydrolysis. In contrast to X-ray, NMR can differentiate the two metal fluorides so that in few cases, the presence of MgF_3^- in the active site was directly shown [56–58]. For more information about such cases and the use of metal fluorides as ATP or phosphate analogues, we refer the reader to the two comprehensive recent reviews [55,58]. Finally, vanadate-containing ATP:Vi or ADP:Vi are used as a transition-state mimic for a variety of proteins. Vanadate is an oxoanion of vanadium which shares structural and chemical similarities with phosphate molecules mimicking the hydrolysis transition state [52,59]. It is known that the simple form of the oxoanion (VO_4^{3-}) can adopt a penta-coordinated, trigonal bipyramidal geometry around the central vanadium in presence of ADP [60]. These properties make the vanadate a phosphate mimic of the transition state for phosphoryl transfer so that vanadate acts as an inhibitor for some ATPases. As previously described by Davies et al. [52], vanadate can be used to mimic phosphoryl transfer, and structures of different protein families including myosin [61,62], dynein [63], kinesin [40], ABC transporters [60,64,65], heat shock protein (Hsp70s) [66], NS3 helicase (dengue virus) [67], nucleoside-diphosphate kinase [68] or F1-ATPase [69] are reported. The main advantage of vanadate is that it can form covalent bonds with the oxygens of phosphate groups from ADP or other ligands [52]. Interestingly, this is not always the case [68], as there are structures where a vanadate is not bound to ADP, but still stabilizes the transition state. It is also noteworthy that vanadate does not work as an inhibitor or as transition-state mimic for all proteins with ATPase activity [70].

Finally, the post-hydrolytic state corresponds to a situation where the nucleotide diphosphate and the previously associated γ-phosphate are separated, but both are still bound to the protein, or, alternatively, where the γ-phosphate is already released, and only ADP is bound to the protein. The post-hydrolytic state where the γ-phosphate is not released can be mimicked not only by an orthophosphate [67,71,72] but also by a sulphate ion, SO_4^{2-} [71,73–75]. Note that sulphate ions have only two ionisable oxygens (with pK_a below 2) [76].

The overall conformational variability of NTP analogues can be seen by overlaying the structures extracted from the PDB and by aligning them on their nucleoside parts (Figure 3). AMPPNP and ATP-γ-S adopt a wider range of conformations (Figure 3A,C) than AMPPCP and ADP:BeF_x (Figure 3B,E), although this allows for a qualitative statement only, since the total numbers of deposited structures in the PDB are different (see Figure 2). AMPPNP and ATP-γ-S thus seem to adapt their conformation to the protein-binding pocket, while for AMPPCP and ADP:BeF_x it may be the protein that adapts. For the transition-state analogues, it is difficult to make the same comparison due to the small number of structures available. However, ADP:AlF_x shows a significant distribution of structures as well (Figure 3F).

In sum, from the eight mainly used analogues for structural studies, five are used to mimic the pre-hydrolytic state: AMPPNP, AMPPCP, ATP-γ-S, AMPCPP and ADP:BeF_x, three to mimic the transition state: ADP:AlF_x, ADP:MgF_x and ADP:V_i, and ADP and ADP:SO_4^{2-} to mimic the post-hydrolytic state. Note that also other NTP analogues exist that differ structurally through the introduction of atoms or groups (e.g., fluorescent probes, biotin groups, etc.) on the base, sugar, or triphosphate regions of the molecule [77–79]. A complete overview is given in reference [77].

3. The DnaB Helicase and the ABC Transporter BmrA

The usefulness of particular ATP mimics for structural studies strongly depends on the nature of the protein of interest, as shown in Figure 2B for the example of DNA helicases and ABC transporters. The two proteins were subject to studies in the last years in our laboratories: the bacterial helicase DnaB from *Helicobacter pylori* [38,80–88] and the ABC transporter BmrA from *Bacillus subtilis* [86,89–92]. In the presence of double-stranded DNA, the DnaB from *Helicobacter pylori* is a double-homo hexamer of

59 kDa monomers with each hexamer moving along its single DNA strand, whereas BmrA from *Bacillus subtilis* is a dimeric membrane protein of 65 kDa monomers. The two proteins are well-characterized ATP-fuelled proteins. In both proteins, the chemical energy released during ATP hydrolysis in the nucleotide-binding domain (NBD) is converted into mechanical work, which, e.g., enables the movement of DnaB along a double-stranded DNA and its unzipping, as well as the transportation of molecules across the membrane by ABC transporters. Both proteins belong to the vast family of P-loop fold NTPases, one of the largest protein superfamilies. In any genome 10–20% of proteins code for P-loop fold domains [93–95]. P-loop fold NTPases are characterized by their signature GxxxxGK [S/T] sequence motif, also known as the Walker A motif [96]. This motif is responsible for binding the triphosphate chain and is often called the P-loop (phosphate-binding loop) motif [97]. In the P-loop fold, the conserved Lys residue forms hydrogen bonds with the β- and γ-phosphate groups of ATP or GTP. Another conserved motif, known as the Walker B motif, is composed of four hydrophobic residues ended by an aspartate residue. The conserved Asp residue stabilizes the metal ion cofactor Mg^{2+} [96].

The *C*-terminal NBD of DnaB belongs to the superfamily 4 (SF4) of helicases, which in turn belongs to the class "RecA and F_1/F_O-related ATPases" (hereafter abbreviated as RecA/F_1-related ATPases) of P-loop old NTPases. The ABC transporter BmrA belongs to a separate class of ABC transporters [93–95]. Both the RecA/F_1-related ATPases and ABC transporters belong to the ASCE division of P-loop fold NTPases. The members of this division are characterized by an additional β-strand in the P-loop fold and a catalytic glutamate (E) residue next to the attacking water molecule [94,95,98]. The glutamate residue stabilizes the catalytic water molecule and, perhaps, operates as a catalytic base for ATP hydrolysis [99].

To avoid a futile NTP hydrolysis, P-loop fold NTPases are initiated before each turnover by activating moieties provided either by other proteins or by domains of the same protein [100–104]. The activating moiety interacts with the triphosphate chain and triggers the hydrolysis. The ATP hydrolysis in DnaB is induced by an interaction with an arginine residue that is provided by the neighbouring subunit of the same oligomer [105,106]. In ABC transporters, one of the NBDs is believed to activate hydrolysis within the active site in the other NBD by providing a signature LSGGQ motif [64,106].

Two analogues were mainly used in structural studies of helicases and ABC transporters (Figure 2): AMPPNP and ATP-γ-S, which both mimic the pre-hydrolytic state. The transition state is mainly mimicked by ADP:AlF_x for the helicases, and ADP:Vi for the ABC transporters. Regarding the literature, this state is underrepresented compared to the pre-hydrolytic state.

We here gather information from published experiments, as well as present complementary original data, in order to give a compilation of ATP analogues and their mimicking power for the two proteins DnaB and BmrA, as assessed by magnetic-resonance methods, namely NMR and EPR.

4. Results and Discussion

4.1. The Pre-Hydrolytic State Mimicked by AMPPCP, AMPPNP and ATP-γ-S

In order to characterize the pre-hydrolytic state, we first investigated DnaB and BmrA in the presence of AMPPNP, AMPPCP, and ATP-γ-S. It however appeared that ATP-γ-S was completely hydrolysed during the rotor filling by BmrA (one hour of filling) and DnaB (overnight filling), as monitored by ^{31}P solid-state NMR experiments (see Figure S1), and was thus of no further use. We therefore focused on AMPPNP and AMPPCP. Since a major function of the DnaB helicase is to bind to DNA, protein samples were also prepared with the ATP analogue and single-stranded DNA (here a DNA-fragment of 20 thymidine nucleotides abbreviated as $(dT)_{20}$). The presence of three signals in the 1D CP ^{31}P NMR spectrum (Figure 4A, left panel) indicates binding of the triphosphate AMPPCP to DnaB. However, the resonances of the phosphorus α and β are rather broad. This broadening might indicate inhomogeneities in the binding site in the environment of the ligand, or chemical-exchange broadening effects. In contrast, in the presence of DNA and AMPPCP, the ^{31}P resonances in the 1D CP

^{31}P spectrum are very sharp (Figure 4B, left panel). This indicates that the DnaB:DNA complex fixes AMPPCP with high homogeneity.

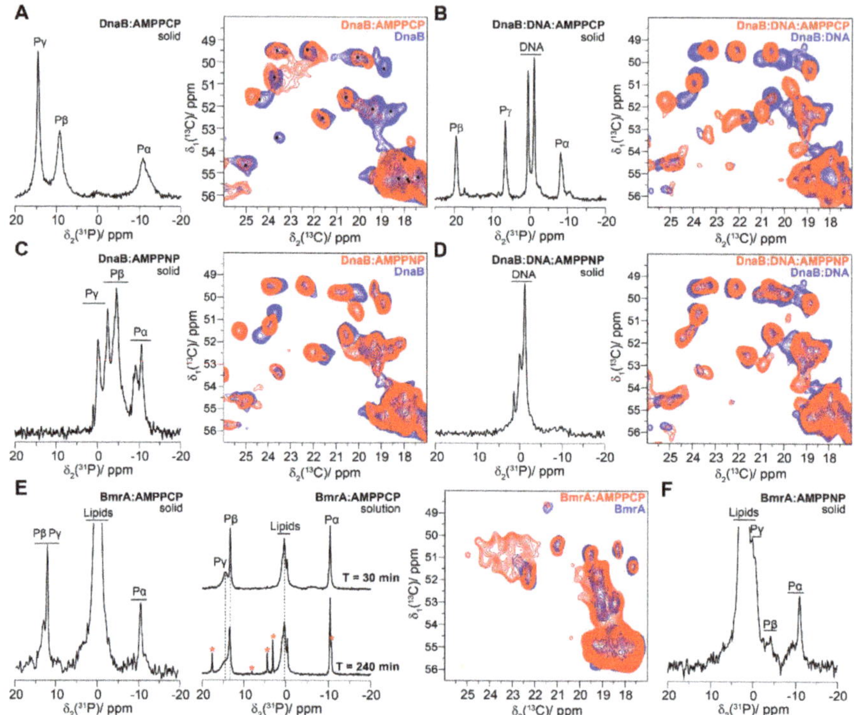

Figure 4. Pre-hydrolytic states using the systems DnaB and BmrA. ^{31}P one dimensional spectrum cross-polarization (1D CP) spectrum (left panel) and the alanine region of ^{13}C-^{13}C-DARR spectra overlay (right panel) of DnaB (blue) and DnaB:AMPPCP (red) (**A**), DnaB:DNA (blue) and DnaB:DNA:AMPPCP (red) (**B**), DnaB (blue) and DnaB:AMPPNP (red) (**C**), DnaB:DNA (blue) and DnaB:DNA:AMPPNP (red) (**D**). ^{31}P 1D CP spectrum (left panel), ^{31}P 1D spectrum (middle panel) spectrum and overlay of the alanine region of ^{13}C-^{13}C-DARR spectra (right panel) of BmrA:AMPPCP (red) and BmrA (blue) (**E**). Red stars indicate hydrolysis products of AMPPCP. ^{31}P 1D CP spectrum of BmrA:AMPPNP (**F**). Spectra in (**A**) and (**B**) were adapted from Wiegand et al. 2019 [82] (http://creativecommons.org/licenses/by/4.0/.), spectra (**C**) and (**D**) were adapted with permission from Wiegand et al. 2016 [38]. Panels (**E**) and (**F**) on BmrA represent original data; Red stars (*) indicate hydrolysis products of AMPPCP.

The 2D ^{13}C-^{13}C DARR experiments recorded on DnaB:AMPPCP show not only chemical-shift perturbations when compared to the apo protein, but also dynamic changes, as can be seen in the extract of the alanine region (Figure 4A, right panel) by the disappearance of resonances, which could be assigned to the *N*-terminal domain [82], which is important for binding the DnaG primase within the primosome. As illustrated by the equivalent 2D ^{13}C-^{13}C DARR experiment on the DNA-bound DnaB (Figure 4B, right panel), the binding of AMPPCP induces stronger CSPs due to larger conformational changes of the protein, but no dynamic effects of the *N*-terminal domain were observed.

In principle, AMPPNP and AMPPCP should have a similar effect on DnaB, as both should induce the pre-hydrolytic state. However, it is clear from the NMR spectra that the effects of these two analogues are very different. First, as highlighted by the 1D CP ^{31}P spectrum (Figure 4C, left panel), the presence of multiple resonances from the phosphate groups of AMPPNP indicates several structurally slightly different bound AMPPNP molecules. Interestingly, the 2D ^{13}C-^{13}C DARR spectrum reveals that the disappearance of the *N*-terminal domain resonances upon binding of AMPPCP is not observed in case of AMPPNP (Figure 4C, right panel). Also, we had observed that in presence of DNA, all AMPPNP is hydrolysed by the helicase [38]. Consequently, as shown by Figure 4D right panel, no AMPPNP is bound to the protein when DNA binds to the helicase, and the 2D ^{13}C-^{13}C DARR spectrum of DnaB in the presence of AMPPNP looks highly similar to DnaB without the analogue, which is not detected in the ^{31}P spectra either (Figure 4D left panel).

BmrA also binds AMPPCP, as shown by the 1D CP ^{31}P spectrum (Figure 4E, left panel). However, the rate of AMPPCP hydrolysis is much higher, compared to DnaB, and degradation products of AMPPCP can be observed already four hours after the rotor filling in the supernatant of the NMR rotor, as shown in Figure 4E right panel (see red stars in the Figure). We recorded a 2D ^{13}C-^{13}C DARR experiment of BmrA:AMPPCP (two days of acquisition), and the spectrum is virtually the same as the one of BmrA in the apo state. Possibly, AMPPCP has been rapidly hydrolysed, and an insufficient amount of AMPPCP only remained bound on BmrA. Similar to AMPPCP, AMPPNP binds to BmrA (Figure 4F), but was also rapidly hydrolysed (data not shown). The analysis of ^{31}P NMR spectra for protein samples containing lipids or DNA is more difficult due to the overlap between the ^{31}P γ- and β-phosphate signals from AMPPNP and those from lipid/DNA.

To overcome the hydrolysis problem with BmrA and to obtain a snapshot of the protein in its pre-hydrolytic state, we used an alternative approach, which is based on using mutant forms of the protein, which do bind ATP, but do not hydrolyse it. For this, catalytic residue/s can be mutated in order to make the protein inactive; still, one must take care that the protein retains its native fold. For BmrA, and also for other ABC transporters, it was shown that the mutation of the catalytic glutamate (E504 in BmrA) does not significantly affect the conformational change occurring upon nucleotide binding [23,99]. In contrast, the protein cannot achieve the pre-hydrolytic conformation when the nucleotide-binding Lys residue of the Walker A motif is mutated, here K380A [23,99]. We incubated the mutant E504A with ATP, and then sedimented it for analysis in the solid-state NMR rotor. While E504A is not completely inactive, it displays a very low ATPase activity (but still even crystals were obtained recently, PDB accession code 6R72, and a cryo-EM based structure was reported, PDB accession code 6R81) when compared to K380A, used as a fully inactive control (Figure 5A). After 40 h, only 50% of ATP is consumed, which allowed for the acquisition of 1D and 2D solid-state NMR experiments. The resulting 1D ^{31}P CP spectrum displays three narrow peaks corresponding to the three phosphate groups from the ATP bound to the protein (Figure 5B). The 2D ^{13}C-^{13}C DARR spectrum displays CSPs and peaks appearing, both induced by the conformational and dynamic changes in the protein as a consequence of ATP binding (Figure 5C) [92].

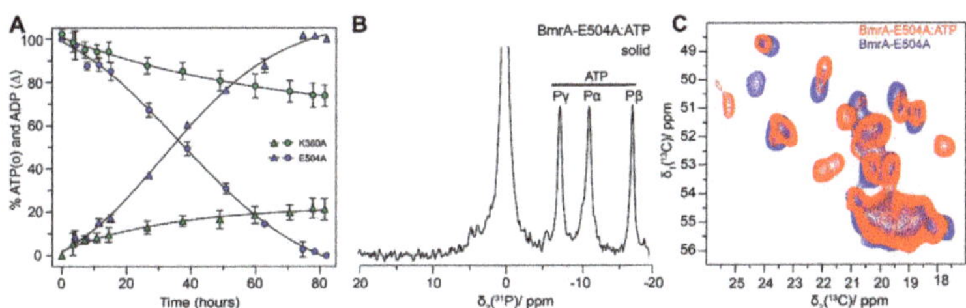

Figure 5. Pre-hydrolytic states using the system BmrA-E504A (catalytically inhibited). Percentage of ADP (Δ) and ATP (○) in the presence of the mutant BmrA-E504A (symbol filled in blue) or of the mutant, which does not bind the nucleotide, BmrA-K380A (symbol filled in green). BmrA-K380A was chosen as a negative control in order to exclude the possibility of an ATPase contaminant in the sample (**A**). ^{31}P 1D CP spectrum of BmrA-E504A:ATP (**B**). The overly of the alanine region of ^{13}C-^{13}C-DARR spectra of BmrA-E504A (blue) and BmrA-E504A:ATP (red) (**C**). Results in panels (**A**) and (**B**) are original, and spectra in (**C**) were adapted from Lacabanne et al. 2019 [92] (http://creativecommons.org/licenses/by/4.0/.).

To summarize, our data show that analysis of the pre-hydrolytic state is difficult both for DnaB and BmrA, since first the corresponding ATP mimics do not behave in a homogenous manner, i.e., analogues which should yield similar states lead to different NMR spectra, and second, most popular analogues are actually hydrolysed by the helicase in presence of DNA, as well as by the ABC transporter.

With respect to the first point, the intriguing observation that the AMPPCP- and AMPPNP-induced pre-hydrolytic states show conformational differences might be linked to the proposition that one can further differentiate each pre-hydrolytic mimic, as discussed by Ogawa et al., and assign the different mimics to specific steps therein: ATP-γ-S for the initial pre-hydrolysis state, AMPPCP for the pre-isomerization state, ADP:BeF_x for the middle pre-hydrolysis state and AMPPNP for the late pre-hydrolysis state [107]. It is difficult to establish a similar statement for DnaB, as one can also explain these differences by the fact that these analogues can behave differently from ATP in terms of their chemical properties: as examples for AMPPNP the oxygen, a hydrogen bond acceptor, is replaced by an NH_2 group, a possible hydrogen bond donor; AMPPCP has one oxygen atom less than ATP.

With respect to the second point, in the DnaB-DNA complex, only AMPPCP resisted to hydrolysis, and was the best choice to study DnaB and its DNA complex. It was however, rapidly hydrolysed in BmrA, which might be caused by the very high ATPase activity of BmrA, which is with an activity of 6.5 $\mu mol \cdot min^{-1} \cdot mg^{-1}$ one of the most active ABC transporters (one to three orders of magnitude higher than typical ABC transporters) [108]. Amongst AMPPNP and ATP-γ-S, which are the most used pre-hydrolytic state analogues for ABC transporters and helicases (see Figure 3B), neither proved useful here. Alternative strategies using mutant forms were successful to analyse a pre-hydrolytic mimic of the protein and presents a valuable alternative when ATP analogues fail to mimic the pre-hydrolysis states.

4.2. *The Transition-State Analogues ATP/ADP:Vi and Aluminium Fluorides (ADP:AlF_x)*

In order to investigate the transition states of BmrA and DnaB, we used the solid-state NMR techniques already described above, and also complemented them by EPR (Figure 1). The conformation of DnaB in the presence of ADP:Vi was compared with DnaB apo (Figure 6A) and DnaB in the presence of ADP only (Figure 6B). We also studied the protein with ADP and DNA, in the presence or absence of vanadate (Figure 6C).

The ^{13}C-^{13}C 2D DARR spectra of DnaB apo and DnaB:ADP:Vi display a few shifting resonances upon binding of the nucleotide (Figure 6A). However, the comparison of DnaB:ADP with and without vanadate shows that the NMR fingerprints of both samples are actually highly similar (Figure 6B), indicating that vanadate did not bind to the NBD and did not induce significant conformational changes. In contrast, when DNA is added to both samples, the NMR spectra of DnaB:ADP:Vi+DNA are different from the ones in the absence of DNA (DnaB:ADP:Vi and DnaB:ADP), with significant CSPs, but the most obvious CSPs are observed for the complex DnaB:ADP:DNA (Figure 6C, left panel). Since these two samples behave differently, a ^{31}P NMR spectrum was recorded to probe the bound ATP-mimics. The 1D ^{31}P-CP spectrum of DnaB:ADP:DNA displays two phosphorus peaks assigned to DNA (two DNA nucleotides bind to one DnaB monomer leading to two different phosphate binding environments [82]), and four peaks which can be assigned to bound ADP [88] (Figure 6C, right panel). $P\alpha$ and $P\beta$ correspond to the DnaB:ADP complex in the absence of DNA, and $P\alpha'$ and $P\beta'$ to the DnaB:ADP:DNA complex, indicating an insufficient DNA concentration to saturate the protein completely with DNA. However, the 1D ^{31}P-CP spectrum of DnaB:ADP:Vi:DNA (Figure 6C, right panel) shows only one population of ADP, with ^{31}P chemical-shift values similar to the DnaB:ADP complex, and a reduced intensity of the peaks assigned to the DNA. One can conclude from these spectra that the presence of vanadate actually inhibits the binding of DNA to DnaB.

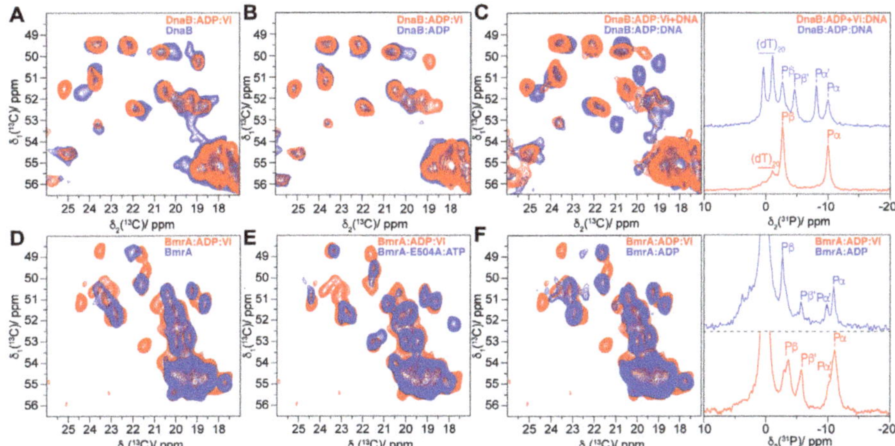

Figure 6. Comparison of the effect of vanadate on BmrA and DnaB. Alanine region from ^{13}C-^{13}C-DARR spectra of DnaB:ADP:Vi overlaid with DnaB apo (**A**), DnaB:ADP (**B**). Alanine region from ^{13}C-^{13}C-DARR (**C**, left panel) and 1D ^{31}P-CP (**C**, right panel) spectra of DnaB:ADP:Vi;DNA overlaid with DnaB:ADP:DNA. Alanine region from ^{13}C-^{13}C-DARR spectra of BmrA:ADP:Vi overlaid with BmrA apo (**D**), BmrA-E504A:ATP (**E**) and BmrA:ADP (left panel) complemented with the 1D ^{31}P spectrum (**F**, right panel). The resonance peak from the lipids (0 ppm) was cut, the separation between the two spectra is indicated by a dashed line. The blue spectrum in (**A**) was adapted from Wiegand et al. 2019 and spectra (**D**) and (**E**) were adapted from Lacabanne et al. 2019 [82,92] (http://creativecommons.org/licenses/by/4.0/.). The red spectra in (**A**) and (**C**), as well as the spectra in (**F**) are original data.

For the ABC transporter BmrA, one must say beforehand that conformational changes are not observed in BmrA upon incubating with vanadate and ADP. The protein requires vanadate and ATP instead of ADP, since ATP hydrolysis is required to induce the conformational changes. The inorganic phosphate is then exchanged by a vanadate anion, and Vi and ADP remain bound. The comparison between the ^{13}C-^{13}C DARR spectra of BmrA apo and BmrA:ATP:Vi shows both CSPs and new appearing peaks (Figure 6D). The presence of additional signals appearing in the protein spectra is indicative of a decrease of the flexibility of the corresponding protein residues [92]. When compared to the pre-hydrolytic state (obtained with BmrA-E504A:ATP), the new peaks appearing in the spectra overlay to a large extent (Figure 6E), indicating that these residues show a similar conformation. Some differences with respect to the pre-hydrolytic state can be observed, which can be associated to the addition of vanadate. To highlight the effect of vanadate, the spectrum of BmrA:ADP:Vi was compared to BmrA:ADP only (Figure 6F). This revealed the presence of new peaks, but only minor CSPs. The appearing peaks can serve as the fingerprint pattern that allows to distinguish the pre-hydrolytic and transition states, while the CSPs serve as the fingerprint pattern reflecting the kind of nucleotide bound.

A 1D ^{31}P-CP NMR experiment can yield complementary information about the bound ATP-mimics (Figure 6F, right panel). The ^{31}P spectrum of BmrA:ADP shows the presence of two populations of ADP (labeled Pα, Pβ and Pα', Pβ'), and the presence of vanadate induces ^{31}P chemical-shift changes for BmrA which were not observed for DnaB [12]. In case of BmrA:ADP:Vi, two populations of Pβ can be clearly distinguished and also for Pα, but less significantly (Pβ of ADP with vanadate has a different chemical shift than Pβ of ADP without vanadate). It is known that the trapping of one nucleotide during the transition state (in presence of vanadate) is possible while the second nucleotide can be poorly bound. This property has been observed for several ABC transporters (p-gp [109]; BmrA [99]; LmrA [110]; Maltose transporter [53]) suggesting an asymmetry of the NBDs [111].

In order to gain additional insight into whether vanadate binding occurred or not, we performed EDNMR experiments. This approach can be used to detect the ^{51}V nucleus (I = 7/2) in proteins in which the Mg^{2+} has been replaced by the EPR-active Mn^{2+} metal ion [16,112] in the nucleotide-binding sites, as sketched in Figure 1. The experiment detects the hyperfine couplings of the unpaired electrons of Mn^{2+} to the nuclei in the vicinity. We applied this both to the ABC transporter and the DnaB helicase. One should mention that it was shown by biochemical investigations for both proteins that upon substitution of Mg^{2+} by Mn^{2+}, their biological function is maintained [86,113]. Figure 7 shows the resulting EDNMR spectrum for the BmrA:ADP:Vi complex (shown in red) with an intense resonance for ^{51}V (for the echo-detected field-swept EPR spectra see Figure S2). In the absence of protein in the sample (black line) the spectrum only shows a ^{51}V peak with very low intensity assigned to vanadate in solution. Unresolved couplings to ^{23}Na would appear at very similar frequencies. We thus conclude that vanadate binds to the NBD in the case of BmrA.

For DnaB, no ^{51}V peak can be observed in the EDNMR spectrum, indicating that no vanadate is found in the vicinity of Mn^{2+} (Figure 7B). The EDNMR spectrum indeed shows the same profile for DnaB in the presence of nucleotide with vanadate (red line) and without (black line). We can thus exclude the presence of vanadate in the NBD of the protein. However, as shown previously in Figure 6B, some spectral differences (mainly appearing peaks upon ADP:Vi incubation) can be noticed when DnaB:ADP:Vi was compared to DnaB:ADP. In other words, these experiments do not allow to exclude that vanadate might bind at another location than in the NBD.

We thus used a complementary experiment which can directly detect ^{51}V using solid-state NMR. ^{51}V has been intensively studied by solid-state NMR due to its rather small nuclear quadrupole moment and its high sensitivity [114–116]. Vanadate has also been studied in biological systems using solution-state NMR [117,118] and solid-state NMR [119]. Figure 7C shows the ^{51}V MAS spectrum of DnaB:ADP:Vi recorded at two different MAS spinning frequencies of 17 and 19 kHz. By measuring at two different MAS frequencies, the central transition (|−1/2> ↔ |+1/2>, to first order free from quadrupole interaction, can be distinguished from the spinning-sideband positions resulting from first-order quadrupolar interaction (a superposition of the remaining single-quantum transitions, marked by asterisks). The presence of the first order quadrupolar coupling sideband pattern already points to immobilized ^{51}V species. We can distinguish two resonances at around −600 ppm (−604 ppm and −618 ppm) and two further vanadate species bound to DnaB at −533 and −681 ppm (Figure 7C). To assign those resonances, a spectrum of the not immobilized (the supernatant) ^{51}V was recorded and assigned (Figure 7D). The resonances of the ^{51}V MAS spectrum can be assigned by comparison with the solution-state spectrum of the supernatant as follows: VO_4^{3-} (V1), $V_2O_7^{4-}$ (V2), $V_4O_{12}^{4-}$ (V4) and $V_5O_{15}^{5-}$ (V5) [120,121]. We can exclude that the peaks corresponding to the immobilized phase peaks result from the precipitation of vanadate, since with an initial orthovanadate concentration of 5 mM (0.92 g L^{-1}) at pH 6, we are two orders of magnitude below the solubility limit. The detected signal thus must stem from DnaB-bound vanadate, which might be related to the observation that addition of vanadate interferes with DNA binding to DnaB (Figure 6C).

To sum up, vanadate is a reasonable ATP-transition-state mimic for the ABC transporter BmrA. The transporter is trapped, most likely in its outward-facing state, when binding ADP:Vi. The transition state is characterized by a characteristic fingerprint in the NMR ^{13}C-^{13}C DARR spectrum, and vanadate is indeed present in the vicinity of the metal ion. In contrast, ADP:Vi is not a suitable ATP-transition-state mimic for the helicase DnaB. Indeed, solid-state NMR and EPR experiments reveal that vanadate does not bind to the NBD together with the nucleotide. Instead, vanadate is bound elsewhere to DnaB, most likely in an unspecific manner. ADP:Vi strongly inhibits binding of DNA, suggesting that they share the same binding site on DnaB, and that vanadate outcompetes DNA.

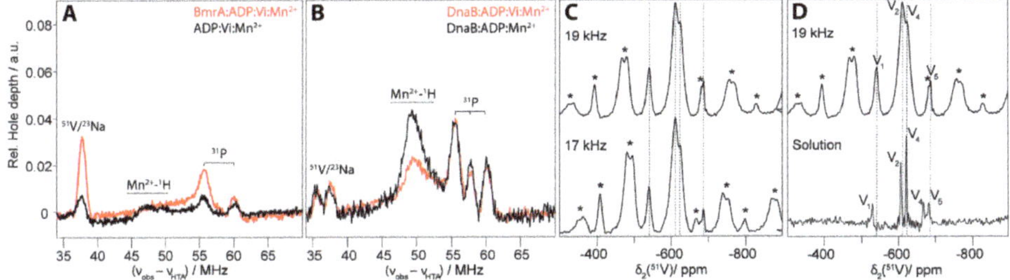

Figure 7. Localisation of the vanadate ion using NMR and EPR experiments. Background corrected 94 GHz EDNMR spectra of BmrA (**A**) and DnaB (**B**) incubated with Mn^{2+} and ADP, with and without vanadate. For ^{31}P, a doublet due to the hyperfine coupling to ^{31}P is observed (~4 MHz), as well as a singlet not assigned so far [122]. ^{51}V spectra of DnaB:ADPMg:DNA:Vi recorded at a MAS frequency of 17 and 19 kHz (**C**). ^{51}V spectra of DnaB:ADPMg:DNA:Vi overlaid with the solution-state spectrum of ADPMg:Vi (**D**). Central transitions are indicated with dashed lines whereas spinning sidebands are indicated with a black star. All panels represent original data.

4.3. Aluminium Fluorides (AlF_X) as Transition-State Mimic

AlF_x is the most frequently used transition-state analogue (Figure 2A), although the pH-dependence of its formation imposes certain limitations to it. At pH ≥ 5 (depending also on the concentration and the anions in the solution), Al^{3+} starts to form an aluminium hydroxide complex, $Al(OH)_3$, which is insoluble. However, the presence of an excess of fluoride shifts the pH upon which $Al(OH)_3$ formation occurs to a higher value. We calculated the concentrations of the different species of aluminium under the conditions used (6 mM of $AlCl_3$ and 30 mM NH_4F) as a function of the pH-value (Figure 8). In our case, the formation of $Al(OH)_3$ starts at pH 7, and almost all Al^{3+} precipitates as $Al(OH)_3$ at pH ≥ 8. The amount of formed AlF_X is thus not sufficient to induce the protein:AlF_x complex formation. At the same time, fluorides present in the solution can form a complex with Mg^{2+} generating the transition-state analogue MgF_3^-. This effect was followed and confirmed by ^{19}F NMR for the conversion of a protein:ADP:AlF_4^- complex to a protein:ADP:MgF_3^- complex by increasing the pH [56]. Moreover, as pointed out above, MgF_3^- and AlF_3 are structurally very similar and some structures comprising AlF_3 as transition state mimic are in reality MgF_3^- because they were obtained at pH ≥ 8 [56–58].

For DnaB, the DnaB:ADP:AlF_x complex can easily be prepared at a pH of 6, since the protein is stable at this pH value. In the presence of the transition-state analogue, the 1D CP ^{31}P spectrum displays two very narrow resonances assigned to the Pα and Pβ of ADP in complex with AlF_x (Figure 9A, left panel). Note a minor amount of DnaB:ADP in the sample. The 2D ^{13}C-^{13}C DARR spectrum of DnaB:ADP:AlF_x displays strong CSPs attributed to conformational changes of the protein (Figure 9A, right panel). While we noticed that the use of vanadate inhibits the binding of DNA, DNA clearly binds to DnaB in the presence of AlF_x, as shown by Figure 9B, left panel. Fluorescence anisotropy measurements revealed that the affinity for DNA-binding is even the highest in the presence of ADP:AlF_x compared to the other ATP-mimics used [82]. The 2D ^{13}C-^{13}C DARR spectrum of the sample in presence of DNA reveals that several peaks, which belong to the *N*-terminal domain, are again missing, indicating a change in the dynamics of the protein, as was already observed for DnaB:AMPPCP without DNA.

The case of BmrA is more complex, since the optimal pH for sample preparation lies at 8. For optimal use of AlF_x, the pH would need to be lowered, but we observed this to result in poor (e.g., strongly broadened) spectra. Nevertheless, we explored this further, and in order to test the pH dependency, BmrA, in the presence of ATP, was incubated with 6 mM of $AlCl_3$ and 30 mM NH_4F at pH 8, 7.5 and 7, and a 1D CP ^{31}P NMR was recorded for all three conditions (Figure 9C, left panel). The 1D CP ^{31}P NMR spectrum at pH 8 shows that ATP/ADP is abundantly co-precipitated with $Al(OH)_3$ which makes the 1D spectrum difficult to analyse due to a broad and rather unstructured resonance of this amorphous species (Figure 9C, left panel). As expected, the fraction of ATP/ADP co-precipitated with $Al(OH)_3$ decreases with decreasing pH-values. While at pH 7 the precipitation of $Al(OH)_3$ is still visible in the ^{31}P NMR spectrum, one can compare it to BmrA:ADP:Vi, which shows that both spectra overlay with only few minor differences (Figure 9D, right panel). This indicates that the conformation is highly similar to the one observed with vanadate. A 2D ^{13}C-^{13}C DARR spectrum recorded on the pH 7 sample (Figure 9C, right panel) confirms this, as the resonances largely superimpose.

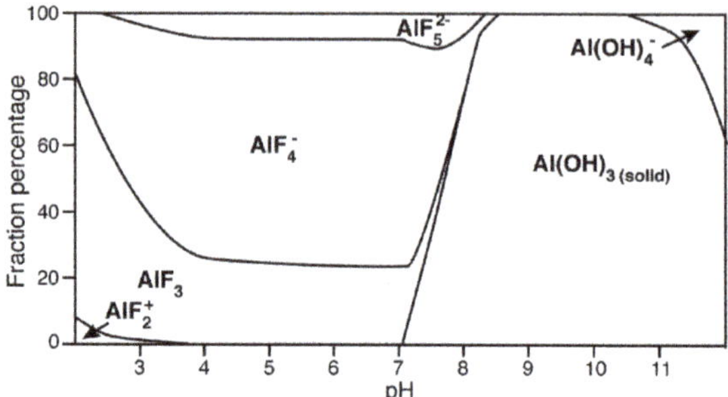

Figure 8. The different species of AlF_x at different pH. The diagram was generated using ChemEQL [123], which calculates chemical speciation and equilibria. Concentrations used were 6 mM $AlCl_3$ and 30 mM NH_4F.

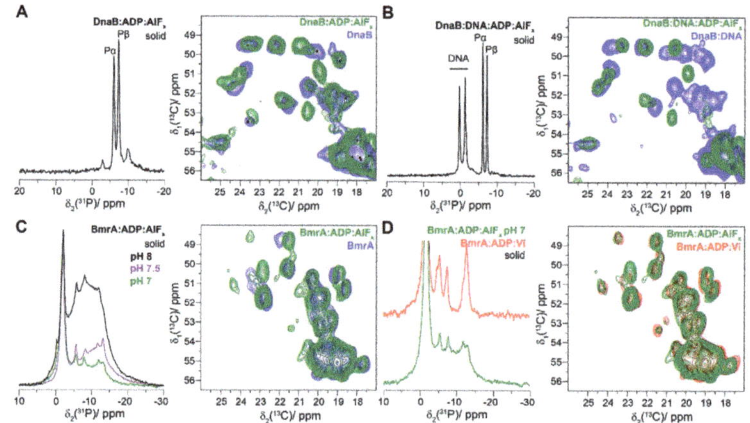

Figure 9. Comparison of the metal fluoride AlF_x on the systems BmrA and DnaB. 1D CP of BmrA:ADP:AlF_x at different pH (**A**, right panel) and ^{13}C-^{13}C-DARR spectra of the alanine region of BmrA:ADP:AlF_x^- (green) and BmrA (**A**, left panel, blue). ^{31}P 1D CP spectra of BmrA:ADP:AlF_x at pH 7 overlaid with BmrA:ADP:Vi (**B**, right panel) and ^{13}C-^{13}C-DARR spectra of the alanine region of BmrA:ADP:AlF_x and BmrA:ADP:Vi (**B**, left panel). ^{31}P 1D CP spectrum (left panel) and ^{13}C-^{13}C-DARR the alanine region spectra overlay (right panel) of DnaB and DnaB:ADP:AlF_x (**C**), DnaB:DNA and DnaB:DNA:ADP:AlF_x (**D**). Spectra (**A**) and (**B**) were adapted from Wiegand et al. 2019 [82] (http://creativecommons.org/licenses/by/4.0/.). Panels (**C**) and (**D**) present original data.

To summarize, the use of AlF_x as a transition analogue heavily depends on the optimal pH value of the protein. Indeed, biological systems are principally studied at pH 5–9, and it is important to take into account the formation and precipitation of $Al(OH)_3$ at pH $\geq$ 7 under our conditions. The spectra of BmrA at pH $\geq$ 7 well illustrate the consequences of the use of AlF_x in alkaline conditions. The protein actually adopts a similar conformation as in the presence of vanadate, but high amounts of amorphous species are detected. In contrast, DnaB at pH 6 shows high affinity to AlF_x which induces substantial conformational changes; and also DNA binding is not affected.

4.4. The Post-Hydrolytic State Induced by ADP

The last state in the ATP hydrolysis cycle is the post-hydrolytic state, where ADP is still bound to the protein and the inorganic phosphate (previously γ-phosphate) is released from the binding pocket. This state is well mimicked by the addition of ADP. We used the ^{31}P and ^{13}C experiments described above to characterize BmrA and DnaB in the presence of ADP.

The ^{31}P CP spectrum for DnaB:ADP is shown in Figure 10A (left panel). The spectrum displays two sharp peaks, which indicates a good homogeneity of the sample. The overlay of the 2D DARR spectra DnaB:ADP and DnaB apo (Figure 10A, right panel) reveals CSPs and also the disappearance of *N*-terminal domain peaks, indicating conformational changes and an increase in the dynamics of the protein.

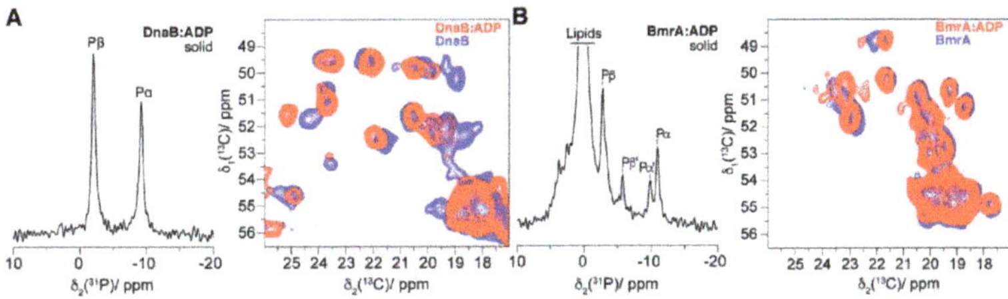

Figure 10. Generation of the post-hydrolytic state using ADP. (**A**) ^{31}P 1D CP spectrum (left panel) and ^{13}C-^{13}C-DARR the alanine region spectra overlay (right panel) of DnaB:ADP/DnaB apo; (**B**) ^{31}P 1D CP spectrum (left panel) and ^{13}C-^{13}C-DARR the alanine region spectra overlay (right panel) of BmrA:ADP/BmrA. Spectra (**A**) were adapted from Wiegand et al. 2019 [82] (http://creativecommons.org/licenses/by/4.0/.), and spectra in panel (**B**) are original data.

In contrast, the conformational changes of the ABC transporter BmrA are minor between the presence and absence of ADP. First of all, the ^{31}P CP spectrum shows the presence of two populations of ADP as identified by peak doubling, labeled Pα, Pβ, and Pα′, Pβ′ (Figure 10B, left panel). These two populations are the result of two different binding modes of ADP to the protein. However, since the intensity of the Pα′, Pβ′ peaks is 50% lower than Pβ, Pα, there is less Pα′, Pβ′ bound to the protein than Pβ, Pα. This is reminiscent to the pattern that was observed with vanadate (Figure 6F). Unspecific binding of ADP to the protein can explain this observation. Secondly, the overlay of the 2D DARR spectra of BmrA:ADP and BmrA apo displays few CSPs compared to what we observed in the presence of other ATP analogues. In contrast to DnaB, the binding of ADP does not induce large conformational or dynamics changes in the protein, and binding of ADP to BmrA seems to be very weak.

4.5. Structural Considerations

It seems worthy to compare the NMR data with structural information as available for proteins that are closely related to DnaB of *Helicobacter pylori* and BmrA of *Bacillus subtilis*, respectively.

BmrA and its structural counterpart. The most suited for comparison appears to be the set of structures that shows the maltose ABC-transporter of *E. coli* (MBP-MalFGK2) in the outward-facing conformation, with two interacting NBDs and in the presence of AMPPNP (PDB 3RLF), ADP:BeF_3 (PDB 3PUX), ADP:Vi (PDB 3PUV), and ADP:AlF_4^- (PDB 3PUW) [64]. The collection of these different structures shapes the view of the transport cycle [122]. Chen and Oldham noticed that, despite the different ATP-analogues used, all residues within the NBD are essentially superimposable. However, structural differences between the pre-hydrolytic state (AMPPNP) and the transition state (ADP:Vi and ADP:AlF_4^-) are (i) the distance between the γ-phosphate or the mimicked γ-phosphate by the analogues and the bridging oxygen of the β-phosphate and (ii) the presence of a water molecule, essential for the ATP hydrolysis, only in the transition state. Although the transmembrane part of the maltose transporter essentially differs from that of BmrA, the NBD homodimers of the two proteins are relatively similar (RMSD of 1.7 Å from the alignment of MBP-MalFGK2:AMPPNP, PDB 3RLF, with BmrAE504A:ATP, PDB 6R72). The two NBDs differ mainly in their ATPase activity. The ATPase activity of MBP-MalFGK2 is one order of magnitude lower than BmrA [124]. For BmrA, a major conformational transition between the open (inward-facing) and closed (outward-facing) conformation was for example experimentally demonstrated by hydrogen/deuterium exchange (HDX) coupled to mass spectrometry [125] and NMR spectroscopy [92]. It is believed that the protein adopts the closed conformation, with interacting NBDs, upon substrate binding. Generally, in membrane transporters, the energies of their sub-conformations should be close to each other and should essentially depend on the protein environment. The NMR spectra of BmrA in the presence of ADP:Vi and ADP:AlF_4^- might be taken as reporters of the enzyme transition state; in the presence of AlF_4^-, the crystal structure of the maltose transporter shows a classical picture with the catalytic water molecule in the apical attack position (Figure 11A). Even in the presence of AlF_4^-, the ^{31}P 1D CP spectra give two signals for the α- and β-phosphates (Figure 9D), respectively, which points to a certain nonequivalence of the two substrate-binding sites in the two similar NBDs. This finding might indicate that the two catalytic sites operate not simultaneously but sequentially.

DnaB of Helicobacter pylori and its structural counterparts. The conformation of DnaB, as could be judged from the 2D ^{13}C-^{13}C DARR spectra, essentially depends on the nature of the analogue used, which matches the great structural variability reported for DnaB from other bacteria and their viral homologues [126–132]. Depending on the presence of substrate analogues and their nature, the SF4 helicase subunits can either form rings of distinct shapes [126–130] or arrange themselves as a hexameric ladder along a DNA strand [131,132]. The latter type of the structure was reported for DnaB from *Bacillus stearothermophilus* (currently *Geobacillus stearothermophilus*), which was crystalized, in the presence of a DNA strand, with GDP:AlF_4^- in five of its six catalytic sites [131] (see Figure 11B). In this structure, each monomer of DnaB interacts in a similar way with two nucleotides of DNA; together, the subunits make a kind of a spiral ladder. It is noteworthy, that the position of AlF_4^- in the structure of *Geobacillus stearothermophilus* DnaB (Figure 11B) differs from that in other P-loop fold NTPases. No catalytic water molecule is present apically to the plane of AlF_4^- (see Figure 11A as a typical example), and the position of the AlF_4^- moiety does not correspond to that of the γ-phosphate group (see Figure 11C,F). Interestingly, the NMR data on DnaB from *Helicobacter pylori* discussed herein point to a full occupation of all six NBDs and a rather high symmetry in the oligomer [82] as it potentially could be achieved by more flat conformations of the helicase hexamer, as reported for several DnaB proteins, including the one from *Geobacillus stearothermophilus*, which were crystallized in the absence of AlF_4^- [127,128]. Whether the physiological shape of the DnaB ring is flat or spiral has to be established yet.

Figure 11D,E show the structures of the NBD of the ABC transporter MBP-MalFGK2 (Figure 11D) and the gene 4 helicase from bacteriophage T (Figure 11E) complexed with the pre-hydrolytic ATP analogue AMPPNP. The overlay of the structures (Figure 11F) shows a similar conformation of the bound phosphate chain of the ATP mimic. Although quite similar enzymes seem to bind ATP mimics in a similar way, they might behave differently to the huge number of ATP analogues available and solid-state NMR seems to be the method-of-choice to address such different behaviors.

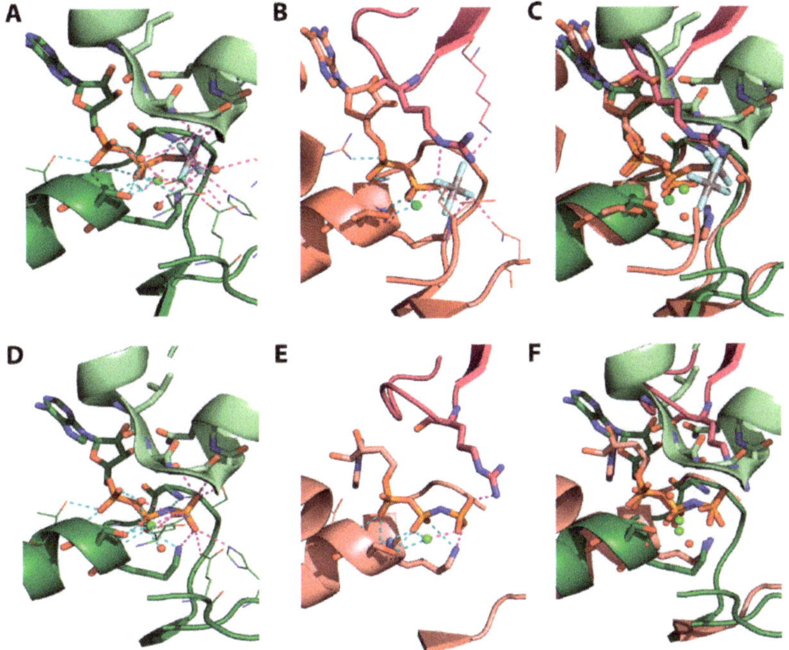

Figure 11. Structural comparison of nucleotide-binding sites in SF4 helicases and ABC-transporters. Different protein subunits are colored in different shades of the same color. Mg^{2+} or Ca^{2+} ions are shown as green spheres; water molecules are shown as red spheres; hydrogen bonds and metal interactions involving α- and β-phosphates are shown as cyan dashes, interactions with γ-phosphate or its fluoroaluminate complex mimic shown as magenta dashes. Nucleotide analogue, P-loop motif residues and activating residues (Arg residue or LSGGQ motif) are shown as thick sticks, other interacting amino acid residues are shown as thin sticks. Enzymes complexed with NDP:AlF_4^- (**A–C**). Maltose/maltodextrin import ATP-binding protein MalK from *Escherichia coli K-12* (PDB ID 3PUW, chain B) (**A**), Replicative helicase DnaB from *Geobacillus stearothermophilus* (PDB ID 4ESV, chain E) (**B**), Structures 4ESV and 3PUW superimposed by phosphate chain and ribose atoms of NDP moieties (**C**); Enzymes complexed with the slowly hydrolyzable ATP analogue AMPPNP (**D–F**). Maltose/maltodextrin import ATP-binding protein MalK from *Escherichia coli K-12*, (PDB ID 3RLF, chain A) (**D**), Gene 4 Ring Helicase from Escherichia phage T7 (PDB ID 1E0J, chain A) (**E**), Structures 1E0J and 3RLF superimposed by phosphate chain (**F**).

5. Conclusions

We herein reviewed magnetic-resonance approaches (in combination with additional data) to provide information at the atomic level on the binding of mimics of the different ATP forms present during the hydrolysis cycle. We investigated this for two ATP-fuelled proteins, an ABC transporter and a DNA helicase (Table 2), both driven by an ATPase motor domain. We showed that the ATP analogues mainly used for structural studies for such systems, AMPPNP and ATP-γ-S, are not suitable for the systems studied here, since both are hydrolysed by the proteins. Furthermore, we show that analogues which should induce the same state in the hydrolysis cycle can fail to do so, since they result in different conformations. We also discuss that some analogues can interfere with protein function, such as DNA binding for DnaB. NMR, and also EPR, are sensitive tools to assess the impact of different analogues for a given protein, a need that arises through the observation that they can have widely differing effects on different proteins. NMR spectroscopy could be of help in tracing minor differences both in the overall protein conformation and in the state of the phosphate groups. Here we showed that solid-state NMR enabled revealing notable differences in the structural properties of closely related P-loop fold NTPases, namely the SF4 DnaB helicase and BmrA ABC-transporter, which both belong to the same division of ASCE-NTPases.

Table 2. Summary of the ATP analogue efficiencies for the two protein systems BmrA and DnaB. ++ indicates high efficiency, + moderate efficiency, - low efficiency and – not efficient.

Analogue	State Being Mimicked	Suitability		Comments
		DnaB	BmrA	
AMPPNP	Pre-hydrolytic	-	–	Hydrolysed by DnaB in the presence of DNA Hydrolysed by BmrA
AMPPCP	Pre-hydrolytic	++	–	Rigidifies DnaB in the presence of DNA Hydrolysis observed
ATP-γ-S	Pre-hydrolytic	–	–	Hydrolysed by BmrA and DnaB
V_i	Transition-state	–	++	Inhibits the DNA binding in DnaB Provides the transition state for BmrA
AlF_x	Transition-state	++	+	Provides the transition state for DnaB and BmrA Starts to precipitate at pH ≥7
ADP	Post-hydrolytic	++	++	Provides the post-hydrolysis state in both systems

Supplementary Materials: The following are available online [133–137].

Author Contributions: D.L. and R.C. prepared the samples. D.L. and T.W. performed the NMR experiments. N.W. and D.K. recorded the EPR experiments. M.I.K. and A.Y.M. analysed the protein structures. D.L., T.W., M.I.K., N.W., D.K., A.Y.M., B.H.M. and A.B. analysed the data. All authors contributed to the writing of the manuscript. D.L., T.W., B.H.M. and A.B. designed the research, and B.H.M. and A.B. supervised the project. All authors have read and agreed to the published version of the manuscript.

Funding: This work was supported by the ETH Career SEED-69 16-1 (T.W.) and the ETH Research Grant ETH-43 17-2 (T.W.), an ERC Advanced Grant (B.H.M., grant number 741863, FASTER), by the Swiss National Science Foundation (B.H.M., grant number 200020_159707 and 200020-188711), the French Agence Nationale de Recherche (A.B., ANR-14-CE09-0024B, ANR-19-CE11-0023), the LABEX ECOFECT (A.B., ANR-11-LABX-0048) within the Université de Lyon program Investissements d'Avenir (A.B., ANR-11-IDEX-0007), the EvoCell Program of the Osnabrueck University (M.I.K.) and a grant from the Russian Science Foundation (A.Y.M., 17-14-01314).

Acknowledgments: We thank Gunnar Jeschke for his comments on the manuscript. D.L. and T.W. acknowledge helpful discussion with Marco E. Weber.

Conflicts of Interest: The authors declare no conflict of interest.

References

1. Manna, R.N.; Dutta, M.; Jana, B. Mechanistic study of the ATP hydrolysis reaction in dynein motor protein. *Phys. Chem. Chem. Phys.* **2020**, *22*, 1534–1542. [CrossRef] [PubMed]
2. Mächtel, R.; Narducci, A.; Griffith, D.A.; Cordes, T.; Orelle, C. An integrated transport mechanism of the maltose ABC importer. *Res. Microbiol.* **2019**, *170*, 321–337. [CrossRef] [PubMed]
3. Prieß, M.; Göddeke, H.; Groenhof, G.; Schäfer, L.V. Molecular mechanism of ATP hydrolysis in an ABC transporter. *ACS Central Sci.* **2018**, *4*, 1334–1343. [CrossRef] [PubMed]
4. Brosh, R.M.; Matson, S.W. History of DNA Helicases. *Genes* **2020**, *11*, 255. [CrossRef]
5. Ma, W.; Schulten, K. Mechanism of substrate translocation by a ring-shaped ATPase motor at millisecond resolution. *J. Am. Chem. Soc.* **2015**, *137*, 3031–3040. [CrossRef]
6. Dittrich, M.; Schulten, K. PcrA Helicase, a prototype ATP-driven molecular motor. *Structure* **2006**, *14*, 1345–1353. [CrossRef]
7. Grigorenko, B.L.; Rogov, A.V.; Topol, I.A.; Burt, S.K.; Martinez, H.M.; Nemukhin, A.V. Mechanism of the myosin catalyzed hydrolysis of ATP as rationalized by molecular modeling. *Proc. Natl. Acad. Sci. USA* **2007**, *104*, 7057–7061. [CrossRef]
8. Davidson, R.B.; Hendrix, J.; Geiss, B.J.; Mccullagh, M. Allostery in the dengue virus NS3 helicase: Insights into the NTPase cycle from molecular simulations. *PLoS Comput. Biol.* **2018**, *14*, e1006103. [CrossRef]
9. Geourjon, C.; Orelle, C.; Steinfels, E.; Blanchet, C.; Deléage, G.; Di Pietro, A.; Jault, J.M. A common mechanism for ATP hydrolysis in ABC transporter and helicase superfamilies. *Trends Biochem. Sci.* **2001**, *26*, 539–544. [CrossRef]

10. Gardiennet, C.; Schütz, A.K.; Hunkeler, A.; Kunert, B.; Terradot, L.; Böckmann, A.; Meier, B.H. A sedimented sample of a 59 kDa dodecameric helicase yields high-resolution solid-state NMR spectra. *Angew. Chem. Int. Ed.* **2012**, *51*, 7855–7858. [CrossRef]
11. Wiegand, T.; Lacabanne, D.; Torosyan, A.; Boudet, J.; Cadalbert, R.; Allain, F.H.-T.; Meier, B.H.; Böckmann, A. Sedimentation yields long-term stable protein samples as shown by solid-state NMR. *Front. Mol. Biosci.* **2020**, *7*, 17. [CrossRef] [PubMed]
12. Wiegand, T. A solid-state NMR tool box for the investigation of ATP-fueled protein engines. *Prog. Nucl. Magn. Reson. Spectrosc.* **2020**, *117*, 1–32. [CrossRef] [PubMed]
13. Takegoshi, K.; Nakamura, S.; Terao, T. 13C-13C polarization transfer by resonant interference recoupling under magic-angle spinning in solid-state NMR. *Chem. Phys. Lett.* **1999**, *307*, 295–302. [CrossRef]
14. Takegoshi, K.; Nakamura, S.; Terao, T. 13C-1H dipolar-assisted rotational resonance in magic-angle spinning NMR. *Chem. Phys. Lett.* **2001**, *344*, 631–637. [CrossRef]
15. Böckmann, A.; Gardiennet, C.; Verel, R.; Hunkeler, A.; Loquet, A.; Pintacuda, G.; Emsley, L.; Meier, B.H.; Lesage, A. Characterization of different water pools in solid-state NMR protein samples. *J. Biomol. NMR* **2009**, *45*, 319–327. [CrossRef]
16. Cox, N.; Lubitz, W.; Savitsky, A. W-band ELDOR-detected NMR (EDNMR) spectroscopy as a versatile technique for the characterisation of transition metal-ligand interactions. *Mol. Phys.* **2013**, *111*, 2788–2808. [CrossRef]
17. Cox, N.; Nalepa, A.; Lubitz, W.; Savitsky, A. ELDOR-detected NMR: A general and robust method for electron-nuclear hyperfine spectroscopy? *J. Magn. Reson.* **2017**, *280*, 63–78. [CrossRef]
18. Goldfarb, D. ELDOR-Detected NMR. In *eMagRes*; Wiley: Hoboken, NJ, USA, 2017; Volume 563, pp. 101–114.
19. Schosseler, P.; Wacker, T.; Schweiger, A. Pulsed ELDOR detected NMR. *Chem. Phys. Lett.* **1994**, *224*, 319–324. [CrossRef]
20. Giannoulis, A.; Feintuch, A.; Barak, Y.; Mazal, H.; Albeck, S.; Unger, T.; Yang, F.; Su, X.-C.; Goldfarb, D. Two closed ATP- and ADP-dependent conformations in yeast Hsp90 chaperone detected by Mn(II) EPR spectroscopic techniques. *Proc. Natl. Acad. Sci. USA* **2019**, *117*, 395–404. [CrossRef]
21. Flowers, S.; Biswas, E.E.; Biswas, S.B. Conformational dynamics of DnaB helicase upon DNA and nucleotide-binding: Analysis by intrinsic tryptophan fluorescence quenching. *Biochemistry* **2003**, *42*, 1910–1921. [CrossRef]
22. Robson, A.; Booth, A.E.; Gold, V.A.; Clarke, A.R.; Collinson, I. A Large conformational change couples the ATP binding site of SecA to the SecY protein channel. *J. Mol. Biol.* **2007**, *374*, 965–976. [CrossRef] [PubMed]
23. Orelle, C.; Gubellini, F.; Durand, A.; Marco, S.; Levy, D.; Gros, P.; Di Pietro, A.; Jault, J.-M. Conformational change induced by ATP binding in the multidrug ATP-binding cassette transporter BmrA. *Biochemistry* **2008**, *47*, 2404–2412. [CrossRef] [PubMed]
24. Kühner, S.; Fischer, S. Structural mechanism of the ATP-induced dissociation of rigor myosin from actin. *Proc. Natl. Acad. Sci. USA* **2011**, *108*, 7793–7798. [CrossRef] [PubMed]
25. Wikström, M. Biophysical and structural aspects of bioenergetics. *R. Soc. Chem.* **2007**, *1*, 414.
26. Chen, B.; Doucleff, M.; Wemmer, D.E.; De Carlo, S.; Huang, H.H.; Nogales, E.; Hoover, T.R.; Kondrashkina, E.; Guo, L.; Nixon, B.T. ATP Ground-and transition states of bacterial enhancer binding AAA+ ATPases support complex formation with their target protein, σ54. *Structures* **2007**, *15*, 429–440. [CrossRef]
27. Chabre, M. Aluminofluoride and beryllofluoride complexes: A new phosphate analogs in enzymology. *Trends Biochem. Sci.* **1990**, *15*, 6–10. [CrossRef]
28. Yount, R.G.; Babcock, D.; Ballantyne, W.; Ojala, D. Adenylyl imidodiphosphate, an adenosine triphosphate analog containing a P-N-P linkage. *Biochemistry* **1971**, *10*, 2484–2489. [CrossRef]
29. Myers, T.C.; Nakamura, K.; Flesher, J.W. Phosphonic acid analogs of nucleoside phosphates. I. The synthesis of 5″-adenylyl methylenediphosphonate, a phosphonic acid analog of ATP. *J. Am. Chem. Soc.* **2002**, *85*, 3292–3295. [CrossRef]
30. Goody, R.S.; Eckstein, F. Thiophosphate analogs of nucleoside di- and triphosphates. *J. Am. Chem. Soc.* **1971**, *93*, 6252–6257. [CrossRef]
31. Mannherz, H.G.; Brehme, H.; Lamp, U. Depolymerisation of F-actin to G-actin and its repolymerisation in the presence of analogs of adenosine triphosphate. *JBIC J. Biol. Inorg. Chem.* **1975**, *60*, 109–116. [CrossRef]

32. Watanabe, F.; Hashimoto, T.; Tagawa, K. Energy-independent protection of the oxidative phosphorylation capacity of mitochondria against anoxic damage by ATP and its nonmetabolizable analogs1. *J. Biochem.* **1985**, *97*, 1229–1234. [CrossRef] [PubMed]
33. Suzuki, Y.; Shimizu, T.; Morii, H.; Tanokura, M. Hydrolysis of AMPPNP by the motor domain of ncd, a kinesin-related protein. *FEBS Lett.* **1997**, *409*, 29–32. [CrossRef]
34. Olesen, C.; Picard, M.; Winther, A.-M.L.; Gyrup, C.; Morth, J.P.; Oxvig, C.; Møller, J.V.; Nissen, P. The structural basis of calcium transport by the calcium pump. *Nat. Cell Biol.* **2007**, *450*, 1036–1042. [CrossRef] [PubMed]
35. Siarheyeva, A.; Liu, R.; Sharom, F.J. Characterization of an asymmetric occluded state of P-glycoprotein with two bound nucleotides. *J. Biol. Chem.* **2010**, *285*, 7575–7586. [CrossRef] [PubMed]
36. Ferguson, A.D.; Sheth, P.R.; Basso, A.D.; Paliwal, S.; Gray, K.; Fischmann, T.O.; Le, H.V. Structural basis of CX-4945 binding to human protein kinase CK2. *FEBS Lett.* **2011**, *585*, 104–110. [CrossRef]
37. Timachi, M.H.; Hutter, C.A.; Hohl, M.; Assafa, T.; Böhm, S.; Mittal, A.; Seeger, M.A.; Bordignon, E. Exploring conformational equilibria of a heterodimeric ABC transporter. *eLife* **2017**, *6*, 257. [CrossRef]
38. Wiegand, T.; Cadalbert, R.; Gardiennet, C.; Timmins, J.; Terradot, L.; Böckmann, A.; Meier, B.H. Monitoring ssDNA binding to the DnaB helicase from *helicobacter pylori* by solid-state NMR spectroscopy. *Angew. Chem. Int. Ed.* **2016**, *55*, 14164–14168. [CrossRef]
39. Kung, G.; Runquist, J.A.; Miziorko, H.M.; Harrison, D.H.T. Identification of the allosteric regulatory site in bacterial phosphoribulokinase. *Biochemistry* **1999**, *38*, 15157–15165. [CrossRef]
40. Nitta, R.; Kikkawa, M.; Okada, Y.; Hirokawa, N. KIF1A Alternately uses two loops to bind microtubules. *Science* **2004**, *305*, 678–683. [CrossRef]
41. Reddy, M.C.M.; Palaninathan, S.K.; Shetty, N.D.; Owen, J.L.; Watson, M.D.; Sacchettini, J.C. High resolution crystal structures of Mycobacterium tuberculosis Adenosine Kinase. *J. Biol. Chem.* **2007**, *282*, 27334–27342. [CrossRef]
42. Shintre, C.A.; Pike, A.C.W.; Li, Q.; Kim, J.-I.; Barr, A.J.; Goubin, S.; Shrestha, L.; Yang, J.; Berridge, G.; Ross, J.; et al. Structures of ABCB10, a human ATP-binding cassette transporter in apo- and nucleotide-bound states. *Proc. Natl. Acad. Sci. USA* **2013**, *110*, 9710–9715. [CrossRef] [PubMed]
43. Hassett, A.; Blaettler, W.; Knowles, J.R. Pyruvate kinase: Is the mechanism of phospho transfer associative or dissociative? *Biochemistry* **1982**, *21*, 6335–6340. [CrossRef]
44. Admiraal, S.J.; Herschlag, D. Mapping the transition state for ATP hydrolysis: Implications for enzymatic catalysis. *Chem. Biol.* **1995**, *2*, 729–739. [CrossRef]
45. Kamerlin, S.C.L.; Florián, J.; Warshel, A. Associative versus dissociative mechanisms of phosphate monoester hydrolysis: On the interpretation of activation entropies. *ChemPhysChem* **2008**, *9*, 1767–1773. [CrossRef]
46. Yang, Y.; Cui, Q. The hydrolysis activity of adenosine triphosphate in myosin: A Theoretical analysis of anomeric effects and the nature of the transition state. *J. Phys. Chem. A* **2009**, *113*, 12439–12446. [CrossRef] [PubMed]
47. Prasad, B.R.; Plotnikov, N.V.; Warshel, A. Addressing open questions about phosphate hydrolysis pathways by careful free energy mapping. *J. Phys. Chem. B* **2012**, *117*, 153–163. [CrossRef]
48. Knowles, J.R. Enzyme-catalyzed phosphoryl transfer reactions. *Annu. Rev. Biochem.* **1980**, *49*, 877–919. [CrossRef]
49. Sondek, J.; Lambright, D.G.; Noel, J.P.; Hamm, H.E.; Sigler, P.B. GTPase mechanism of G proteins from the 1.7-Å crystal structure of transducin α-GDP AIF−4. *Nat. Cell Biol.* **1994**, *372*, 276–279. [CrossRef]
50. Wittinghofer, A. Signaling mechanistics: Aluminum fluoride for molecule of the year. *Curr. Biol.* **1997**, *7*, R682–R685. [CrossRef]
51. Graham, D.L.; Lowe, P.N.; Grime, G.W.; Marsh, M.; Rittinger, K.; Smerdon, S.J.; Gamblin, S.J.; Eccleston, J.F. MgF_3- as a transition state analog of phosphoryl transfer. *Chem. Biol.* **2002**, *9*, 375–381. [CrossRef]
52. Davies, D.R.; Hol, W.G. The power of vanadate in crystallographic investigations of phosphoryl transfer enzymes. *FEBS Lett.* **2004**, *577*, 315–321. [CrossRef] [PubMed]
53. Sharma, S.; Davidson, A.L. Vanadate-induced trapping of nucleotides by purified maltose transport complex requires ATP hydrolysis. *J. Bacteriol.* **2000**, *182*, 6570–6576. [CrossRef] [PubMed]
54. Loo, T.W.; Clarke, D.M. Vanadate trapping of nucleotide at the ATP-binding sites of human multidrug resistance P-glycoprotein exposes different residues to the drug-binding site. *Proc. Natl. Acad. Sci. USA* **2002**, *99*, 3511–3516. [CrossRef] [PubMed]

55. Jin, Y.; Richards, N.G.; Waltho, J.P.; Blackburn, G.M. Metal fluorides as analogues for studies on phosphoryl transfer enzymes. *Angew. Chem. Int. Ed.* **2017**, *56*, 4110–4128. [CrossRef] [PubMed]
56. Jin, Y.; Cliff, M.J.; Baxter, N.J.; Dannatt, H.R.W.; Hounslow, A.M.; Bowler, M.W.; Blackburn, G.M.; Waltho, J.P. Charge-balanced metal fluoride complexes for protein kinase a with adenosine diphosphate and substrate peptide SP20. *Angew. Chem. Int. Ed.* **2012**, *51*, 12242–12245. [CrossRef]
57. Baxter, N.J.; Blackburn, G.M.; Marston, J.P.; Hounslow, A.M.; Cliff, M.J.; Bermel, W.; Williams, N.H.; Hollfelder, F.; Wemmer, A.D.E.; Waltho, J.P. Anionic charge is prioritized over geometry in aluminum and magnesium fluoride transition state analogs of phosphoryl transfer enzymes. *J. Am. Chem. Soc.* **2008**, *130*, 3952–3958. [CrossRef]
58. Jin, Y.; Molt, R.W.; Blackburn, G.M. Metal fluorides: Tools for structural and computational analysis of phosphoryl transfer enzymes. *Top Curr Chem (Cham)* **2017**, *375*, 1–31. [CrossRef]
59. Akabayov, S.R.; Akabayov, B. Vanadate in structural biology. *Inorganica Chim. Acta* **2014**, *420*, 16–23. [CrossRef]
60. Chen, J.; Sharma, S.; Quiocho, F.A.; Davidson, A.L. Trapping the transition state of an ATP-binding cassette transporter: Evidence for a concerted mechanism of maltose transport. *Proc. Natl. Acad. Sci. USA* **2001**, *98*, 1525–1530. [CrossRef]
61. Münnich, S.; Taft, M.H.; Manstein, D.J. Crystal structure of human myosin 1c-The Motor in GLUT4 exocytosis: Implications for Ca^{2+} regulation and 14-3-3 binding. *J. Mol. Biol.* **2014**, *426*, 2070–2081. [CrossRef]
62. Chinthalapudi, K.; Heissler, S.M.; Preller, M.; Sellers, J.R.; Manstein, D.J. Mechanistic insights into the active site and allosteric communication pathways in human nonmuscle myosin-2C. *eLife* **2017**, *6*, e32742. [CrossRef] [PubMed]
63. Schmidt, H.; Zalyte, R.; Urnavicius, L.; Carter, A.P. Structure of human cytoplasmic dynein-2 primed for its power stroke. *Nat. Cell Biol.* **2015**, *518*, 435–438. [CrossRef] [PubMed]
64. Oldham, M.L.; Chen, J. Snapshots of the maltose transporter during ATP hydrolysis. *Proc. Natl. Acad. Sci. USA* **2011**, *108*, 15152–15156. [CrossRef] [PubMed]
65. Hofmann, S.; Januliene, D.; Mehdipour, A.R.; Thomas, C.; Stefan, E.; Brüchert, S.; Kuhn, B.T.; Geertsma, E.R.; Hummer, G.; Tampé, R.; et al. Conformation space of a heterodimeric ABC exporter under turnover conditions. *Nat. Cell Biol.* **2019**, *571*, 580–583. [CrossRef]
66. Jiang, J.; Maes, E.G.; Taylor, A.B.; Wang, L.; Hinck, A.P.; Lafer, E.M.; Sousa, R. Structural basis of J Cochaperone binding and regulation of Hsp70. *Mol. Cell* **2007**, *28*, 422–433. [CrossRef]
67. Luo, D.; Xu, T.; Watson, R.P.; Scherer-Becker, D.; Sampath, A.; Jahnke, W.; Yeong, S.S.; Wang, C.H.; Lim, S.P.; Strongin, A.; et al. Insights into RNA unwinding and ATP hydrolysis by the flavivirus NS3 protein. *EMBO J.* **2008**, *27*, 3209–3219. [CrossRef]
68. Dumais, M.; Davies, D.R.; Lin, T.; Staker, B.L.; Myler, P.J.; Van Voorhis, W.C. Structure and analysis of nucleoside diphosphate kinase from Borrelia burgdorferi prepared in a transition-state complex with ADP and vanadate moieties. *Acta Crystallogr. F Struct. Biol. Commun.* **2018**, *74*, 373–384. [CrossRef]
69. Chen, C.; Saxena, A.K.; Simcoke, W.N.; Garboczi, D.N.; Pedersen, P.L.; Ko, Y.H. Mitochondrial ATP synthase. Crystal structure of the catalytic F1 unit in a vanadate-induced transition-like state and implications for mechanism. *J Biol Chem* **2006**, *281*, 13777–13783. [CrossRef]
70. Luo, D.; Nakazawa, M.; Yoshida, Y.; Cai, J.; Imai, S. Effects of three different $Ca(_2+)$ pump ATPase inhibitors on evoked contractions in rabbit aorta and activities of $Ca(_2+)$ pump ATPases in porcine aorta. *Gen. Pharmacol. Vasc. Syst.* **2000**, *34*, 211–220. [CrossRef]
71. Drakou, C.E.; Malekkou, A.; Hayes, J.M.; Lederer, C.W.; Leonidas, D.; Oikonomakos, N.G.; Lamond, A.I.; Santama, N.; Zographos, S. hCINAP is an atypical mammalian nuclear adenylate kinase with an ATPase motif: Structural and functional studies. *Proteins: Struct. Funct. Bioinform.* **2011**, *80*, 206–220. [CrossRef]
72. He, C.; Chen, J.; Wang, H.; Wan, Y.; Zhou, J.; Dan, Z.; Zeng, Y.; Xu, W.; Zhu, Y.; Huang, W.; et al. Crystal structures of rice hexokinase 6 with a series of substrates shed light on its enzymatic mechanism. *Biochem. Biophys. Res. Commun.* **2019**, *515*, 614–620. [CrossRef] [PubMed]
73. Ho, M.-C.; Shi, W.; Rinaldo-Matthis, A.; Tyler, P.C.; Evans, G.B.; Clinch, K.; Almo, S.C.; Schramm, V.L. Four generations of transition-state analogues for human purine nucleoside phosphorylase. *Proc. Natl. Acad. Sci. USA* **2010**, *107*, 4805–4812. [CrossRef] [PubMed]
74. Menz, R.; Walker, J.E.; Leslie, A.G. Structure of bovine mitochondrial F1-ATPase with nucleotide bound to all three catalytic sites. *Cell* **2001**, *106*, 331–341. [CrossRef]

75. Dinescu, A.; Bhansali, V.S.; Cundari, T.R.; Luo, J.-L.; Anderson, M.E. Function of conserved residues of human glutathione synthetase. *J. Biol. Chem.* **2004**, *279*, 22412–22421. [CrossRef] [PubMed]
76. Hunter, T. Why nature chose phosphate to modify proteins. *Philos. Trans. R. Soc. B: Biol. Sci.* **2012**, *367*, 2513–2516. [CrossRef]
77. Bagshaw, C. ATP analogues at a glance. *J. Cell Sci.* **2001**, *114*, 459–460.
78. Elphick, L.M.; Lee, S.E.; Gouverneur, V.; Mann, D.J. Using chemical genetics and ATP analogues to dissect protein kinase function. *ACS Chem. Biol.* **2007**, *2*, 299–314. [CrossRef]
79. Wiberg, K. Application of the pople-santry-segal CNDO method to the cyclopropylcarbinyl and cyclobutyl cation and to bicyclobutane. *Tetrahedron* **1968**, *24*, 1083–1096. [CrossRef]
80. Gardiennet, C.; Wiegand, T.; Bazin, A.; Cadalbert, R.; Kunert, B.; Lacabanne, D.; Gutsche, I.; Terradot, L.; Meier, B.H.; Böckmann, A. Solid-state NMR chemical-shift perturbations indicate domain reorientation of the DnaG primase in the primosome of Helicobacter pylori. *J. Biomol. NMR* **2016**, *64*, 189–195. [CrossRef]
81. Keller, K.; Wiegand, T.; Cadalbert, R.; Meier, B.H.; Böckmann, A.; Jeschke, G.; Yulikov, M. High-spin Metal Centres in Dipolar EPR Spectroscopy. *Chim. Int. J. Chem.* **2018**, *72*, 216–220. [CrossRef]
82. Wiegand, T.; Cadalbert, R.; Lacabanne, D.; Timmins, J.; Terradot, L.; Böckmann, A.; Meier, B.H. The conformational changes coupling ATP hydrolysis and translocation in a bacterial DnaB helicase. *Nat. Commun.* **2019**, *10*, 1–11. [CrossRef] [PubMed]
83. Wiegand, T.; Cadalbert, R.; von Schroetter, C.; Allain, F.H.T.; Meier, B.H. Segmental isotope labelling and solid-state NMR of a 12 × 59 kDa motor protein: Identification of structural variability. *J. Biomol. NMR* **2018**, *71*, 237–245. [CrossRef] [PubMed]
84. Wiegand, T.; Gardiennet, C.; Cadalbert, R.; Lacabanne, D.; Kunert, B.; Terradot, L.; Böckmann, A.; Meier, B.H. Variability and conservation of structural domains in divide-and-conquer approaches. *J. Biomol. NMR* **2016**, *65*, 79–86. [CrossRef] [PubMed]
85. Wiegand, T.; Gardiennet, C.; Ravotti, F.; Bazin, A.; Kunert, B.; Lacabanne, D.; Cadalbert, R.; Güntert, P.; Terradot, L.; Böckmann, A.; et al. Solid-state NMR sequential assignments of the *N*-terminal domain of HpDnaB helicase. *Biomol. NMR Assign.* **2015**, *10*, 13–23. [CrossRef] [PubMed]
86. Wiegand, T.; Lacabanne, D.; Keller, K.; Cadalbert, R.; Lecoq, L.; Yulikov, M.; Terradot, L.; Jeschke, G.; Meier, B.H.; Böckmann, A. Solid-state NMR and EPR Spectroscopy of Mn^{2+}-substituted ATP-fueled protein engines. *Angew. Chem. Int. Ed.* **2017**, *56*, 3369–3373. [CrossRef]
87. Wiegand, T.; Liao, W.-C.; Ong, T.-C.; Däpp, A.; Cadalbert, R.; Copéret, C.; Böckmann, A.; Meier, B.H. Protein-nucleotide contacts in motor proteins detected by DNP-enhanced solid-state NMR. *J. Biomol. NMR* **2017**, *69*, 157–164. [CrossRef]
88. Wiegand, T.; Schledorn, M.; Malär, A.A.; Cadalbert, R.; Däpp, A.; Terradot, L.; Meier, B.H.; Böckmann, A. Nucleotide binding modes in a motor protein revealed by ^{31}P-and ^{1}H-detected MAS solid-state NMR spectroscopy. *ChemBioChem* **2019**, *21*, 324–330. [CrossRef]
89. Kunert, B.; Gardiennet, C.; Lacabanne, D.; Calles-Garcia, D.; Falson, P.; Jault, J.-M.; Meier, B.H.; Penin, F.; Bã¶ckmann, A.; Böckmann, A. Efficient and stable reconstitution of the ABC transporter BmrA for solid-state NMR studies. *Front. Mol. Biosci.* **2014**, *1*, 5. [CrossRef]
90. Lacabanne, D.; Kunert, B.; Gardiennet, C.; Meier, B.H.; Böckmann, A. Sample preparation for membrane protein structural studies by solid-state NMR. *Adv. Struct. Saf. Stud.* **2017**, *1635*, 345–358. [CrossRef]
91. Lacabanne, D.; Lends, A.; Danis, C.; Kunert, B.; Fogeron, M.-L.; Jirasko, V.; Chuilon, C.; Lecoq, L.; Orelle, C.; Chaptal, V.; et al. Gradient reconstitution of membrane proteins for solid-state NMR studies. *J. Biomol. NMR* **2017**, *69*, 81–91. [CrossRef]
92. Lacabanne, D.; Orelle, C.; Lecoq, L.; Kunert, B.; Chuilon, C.; Wiegand, T.; Ravaud, S.; Jault, J.-M.; Meier, B.H.; Böckmann, A. Flexible-to-rigid transition is central for substrate transport in the ABC transporter BmrA from *Bacillus subtilis*. *Commun. Biol.* **2019**, *2*, 149. [CrossRef]
93. Aravind, L.; Iyer, L.M.; Leipe, D.D.; Koonin, E.V. A novel family of P-loop NTPases with an unusual phyletic distribution and transmembrane segments inserted within the NTPase domain. *Genome Biol.* **2004**, *5*, R30. [CrossRef] [PubMed]
94. Leipe, D.D.; Aravind, L.; Grishin, N.V.; Koonin, E.V. The bacterial replicative helicase DnaB evolved from a RecA duplication. *Genome Res.* **2000**, *10*, 5–16. [PubMed]
95. Leipe, D.D.; Wolf, Y.I.; Koonin, E.V.; Aravind, L. Classification and evolution of P-loop GTPases and related ATPases. *J. Mol. Biol.* **2002**, *317*, 41–72. [CrossRef] [PubMed]

96. Walker, J.; Saraste, M.; Runswick, M.; Gay, N. Distantly related sequences in the alpha- and beta-subunits of ATP synthase, myosin, kinases and other ATP-requiring enzymes and a common nucleotide binding fold. *EMBO J.* **1982**, *1*, 945–951. [CrossRef]
97. Saraste, M.; Sibbald, P.R.; Wittinghofer, A. The P-loop-A common motif in ATP- and GTP-binding proteins. *Trends Biochem. Sci.* **1990**, *15*, 430–434. [CrossRef]
98. Iyer, L.M.; Leipe, D.D.; Koonin, E.V.; Aravind, L. Evolutionary history and higher order classification of AAA + ATPases. *J. Struct. Biol.* **2004**, *146*, 11–31. [CrossRef]
99. Orelle, C.; Dalmas, O.; Gros, P.; Di Pietro, A.; Jault, J.-M. The conserved glutamate residue adjacent to the walker-B motif is the catalytic base for ATP hydrolysis in the ATP-binding cassette transporter BmrA. *J. Biol. Chem.* **2003**, *278*, 47002–47008. [CrossRef]
100. Feniouk, B.A.; Yoshida, M. Regulatory mechanisms of proton-translocating F(O)F (1)-ATP synthase. *Results Probl. in cell Differ.* **2008**, *45*, 279–308. [CrossRef]
101. Wittinghofer, A. Phosphoryl transfer in Ras proteins, conclusive or elusive? *Trends Biochem. Sci.* **2006**, *31*, 20–23. [CrossRef]
102. Wittinghofer, A.; Vetter, I.R. Structure-function relationships of the G Domain, a canonical switch motif. *Annu. Rev. Biochem.* **2011**, *80*, 943–971. [CrossRef] [PubMed]
103. Gerwert, K.; Mann, D.; Kötting, C. Common mechanisms of catalysis in small and heterotrimeric GTPases and their respective GAPs. *Biol. Chem.* **2017**, *398*, 523–533. [CrossRef] [PubMed]
104. Shalaeva, D.N.; Cherepanov, D.A.; Galperin, M.Y.; Golovin, A.V.; Mulkidjanian, A.Y. Evolution of cation binding in the active sites of P-loop nucleoside triphosphatases in relation to the basic catalytic mechanism. *Elife* **2018**, *7*, e37373e. [CrossRef]
105. Wilkens, S. Structure and mechanism of ABC transporters. *F1000Prime Rep.* **2015**, *7*, 14. [CrossRef] [PubMed]
106. Ford, R.C.; Beis, K. Learning the ABCs one at a time: Structure and mechanism of ABC transporters. *Biochem. Soc. Trans.* **2019**, *47*, 23–36. [CrossRef]
107. Ogawa, T.; Saijo, S.; Shimizu, N.; Jiang, X.; Hirokawa, N. Mechanism of catalytic microtubule depolymerization via KIF2-tubulin transitional conformation. *Cell Rep.* **2017**, *20*, 2626–2638. [CrossRef]
108. Steinfels, E.; Orelle, C.; Fantino, J.-R.; Dalmas, O.; Rigaud, J.-L.; Denizot, F.; Di Pietro, A.; Jault, J.-M. Characterization of YvcC (BmrA), a multidrug ABC transporter constitutively expressed in *Bacillus subtilis*. *Biochemistry* **2004**, *43*, 7491–7502. [CrossRef]
109. Tombline, G.; Senior, A.E. The occluded nucleotide conformation of P-glycoprotein. *J. Bioenerg. Biomembr.* **2005**, *37*, 497–500. [CrossRef]
110. Van Veen, H.W.; Margolles, A.; Müller, M.; Higgins, C.F.; Konings, W.N. The homodimeric ATP-binding cassette transporter LmrA mediates multidrug transport by an alternating two-site (two-cylinder engine) mechanism. *EMBO J.* **2000**, *19*, 2503–2514. [CrossRef]
111. Orelle, C.; Jault, J.-M. Structures and transport mechanisms of the ABC efflux pumps. In *Efflux-Mediated Antimicrobial Resistance in Bacteria*; Springer Science and Business Media LLC: Basel, Switzerland, 2016; pp. 73–98.
112. Collauto, A.; Mishra, S.; Litvinov, A.; Mchaourab, H.S.; Goldfarb, D. Direct spectroscopic detection of ATP turnover reveals mechanistic divergence of ABC exporters. *Structure* **2017**, *25*, 1264–1274.e3. [CrossRef]
113. Soni, R.K.; Mehra, P.; Choudhury, N.R.; Mukhopadhyay, G.; Dhar, S.K. Functional characterization of Helicobacter pylori DnaB helicase. *Nucleic Acids Research* **2003**, *31*, 6828–6840. [CrossRef] [PubMed]
114. Lapina, O.; Shubin, A.; Khabibulin, D.; Terskikh, V.V.; Bodart, P.; Amoureux, J.-P. Solid-state NMR for characterization of vanadium-containing systems. *Catal. Today* **2003**, *78*, 91–104. [CrossRef]
115. Fernandez, C.; Bodart, P.; Amoureux, J.-P. Determination of 51V quadrupole and chemical shift tensor orientations in V2O5 by analysis of magic-angle spinning nuclear magnetic resonance spectra. *Solid State Nucl. Magn. Reson.* **1994**, *3*, 79–91. [CrossRef]
116. Rehder, D.; Polenova, T.; Buhl, M. Vanadium-51 NMR. In *Annual Reports on NMR Spectroscopy*; Elsevier BV: Amsterdam, The Netherlands, 2007; Volume 62, pp. 49–114.
117. Aureliano, M.; Tiago, T.; Gândara, R.M.; Sousa, A.; Moderno, A.; Kaliva, M.; Salifoglou, A.; Duarte, R.O.; Moura, J.J.G. Interactions of vanadium(V)-citrate complexes with the sarcoplasmic reticulum calcium pump. *J. Inorg. Biochem.* **2005**, *99*, 2355–2361. [CrossRef]
118. Fenn, A.; Wächtler, M.; Gutmann, T.; Breitzke, H.; Buchholz, A.; Lippold, I.; Plass, W.; Buntkowsky, G. Correlations between 51V solid-state NMR parameters and chemical structure of vanadium (V) complexes as

models for related metalloproteins and catalysts. *Solid State Nucl. Magn. Reson.* **2009**, *36*, 192–201. [CrossRef] [PubMed]

119. Pooransingh-Margolis, N.; Renirie, R.; Hasan, Z.; Wever, R.; Vega, A.J.; Polenova, T. 51V solid-state magic angle spinning NMR spectroscopy of vanadium chloroperoxidase. *J. Am. Chem. Soc.* **2006**, *128*, 5190–5208. [CrossRef]
120. McCann, N.; Wagner, M.; Hasse, H. A thermodynamic model for vanadate in aqueous solution-equilibria and reaction enthalpies. *Dalton Trans.* **2013**, *42*, 2622–2628. [CrossRef] [PubMed]
121. Iannuzzi, M.; Young, T.; Frankel, G.S. Aluminum alloy corrosion inhibition by vanadates. *J. Electrochem. Soc.* **2006**, *153*, B533–B541. [CrossRef]
122. Lewinson, O.; Orelle, C.; Seeger, M.A. Structures of ABC transporters: Handle with care. *FEBS Letters* **2020**. [CrossRef]
123. Müller, B. *ChemEQL Version 3.2*; Swiss Federal Institute of Aquatic Science and Technology: Kastanienbaum, Switzerland, 2015.
124. Bao, H.; Dalal, K.; Cytrynbaum, E.N.; Duong, F. Sequential action of MalE and maltose allows coupling ATP hydrolysis to translocation in the MalFGK2 transporter. *J. Biol. Chem.* **2015**, *290*, 25452–25460. [CrossRef]
125. Mehmood, S.; Domene, C.; Forest, E.; Jault, J.-M. Dynamics of a bacterial multidrug ABC transporter in the inward-and outward-facing conformations. *Proc. Natl. Acad. Sci. USA* **2012**, *109*, 10832–10836. [CrossRef] [PubMed]
126. Singleton, M.R.; Sawaya, M.R.; Ellenberger, T.; Wigley, D.B. Crystal structure of T7 gene 4 ring helicase indicates a mechanism for sequential hydrolysis of nucleotides. *Cell* **2000**, *101*, 589–600. [CrossRef]
127. Bailey, S.; Eliason, W.K.; Steitz, T.A. Structure of hexameric DnaB helicase and its complex with a domain of DnaG primase. *Science* **2007**, *318*, 459–463. [CrossRef] [PubMed]
128. Singleton, M.R.; Dillingham, M.S.; Wigley, D.B. Structure and mechanism of helicases and nucleic acid translocases. *Annu. Rev. Biochem.* **2007**, *76*, 23–50. [CrossRef] [PubMed]
129. Enemark, E.J.; Joshua-Tor, L. On helicases and other motor proteins. *Curr. Opin. Struct. Biol.* **2008**, *18*, 243–257. [CrossRef]
130. Wang, G.; Klein, M.G.; Tokonzaba, E.; Zhang, Y.; Holden, L.G.; Chen, X.S. The structure of a DnaB-family replicative helicase and its interactions with primase. *Nat. Struct. Mol. Biol.* **2007**, *15*, 94–100. [CrossRef]
131. Itsathitphaisarn, O.; Wing, R.A.; Eliason, W.K.; Wang, J.; Steitz, T.A. The hexameric helicase DnaB adopts a nonplanar conformation during translocation. *Cell* **2012**, *151*, 267–277. [CrossRef]
132. Gao, Y.; Cui, Y.; Fox, T.; Lin, S.; Wang, H.; De Val, N.; Zhou, Z.H.; Yang, W. Structures and operating principles of the replisome. *Science* **2019**, *363*, eaav7003. [CrossRef]
133. Gordon, J.A. Use of vanadate as protein-phosphotyrosine phosphatase inhibitor. In *Methods in Enzymology*; Academic Press: San Diego, CA, USA, 1991; Volume 201, pp. 477–482.
134. Gor'kov, P.L.; Witter, R.; Chekmenev, E.Y.; Nozirov, F.; Fu, R.; Brey, W.W. Low-E probe for (19)F-(1)H NMR of dilute biological solids. *J. Magn. Reson.* **2007**, *189*, 182–189. [CrossRef]
135. Fogh, R.; Ionides, J.; Ulrich, E.; Boucher, W.; Vranken, W.; Linge, J.P.; Habeck, M.; Rieping, W.; Bhat, T.N.; Westbrook, J.; et al. The CCPN project: An interim report on a data model for the NMR community. *Nat. Struct. Biol.* **2002**, *9*, 416–418. [CrossRef]
136. Stevens, T.J.; Fogh, R.H.; Boucher, W.; Higman, V.A.; Eisenmenger, F.; Bardiaux, B.; van Rossum, B.J.; Oschkinat, H.; Laue, E.D. A software framework for analysing solid-state MAS NMR data. *J. Biomol. NMR* **2011**, *51*, 437–447. [CrossRef] [PubMed]
137. Vranken, W.F.; Boucher, W.; Stevens, T.J.; Fogh, R.H.; Pajon, A.; Llinas, M.; Ulrich, E.L.; Markley, J.L.; Ionides, J.; Laue, E.D. The CCPN data model for NMR spectroscopy: Development of a software pipeline. *Proteins* **2005**, *59*, 687–696. [CrossRef] [PubMed]

Publisher's Note: MDPI stays neutral with regard to jurisdictional claims in published maps and institutional affiliations.

© 2020 by the authors. Licensee MDPI, Basel, Switzerland. This article is an open access article distributed under the terms and conditions of the Creative Commons Attribution (CC BY) license (http://creativecommons.org/licenses/by/4.0/).

Review

The TRIOBP Isoforms and Their Distinct Roles in Actin Stabilization, Deafness, Mental Illness, and Cancer

Beti Zaharija †, Bobana Samardžija † and Nicholas J. Bradshaw *

Department of Biotechnology, University of Rijeka, 51000 Rijeka, Croatia; beti.zaharija@biotech.uniri.hr (B.Z.); bobana.samardzija@biotech.uniri.hr (B.S.)

* Correspondence: nicholas.b@uniri.hr

† These authors contributed equally to the manuscript.

Received: 16 September 2020; Accepted: 26 October 2020; Published: 27 October 2020

Abstract: The *TRIOBP* (*TRIO* and *F-actin Binding Protein*) gene encodes multiple proteins, which together play crucial roles in modulating the assembly of the actin cytoskeleton. Splicing of the *TRIOBP* gene is complex, with the two most studied TRIOBP protein isoforms sharing no overlapping amino acid sequence with each other. TRIOBP-1 (also known as TARA or TAP68) is a mainly structured protein that is ubiquitously expressed and binds to F-actin, preventing its depolymerization. It has been shown to be important for many processes including in the cell cycle, adhesion junctions, and neuronal differentiation. TRIOBP-1 has been implicated in schizophrenia through the formation of protein aggregates in the brain. In contrast, TRIOBP-4 is an entirely disordered protein with a highly specialized expression pattern. It is known to be crucial for the bundling of actin in the stereocilia of the inner ear, with mutations in it causing severe or profound hearing loss. Both of these isoforms are implicated in cancer. Additional longer isoforms of TRIOBP exist, which overlap with both TRIOBP-1 and 4. These appear to participate in the functions of both shorter isoforms, while also possessing unique functions in the inner ear. In this review, the structures and functions of all of these isoforms are discussed, with a view to understanding how they operate, both alone and in combination, to modulate actin and their consequences for human illness.

Keywords: TRIOBP; cancer; deafness; hearing loss; mental illness; schizophrenia; actin; cytoskeleton; disordered structure; protein aggregation

---

1. Introduction

Actin filaments are one of the key elements of the cytoskeleton, and are vital for processes including cellular motility, neuronal differentiation, and cell–cell junctions. The core of these are composed of filamentous F-actin. These are formed by the polymerization of globular units of G-actin, and fibers can in turn depolymerize back to G-actin again [1]. The correct regulation of this key molecular process therefore impacts upon a wide array of cellular functions, and incorrect regulation is associated with various diseases [2]. Among the regulators of actin discovered in the last few decades are the proteins encoded by the *TRIO* and *F-actin Binding Protein* (*TRIOBP*) locus [3]. The *TRIOBP* gene is subject to complicated alternative splicing (Figure 1a). Multiple long splice variants exist [4,5], of which the longest is *TRIOBP-6*, although the slightly shorter *TRIOBP-5* is more often studied. The majority of published work into TRIOBP proteins, however, has instead focused on the products of two shorter transcripts. Of these, *TRIOBP-1* is transcribed from the 3′ end of the *TRIOBP* gene and encodes a largely structured protein [3] with a ubiquitous expression pattern [4,5]. In contrast, *TRIOBP-4* is transcribed from the 5′ end of the gene and encodes a structurally disordered protein, expressed predominantly in the inner ear and retina [4]. The TRIOBP-1 and TRIOBP-4 proteins share no common amino acid sequence, however, both share most or all of their primary structure with the longer variants (Figure 1b).

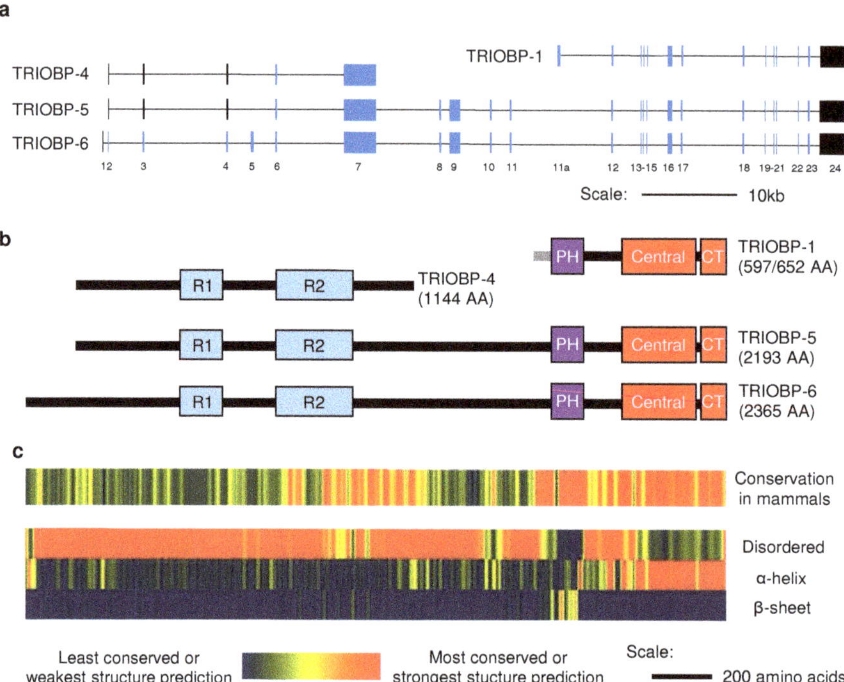

Figure 1. (**a**) Scale schematic of the alternative splicing of *TRIOBP* in humans. Exons (vertical bars) on the four most studied isoforms are shown, with introns represented by horizontal lines. Blue exons are entirely or mainly coding, black exons are entirely or mainly non-coding. Exon numbering is according to Park et al. [6]. (**b**) Scale schematic of the human TRIOBP-1, 4, 5, and 6 proteins with structural regions highlighted: R1, R2: First and second repeat domains, PH: Pleckstrin homology domain, Central: Central coiled coil domain, CT: C-terminal coiled coil domain. The number of amino acids (AA) in humans is also indicated. (**c**) The level of conservation of each section of the TRIOBP-6 amongst mammalian orthologues and predictions of three forms of secondary structure: disordered/unstructured protein, α-helix, and β-sheet. These are displayed as heat maps to scale with the schematic in part (**b**). Conservation determined using AL2CO [7], based on amino acid sequences of TRIOBP-6 (or similar splice variants) from 57 different mammalian genera. These were identified using BLAST (reference sequence human: TRIOBP-6, NP_001034230.1), aligned with CLUSTAL Omega 1.2.4 [8] and the alignment was then manually curated. Secondary structure predictions were made using PSIPDRED 4.0 and DISOPRED3 [9–11] with protein analyzed in three overlapping sections. All results were averaged over an 11 amino acid sliding window for clarity. The N-terminal 61 amino acids of TRIOBP-1 from exon 11a that are not present in TRIOBP-6 were not evaluated here, but were previously predicted to be disordered with comparatively poor conservation [12].

2. TRIOBP-1: A Structured Protein Implicated in Mental Illness and Cancer

2.1. The Structure of TRIOBP-1

TRIOBP-1 consists of two major structured regions: a predicted Pleckstrin homology (PH) domain near the N-terminus and coiled coil domains that make up the C-terminal half of the protein (Figure 1b). These are separated by a linker region of approximately 100 amino acids, referred to as the "mid domain" [13], which is predicted to be intrinsically disordered. Finally, TRIOBP-1 has an optionally translated disordered region at its extreme N-terminus, which is targeted to the nucleus of the cell [12]. This results from the existence of two different potential start codons, 59 amino acids apart from each other, and means that full length TRIOBP-1 can be either 593 or 652 amino acids in length [12]. The 593 amino acid long version of TRIOBP-1 was the first TRIOBP protein to be described, under the name TARA, for TRIO Associated Repeat on Actin [3] (also referred to as TAP68 [14]). This appears to be the more abundant species in many cell culture systems. The 652 amino acid long version may, however, be the principle TRIOBP-1 species in the human heart [15].

The presence of a PH domain near the N-terminus is strongly predicted [3], and it was confirmed that this region of TRIOBP-1 forms a compact folded domain [12]. Its structure has never been studied experimentally, but based on homology with other proteins, it seems to be a fairly typical PH domain with two extended unstructured loops sticking out of it (Figure 2). These loops consist predominantly of polar and charged amino acids. The second, and larger, of these loops is highly conserved in mammals (Figure 1c). The function of the PH domain is currently unknown, however, it likely acts as a protein–protein interaction domain. No interaction of TRIOBP-1 with phosphoinositides has been published, although this cannot be formally discounted.

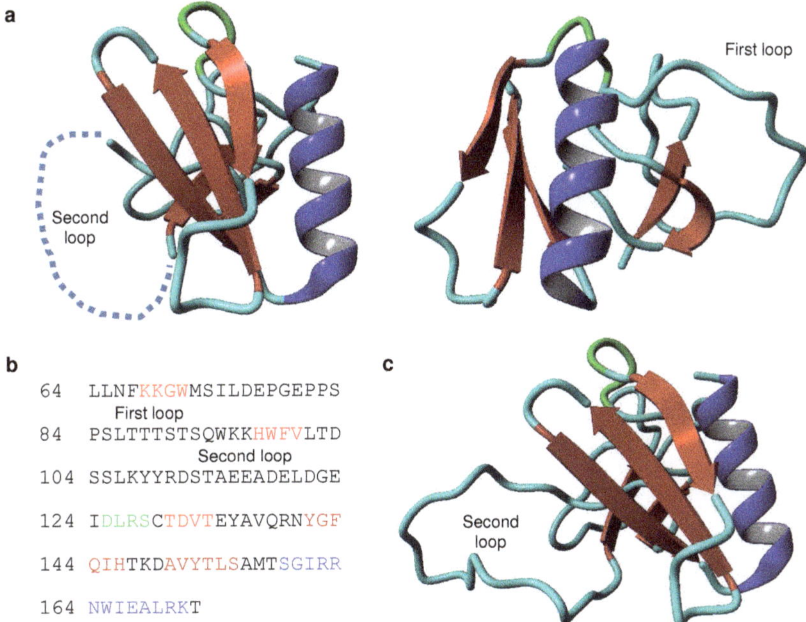

Figure 2. A structural homology model of the PH domain of TRIOBP-1 (amino acids 64–172 of 652 amino acid TRIOBP-1). (**a**) Images of the model, with the first loop region displayed. (**b**) Amino acid sequence of the PH domain, with the first and second loop regions indicated. Coloring corresponds to the secondary structures seen in the molecular images. (**c**) Image of the model including a low quality prediction of the strength of the second loop region. Model generated using MODELLER 9.20 [16], based principally on the structure of the PH domain of DAPP1 (PDB ID: 1FAO), which includes sequence analogous to the first loop region. Shorter sections including the second loop were modeled with additional templates (PDB ID: 2DYN, 2D9Y, 3GOC, and 5YUG). Alignments were generated using CLUSTAL Omega 1.2.4 [8], and then optimized manually. Of the 20 models generated, the one with the lowest objective function score was visualized using YASARA 18.4.24 [17].

The C-terminal half of TRIOBP-1 is highly structured, having long been predicted to consist of coiled-coil (CC) domains (Figure 3a) [3]. Recent predictions suggest there to be approximately six CCs within this section of TRIOBP-1, which separate into two distinct domains: a larger central CC domain and a smaller C-terminal CC domain (Figure 3b) [12,14]. These two domains, and the central CC domain in particular, appear to be involved in many of the functions of TRIOBP-1 in the cell (Figure 3c). While the C-terminal CC domain is monomeric when expressed in isolation, the central CC domain forms an elongated hexamer, seemingly through distinct homodimeric and homotrimeric interactions (Figure 3d) [12]. The central CC domain is therefore responsible for the known oligomerization [22] of the full length TRIOBP-1 protein.

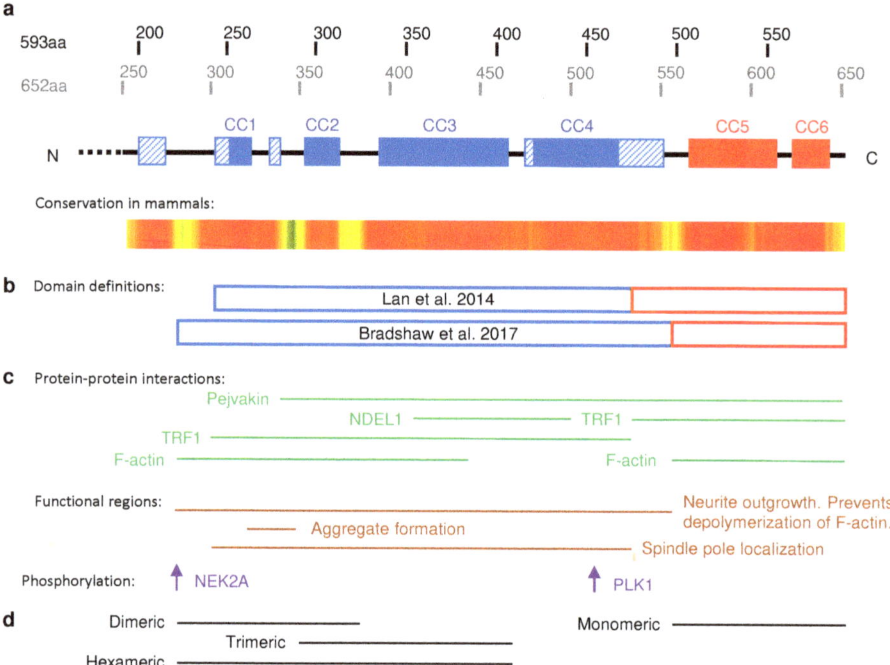

Figure 3. The structure of the coiled-coil regions of TRIOBP-1. All parts of this figure are to scale with each other. (**a**) Locations of predicted coiled-coils (CC). Solid filled boxes represent high confidence predictions, striped boxes represent lower confidence predictions, derived from PSIRPED [9]. CCs are colored based on their predicted inclusion in the central CC domain (blue) or C-terminal CC domain (red). Amino acid numbering from both the 593 amino acid and 652 amino acid TRIOBP-1 proteins are shown. Labeling of CCs is based on Bradshaw et al. [12] and differs from the numbering used by Katsuno et al. [18], who do not count the putative coiled-coil labeled here as CC1 in their numbering. Level of amino acid conservation is displayed using the same calculation and heat map as in Figure 1c. (**b**) Locations of constructs representing the central and C-terminal CC domains from two publications [12,14]. (**c**) Locations of regions of TRIOBP-1 involved in protein–protein interactions and functions [3,12,14,19,20]. Note that some proteins bind more than one region of TRIOBP-1. The only proteins so far reported to bind to TRIOBP-1 outside of these CC regions is TRIO, which binds to the mid domain between the central CC region and the PH domain [13]. The locations of two known phosphorylated residues and their associated kinases are also shown [14,21]. (**d**) Locations of fragments of TRIOBP-1 and the oligomeric states they adopt when expressed in isolation in vitro [12].

In addition to the main TRIOBP-1 species, which are approximately 70 kDa, smaller species have also been detected by western blotting, ranging in size from 45–60 kDa [15,23]. Based on the specificity of the antibodies used, these would be expected to represent the C-terminal 400–540 amino acids of TRIOBP-1, that is, the coiled-coil domains and variable amounts of the unstructured linker region, but not a complete PH domain. An additional splice variant, *TRIOBP-2*, has also been sequenced (annotated in genome assembly hg38), which encode the N-terminal sections of TRIOBP-1 including the PH domain and parts of the central coiled-coil domain. However, to date, this has not been thoroughly characterized.

2.2. TRIOBP-1 as a Regulator of Actin Polymerization

Upon its initial discovery, TRIOBP-1 was noted to adopt a filamentous expression pattern, appearing at 350 nm periodic intervals along the length of actin filaments [3]. Direct interaction between TRIOBP-1 and actin could be demonstrated in vitro, strongly indicating TRIOBP-1 to be an actin-associated protein [3]. Furthermore, TRIOBP-1 co-localizes with, although seemingly does not bind to, two other actin-associated proteins, actinin and myosin II [3]. Knockdown of TRIOBP-1 by siRNA has repeatedly been shown to lower the expression of filamentous F-actin in cell systems [24–26], while over-expression of TRIOBP-1 in cell lines leads to a "cell spreading" phenotype, resulting from excessive F-actin formation [3]. Notably, the central CC domain of TRIOBP-1 is capable of interacting with F-actin and blocking its depolymerization into G-actin [12]. One of the principle cellular functions of TRIOBP-1 therefore appears to be maintaining the existence of F-actin fibers.

In wound healing assays, performed in neuroblastoma cells, overexpression of TRIOBP-1 was seen to increase the rate of cellular migration [19]. This effect was cumulative with that of overexpressing NDEL1 (Nuclear Distribution Element-Like 1, also known as Nudel) [19]. NDEL1 is a key neurodevelopmental protein with links to mental illness, which is more commonly associated with the microtubule cytoskeleton [27]. Nevertheless, NDEL1 directly interacted with TRIOBP-1, binding to the central CC region at approximately the fourth coiled coil, and appeared to work co-operatively with TRIOBP-1 to enhance levels of F-actin [19]. Furthermore, in neuronal systems, TRIOBP-1 appears to recruit two key kinases to NDEL1 [28]. The ensuing phosphorylation events lead to increased F-actin formation, neurite outgrowth, and dendritic arborization [28]. TRIOBP-1 and NDEL1 therefore appear to act synergistically in cell migration and neuronal differentiation.

Another important role of actin is in relation to the receptors that modulate adhesion between the cell and both its extracellular matrix and other cells. The actin cytoskeleton physically links these and provides the basis of mechanical force within the cell that allows it to interact with external stimuli [29]. TRIOBP-1 has been identified in the focal adhesions that link cells to the extracellular matrix, and its expression there is regulated by myosin II [30], which generates tension, leading to maturation of the focal adhesions. TRIOBP-1 is also found at the adhesion junctions between cells [13]. In adhesion junctions of epithelial cells, expression of the crucial transmembrane protein E-cadherin is regulated by the RhoGEF TRIO. TRIOBP-1 binds to TRIO using its mid domain and prevents this effect, leading to increased E-cadherin expression and increased density of actin filaments [13]. It remains to be clarified whether this role of TRIOBP-1 in modulating actin via TRIO is distinct from its effect on actin depolymerization, which seems to occur through direct binding [3,12].

TRIOBP-1 is also found at the adherens junctions in the heart, where it interacts with JCAD (Junctional Protein Associated with Coronary Artery Disease) [31]. Knockdown of either TRIOBP-1 or JCAD in epithelial cells led to reduced F-actin stress fiber formation [31]. TRIOBP-1 also possesses an additional function in the heart through its interaction with the voltage gated ion channel hERG1 (human Ether-à-go-go-Related Gene 1, also known as KCNH2) [15]. In cardiomyocytes, TRIOBP-1 affects expression of hERG, with direct effects on cardiac rapidity, leading the authors to speculate that TRIOBP-1 may function as a bridge between actin filaments and hERG1 in the membrane, linking excitation of the ion channel to cell mobility [15].

2.3. TRIOBP-1 in the Cell Cycle

TRIOBP-1 is essential for correct mitotic progression, with its knockdown in cells leading to multipolar spindle formation [14]. Similar effects are also observed when expression levels of TRIOBP-1 expression were increased, through knockdown of ubiquitin ligase HECTD3 [32]. This suggests that regulation of TRIOBP-1 expression is of significant importance. The most likely mechanism by which TRIOBP-1 affects mitotic progression is through its interaction with TRF1 (Telomere Repeat Factor 1 [22,33]). TRF1 is found at the telomeres of cells, and is involved in both telomere stability and cell cycle regulation. Notably, the localization of TRF1 during mitosis is dependent on that of TRIOBP-1 [14]. The localization of TRIOBP-1 during the cell cycle is itself regulated by two kinases, with PLK1 in particular being required for both its localization in prophase and metaphase, and also for its interaction with TRF1 [14,21]. Strikingly, mutation of the threonine in TRIOBP-1 that is phosphorylated by PLK1 leads to mitotic arrest in prometaphase [21]. Specifically, the chromosomes fail to segregate, highlighting the importance of TRIOBP-1 in this process. While there is some evidence that actin plays a role in mitosis, it remains to be determined whether the function of TRIOBP-1 in mitosis is directly related to its F-actin stabilization effect.

2.4. TRIOBP-1 in Mental Illness

It has recently been suggested that chronic mental illnesses such as schizophrenia, bipolar disorder, and major depression may be caused in part by the accumulation of aggregates of specific proteins in the brains of patients [34,35], in partial analogy to similar insoluble protein deposits in neurodegenerative conditions. In order to detect such proteins, the total insoluble (and aggregated) protein fraction was isolated from the brains of patients with schizophrenia, and used to inoculate a mouse. Monoclonal antibodies were generated from this animal and screened for the ability to specifically recognize the insoluble protein fraction of the patient brain compared to an equivalent preparation from the control brain tissue [36]. One such antibody was found to recognize TRIOBP-1, suggesting it to be present in an aggregated state in the brains of at least a subgroup of patients [23].

TRIOBP-1, but not TRIOBP-4, formed insoluble aggregates when expressed in mammalian cell culture or rodent primary neurons [23]. Subsequent mapping studies determined the central CC region of TRIOBP-1 to be the basis of its aggregation propensity [12]. The critical region for aggregation has now been mapped to a 25 amino acid long loop containing multiple charged amino acids [12]. In addition to 70 kDa full length TRIOBP-1, aggregation is also seen of shorter (45–60 kDa) protein species, representing coiled-coil regions of TRIOBP-1, but without the PH domain [23]. The consequences of TRIOBP-1 aggregation are still being determined, although effects have been seen on neurite outgrowth in cell culture [23]. Structures resembling aggregates have also been seen when TRIOBP-1 is expressed in other tissues [15,20]. Regulation of *TRIOBP* expression and folding may therefore be important for mental health. One such regulatory factor is already known, the ubiquitin ligase HECTD3, which leads to degradation of TRIOBP-1 [32].

Unlike several other proteins that are implicated as aggregating in mental illness [35], TRIOBP-1 is not encoded for by a known genetic risk factor for major mental illness. This may be because the functions of TRIOBP-1 in actin regulation are fundamental to life, and as such, mutations in its (highly conserved) sequence would lead to outcomes more detrimental than those seen in mental illness. Supporting evidence comes from a handful of studies, however. First, in two screens of samples from separate brain banks, levels of *TRIOBP* transcripts were seen to be subtly, but significantly higher in schizophrenia patients than in the controls [37]. Second, a polymorphism in the *NDE1/miR-484* locus, previously associated with schizophrenia in the Finnish population [38], was found to affect the expression of *TRIOBP* transcripts [39,40]. *MiR-484* expression was subsequently shown to lead to increased levels of the TRIOBP-1 protein [40]. Finally, a consanguineous family has been reported who suffer from schizophrenia, epilepsy, and hearing, with linkage to chromosome 22q12.3 q13.3 [41]. It is therefore possible, although not yet verified, that rare variants in *TRIOBP* could be responsible for these phenotypes.

2.5. TRIOBP-1 in Cancer

TRIOBP-1 has been identified in cell lines from a range of different cancers including lung carcinoma [42], glioblastoma [43], esophageal [44], pancreatic [45], prostate, lung, and breast cancer [46].

Studies with glioblastoma showed that TRIOBP (from the specificity of the antibody used: TRIOBP-1, 5, and/or 6) was more abundant in the tumors themselves than in the surrounding tissues [43]. Analysis of existing datasets suggested that it was also over-expressed in classical, mesenchymal, neuronal, and pro-neuronal glioblastoma [43]. Further analysis in glioblastoma cell lines demonstrated that knockdown of TRIOBP-1 (and TRIOBP-5/6) reduced the proliferation and migration of these cells [43].

Another interesting line of research comes from study of the microRNA *miR-3178*, a target of the cancer-suppressing protein SP1, which was shown to have anti-metastatic properties in a mouse model [46]. *MiR-3178* inhibits the expression of *TRIOBP-1* and *5*, as measured at both the transcript and protein levels, through binding to their untranslated 3′ exon. Crucially, while *miR-3178* inhibits the migration and integration of metastatic cells, this effect can be reversed by expression of TRIOBP-1 [46]. Together, there is therefore evidence that TRIOBP-1 affects tumor metastasis through its known roles in actin modulation as well as potentially through its roles in the cell cycle.

2.6. TRIOBP-1 in Other Diseases

While TRIOBP-1 is not generally considered to have a significant role in hearing loss, unlike TRIOBP-4, it should be noted that TRIOBP-1 is expressed in the stereocilia of the inner ear [20]. Here, it binds to the hearing-related protein Pejvakin, with over-expression of TRIOBP-1 causing Pejvakin to form aggregates [20]. There have also been reported missense mutations within *TRIOBP-1* in patients with hearing loss [47,48] (Table 1), however, these would also affect longer splice variants of *TRIOBP*.

3. TRIOBP-4: A Disordered Protein Implicated in Deafness

3.1. The Structure of TRIOBP-4

Human TRIOBP-4 is a 1144 amino acid long protein, which is predicted to be almost entirely disordered, possessing no fixed secondary or tertiary structure [49]. While TRIOBP-4 therefore possesses no folded domains, it has been observed to contain two repeat regions [4], referred to as R1 and R2 (Figure 4a) [49]. The R1 repeat region lies near the center of the protein. In humans, it has a high isoelectic point of 11.7 and consists of six repeats (with slight variations) of the sequence SSPNRTTQRDNPRTPCAQRDNPRA [49]. R2, in humans, consists of five repeats of the sequence VCIGHRDAPRASSPPR (with slight variations), with 30–40 amino acids between each repeat. It lies in the C-terminal half of TRIOBP-4 and has a much lower isoelectric point of 5.4 [49].

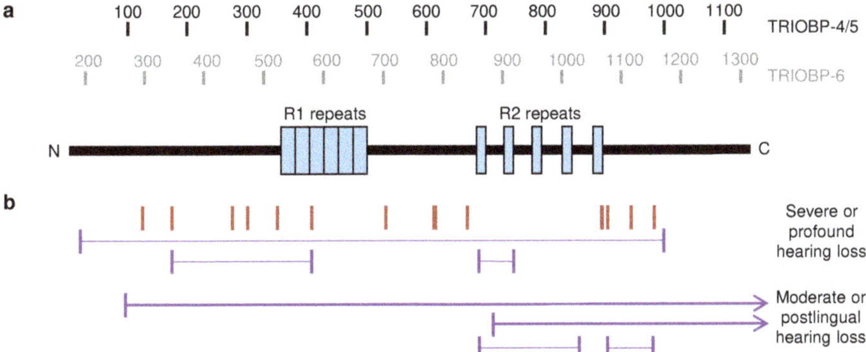

Figure 4. (**a**) The location of the repeats that make up the R1 and R2 regions of TRIOBP-4, with ¸amino acid numbering of both TRIOBP-4/5 and TRIOBP-6. (**b**) The location of frameshift and nonsense mutations from patients with hearing loss. Red bars indicate homozygous mutations, while purple bars joined by dotted lines indicate compound heterozygous mutations. Arrowheads indicate that the other heterozygous mutations lie in a region of *TRIOBP-5/6* that is 3′ of the *TRIOBP-4* open reading frame. Full details of these are in Table 1. All elements of this figure are shown to scale.

3.2. *TRIOBP-4 as an Actin Bundling Protein in the Inner Ear*

TRIOBP-4 binds directly to F-actin, principally through its R1 repeat domain, and is found along the length of filaments [49,50]. R2 shows a considerably weaker, probably hydrophobic interaction to actin [49]. In vitro assays showed TRIOBP-4 molecules to bind actin subunits at a ratio of 1:3–1:4, and that addition of TRIOBP-4 caused actin filaments to become organized into densely packed bundles, which resembled the hair cell rootlets of the inner ear [50].

TRIOBP-4 has a very specialized expression pattern and is highly expressed in the hair cells of the inner ear [50]. These cells perform mechano-electrical transduction from the fluid motion that is induced by sound into neuronal signaling. This occurs through stereocilia, organelles containing an F-actin core, which are anchored into the cuticular plate of hair cells by rootlets, and which pivot in response to fluid motion. TRIOBP-4 is found in the upper sections of these rootlets as well as along the length of the stereocilia themselves in their actin cores [18,50]. TRIOBP-4 is also found in Deiters' cells, which support the hair cells [18]. Normally, stereocilia rootlets would form in the first 16 postnatal days of mice, however, they were not seen to form at all in mice lacking the ability to produce either TRIOBP-4 or the longer isoforms (homozygous deletion of mouse exon 6, equivalent to human exon 7, Figure 1a) [50]. While stereocilia still form, they are considerably less rigid than those of wild type animals, often being found pointing in the wrong direction, and progressively degenerate [50]. These stereocilia still react to mechano-electrical transduction, but no longer have the rigidity required to remain upright and pivot in response to sound [50]. Seemingly as a result of this, these mice are profoundly deaf [50]. It therefore appears that the actin bundling function of TRIOBP-4 is crucial for the formation of the stereocilia rootlet and forming them into tight actin bundles, which are required for their stability and rigidity [18,50].

While the known role of TRIOBP-4 as an acting bundling protein has been largely restricted, so far, to studies in the inner ear, a more general role for it is suggested by two lines of evidence. First, while *TRIOBP-4* does show a very specialized expression pattern, it is not unique to the inner ear, with its transcripts notably being highly expressed in the retina [4]. Second, knockdown of TRIOBP-4 in a pancreatic cancer cell line led to reduced filopodia formation, with TRIOBP-4 seen at actin bundles of these structures [45].

3.3. *TRIOBP-4 in Hearing Loss*

In 2000, details were reported of a Palestinian family with nonsyndromic hereditary deafness, linked to a locus on chromosome 22, which was labeled as DFNB28 [5]. Homozygosity mapping implicated the *TRIOBP* locus, but no mutations were found in *TRIOBP-1*, the only open reading frame of *TRIOBP* known at that time. This directly led to the cloning of the long splice form *TRIOBP-5* and the discovery of a homozygous nonsense mutation within it [5] as well as separate mutations in other families with nonsyndomic deafness [5]. Simultaneously, studies of deafness linked-loci in families from India and Pakistan led to the discovery of a range of other *TRIOBP* mutations as well as cloning of *TRIOBP-4* and *6* [4]. Subsequently, a large number of studies have sequenced the *TRIOBP* gene in families or individuals with severe or profound prelingual hearing loss, revealing a wide range of seemingly pathogenic recessive mutations (Table 1, Figure 4b). These pathogenic mutations tend to be homozygous in patients with deafness, in many instances as a result of consanguinity. Patients have also been found with compound heterozygous expression of two different *TRIOBP* mutations.

The majority of mutations detected to date in patients are either nonsense or frameshift mutations in *TRIOBP-4*, which would lead to the expression of truncated TRIOBP-4 and longer splice variants such as TRIOBP-5 and 6, but with no predicted effect on TRIOBP-1. While many of the mutations lie with the large exon 7 (as in *TRIOBP-6*, Figure 1a), their location in the TRIOBP-4 protein varies considerably (Figure 4b). Many, but not all, of the predicted truncated proteins would still contain the R1 repeat region. Most either lack the R2 region, or would only partially express it. It is likely that these putative truncated proteins would be non-functional and degraded by the proteasome. Deafness in patients with these mutations therefore likely arises through lack of functional TRIOBP-4, which is consistent with the finding that mice lacking TRIOBP-4 (and longer isoforms) are profoundly deaf [50]. An alternative hypothesis would be that loss of the R2 region and/or C-terminal region of TRIOBP-4 would lead to expression of truncated proteins, which could interfere with normal stereocilia function. While the roles of these regions are not well characterized, it is notable that they are among the most highly conserved regions of *TRIOBP-4* in mammals (Figure 1c). In both instances, based on studies in mice [18], it is likely that stereocilia rootlets fail to form in the patients, leading to degeneration of stereocilia and thus hearing loss. Consistent with this, some patients with TRIOBP-4 mutations and hearing loss have had been successfully treated using cochlear implants, which bypass the need for stereocilia [50,51].

While most *TRIOBP* mutations implicated in deafness were found in patients with severe or profound hearing loss detected before speaking, compound heterozygous mutations have been reported in patients with moderate hearing loss or later onset severe hearing loss (Table 1). Notably, many of these patients either possess a mutation that lies 3′ of the *TRIOBP-4* reading frame, affecting longer splice variants only, or else have a mutation near the C-terminus of TRIOBP-4, meaning that the R2 domain would still be intact (Figure 4b). Potentially, the presence of some TRIOBP-4 functionality could therefore explain the milder phenotype, although there are instances of similar C-terminal mutations in patients with severe hearing loss.

Genome wide association studies have also shown that intronic SNP rs58389158 is associated with age related hearing impairment in non-Hispanic white individuals from California [52]. This SNP lies in an intron common to *TRIOBP-4, 5,* and *6*. It is close to, and correlates strongly with, the coding SNP rs5756795, which leads to an F1187I protein variant [52]. The rs58389158 finding was replicated in the UK Biobank [52], in which rs5756795 was also found to be associated with both hearing difficulty and hearing aid use at the genome-wide level [53]. Common sequence variants in *TRIOBP* therefore appear to have an impact on hearing, in addition to rare nonsense and frameshift mutations.

Table 1. Published mutations in *TRIOBP* from individuals and families with hearing loss.

Mutation [1]	Type [2]	Zygosity [3]	Isoforms (Location) [4]	Origin [5] of Proband(s)	Ref (s)
		Severe to Profound Hearing Loss			
p.P191Rfs*50	FS	CHT (p.P1172Cfs*13)	4, 5, 6	South Africa	[54]
p.Q297*	NON	HM	4, 5, 6	India	[4]
p.R347*	NON	HM	4, 5, 6	Palestinian	[5]
		CHT (p.Q581*)		Palestinian	[5]
p.R448*	NON	HM	4, 5, 6	China, Afghan	[51,55]
p.R474*	NON	HM [7]	4, 5, 6	Pakistan [7]	[50]
p.R523*	NON	HM [7]	4, 5, 6	Pakistan [7]	[50]
p.Q581*	NON	HM	4, 5, 6 (R1)	Palestinian	[5]
		CHT (p.R347*)		Palestinian	[5]
		CHT (p.G1019R)		Palestinian	[5]
p.Q740*	NON	HM [7]	4, 5, 6	Pakistan [7]	[50]
p.R785Sfs*50	FS	HM	4, 5, 6	Turkey	[56]
p.R788*	NON	HM	4, 5, 6	Pakistan	[4]
p.R841*	NON	HM	4, 5, 6	Turkey	[54]
p.R861*	NON	CHT (p.R920*)	4, 5, 6 (R2)	China	[57]
p.R920*	NON	CHT (p.R861*)	4, 5, 6 (R2)	China	[57]
p.G1019R	MIS	CHT (p.Q581*)	4, 5, 6 (R2)	Palestinian	[5]
p.I1065V	MIS	CHT (p.R1982H)	4, 5, 6 (R2)	China	[48]
p.R1068*	NON	HM	4, 5, 6 (R2)	Pakistan, Iran	[4,58]
p.D1069fs*12	FS	HM	4, 5, 6 (R2)	India	[4]
p.R1078Pfs*6	FS	HM	4, 5, 6 (R2)	India	[4]
p.R1117*	NON	HM	4, 5, 6	India	[4]
p.E1156*	NON	HM [7]	4, 5, 6	Pakistan [7]	[50]
p.P1172Cfs*13	FS	CHT (p.R191Rfs*50)	4, 5, 6	South Africa	[54]
p.R1982H	MIS	CHT (p.I1065V)	1, 5, 6	China	[48]
p.S2121L	MIS	HM	1, 5, 6 (Centr.)	Iran	[47]

Table 1. *Cont.*

Mutation [1]	Type [2]	Zygosity [3]	Isoforms (Location) [4]	Origin [5] of Proband(s)	Ref (s)
		Moderate or Postlingual Hearing Loss [6]			
p.Q268Lfs*432	FS	CHT (p.G1672*)	4, 5, 6	Poland	[59]
p.R861*	NON	CHT(p.P1030Lfs*183)	4, 5, 6 (R2)	USA, Iran	[54,60]
pR885Afs*120	FS	CHT (p.G1672*)	4, 5, 6	Netherlands	[61]
p.P1030Lfs*183	FS	CHT (p.R861*)	4, 5, 6 (R2)	USA, Iran	[54,60]
p.R1078Pfs*6	FS	CHT (p.L1154Afs*29)	4, 5, 6 (R2)	Netherlands	[61]
p.M1151V	MIS	CHT (p.P1396R)	4, 5, 6	China	[57]
p.L1154Afs*29	FS	CHT (R1078Pfs*6)	4, 5, 6	Netherlands	[61]
p.P1396R	MIS	CHT (p.M1151V)	5, 6	China	[57]
p.G1672*	NON	CHT (p.Q268Lfs*432) CHT (pR885Afs*120)	5, 6	Poland Netherlands	[59] [61]

[1] Amino acid number of human TRIOBP-6, NM_001039141.2 (for amino acid locations in TRIOBP-4 or 5, subtract 172 amino acids). [2] FS: Frameshift, NON: Nonsense, MIS: Missense. [3] HM: Homozygous, HT: Heterozygous, CHT: Compound heterozygous with the mutation indicated. [4] Numbers refer to isoforms, e.g., "5,6" indicates the mutation lies within TRIOBP-5 and TRIOBP-6. Key to locations: R1, R2: first and second repeats of TRIOBP-4, Centr.: Central coiled-coil domain of TRIOBP-1. [5] Country name, unless a more specific ethnicity was stated in the original paper. [6] Or prelingual, but severity not stated. [7] Personal communication of additional details by Prof. Shin-ichiro Kitajiri.

3.4. *TRIOBP-4 in Cancer*

In contrast to the general expression of *TRIOBP-1* and, to a lesser extent, longer *TRIOBP* splice variants in cancer [46], *TRIOBP-4* transcripts were specifically seen to be expressed in a cancer cell line, HPAC [45]. Subsequent analysis found TRIOBP-4 to be upregulated in human pancreatic and, to an extent, breast cancer tissue, but not in prostrate or lung cancer tissue. Knockdown of TRIOBP-4 (and longer variants) in several pancreatic cancer cells lines led to a reduction in cell proliferation [45]. Therefore, it appears that TRIOBP-4 may play a specialized role in pancreatic cancer.

Additionally, a T195I missense mutation in TRIOBP-4 was among several mutations detected in a family with seemingly genetic, gastric, and rectal cancer [62]. Subsequent exome sequencing of additional families with these diseases led to the identification of several additional missense mutations in patients, two of which, A660V and S826L, segregated with disease in families [62]. These would also effect longer *TRIOBP* splice forms, and it remains to be confirmed whether they are pathogenic.

3.5. *TRIOBP-4 Mutations in Other Illnesses*

In addition to hearing loss, rare missense mutations in *TRIOBP-4* (and longer splice variants) have also been detected in a patient with multiple sclerosis (A322S mutation) [63] and in a patient with developmental delay, visual impairment, muscle weakness, hypotonia, clinodactyly, and mild hearing impairment (R1078C mutation) [64].

4. Potential Significance of the Longer Splice Variants TRIOBP-5 and TRIOBP-6

4.1. *The Structure of the Long Splice Variants*

While TRIOBP-1 and 4 share no common amino acid sequence with each other, they do with the longer TRIOBP splice variants. These contain the entire coding sequence of TRIOBP-1 and 4, except for the optionally translated extreme N-terminus of TRIOBP-1 (Figure 1). In human, the longest isoform, TRIOBP-6, is derived from a 24 exon long transcript, of which all but exons 1 and 24 are coding. This leads to a 2365 amino acid peptide, which forms the basis of numbering for all *TRIOBP* putative pathological mutations (Table 1). The majority of biological experiments, however, have instead focused on TRIOBP-5 (also called TRIOBP-3 in some earlier articles), which is a 2193 amino acid protein in humans. TRIOBP-5 is also the longest established isoform in mice. It derives from a transcript lacking exons 1 and 5, and whose open reading frame only begins on exon 6. This is because the Kozak sequence used for *TRIOBP-6* is encoded across exons 1 and 2, and is therefore incomplete in *TRIOBP-5* transcripts. As a result, TRIOBP-5 begins its reading frame at the same point as TRIOBP-4, but is otherwise identical in the amino acid sequence to TRIOBP-6.

TRIOBP-6 has some isoform-specific amino acid sequence at its N-terminus, plus both it and TRIOBP-5 share some of the coding sequence, which lies between the coding exons of *TRIOBP-1* and *5* (Figure 1a,b). These additional sequences are predicted to be predominantly unstructured, with the exception of a possible short stretch of α-helix in the isoform specific N-terminus of TRIOBP-6, and another near the center of the long isoforms (Figure 1c). Neither of these regions show significant sequence similarity to known protein structures. The long variants are therefore predicted to be intrinsically disordered for most of their length, but with the PH domain and coiled-coil domains of TRIOBP-1 at their C-terminal ends. The coiled-coil regions of TRIOBP-5 have been shown to lead to oligomerization, in a similar manner to TRIOBP-1 [12,18]. These proteins also possess multiple actin binding domains, sharing both the R1 repeat of TRIOBP-4 and coiled-coil domains of TRIOBP-1.

4.2. TRIOBP-5 in the Inner Ear and Deafness

While the majority of *TRIOBP* mutations found in patients with profound hearing loss lie within the reading frame of *TRIOBP-4* (Table 1), these would also affect the TRIOBP-5 and 6 proteins. Additionally, patients with moderate and/or progressive hearing loss have been described that possess both a mutation in *TRIOBP-4* and a p.G1672* mutation on the other *TRIOBP* allele, which would affect only the longer splice variants [59,61]. Consistent with this, while mice who lack both TRIOBP-4 and 5 show profound deafness [50], those engineered to express TRIOBP-4, but not 5, instead display a progressive form of deafness [18]. Together, these findings strongly imply that while TRIOBP-4 is essential for prelingual hearing ability, specific loss of the longer splice variants is also required for maintenance of hearing.

TRIOBP-5 is expressed in the same inner ear cell types as TRIOBP-4, with both being found in the stereocilia rootlets [50]. In contrast to TRIOBP-4, however, TRIOBP-5 is predominantly found in the lower parts of the rootlet, below the apical surface [18]. The specific role of TRIOBP-5 in the ear has been studied using various *TRIOBP*-deficient mice. While deletion of both *TRIOBP-1* and *TRIOBP-5* is lethal [50], mice lacking two *TRIOBP-5* specific exons are viable, as are heterozygous mice that can express TRIOBP-1 from one allele and TRIOBP-4 from the other [18]. These *TRIOBP-5*-deficient mice still express TRIOBP-4 in the stereocilia and retain residual hearing for at least 4–8 weeks [18]. This contrasts with the profound deafness of *TRIOBP-4*-deficient mice [50], indicating a unique role of TRIOBP-5, which is also essential for hearing. Detailed analysis of the *TRIOBP-5* knockout mice revealed that stereocilia appear to form normally, but then become increasingly disorganized over time. Specifically, some fuse together or are missing, while others appear thin and fragmented compared to those of wild-type animals [18]. The stereocilia are also seen to be less stiff, and to rotate less freely than wild-type ones [18]. Therefore, while TRIOBP-4 appears to be required to form stereocilia rootlets and elongate them into tight actin bundles (a role indispensable for hearing), TRIOBP-5 instead plays a separate, later role in widening and giving structure to the stereocilia (loss of which leads to progressive hearing loss) [18,50].

Interestingly, this role of TRIOBP-5 in modeling of the rootlets is retained in mice that express incomplete TRIOBP-5 (terminating after the PH domain), however, they do not gain the usual resilience [18]. Such mice may therefore reflect patients with mutations like p.G1672*, who have moderate progressive hearing loss, but not the profound hearing loss associated with mutations in *TRIOBP-4* [18,59,61]. This also implies that the role of TRIOBP-5 in the stereocilia is likely to involve its coiled-coil domains. One possible explanation for this is that these domains interact with Pejkavin, a protein also required for bundling of actin in the inner ear and for hearing, which was seen to interact with this region of TRIOBP-1 [20,65].

It is likely that TRIOBP-6 could also be involved in this process, but this remains untested due to lack of a known murine TRIOBP-6 species.

4.3. Potential Significance for the Long Splice Variants in Other Processes and Diseases

TRIOBP-5 and/or 6 are known to be expressed in the brain alongside TRIOBP-1 [4,5], and TRIOBP-5 exogenously expressed in neurons forms aggregates similar to those of TRIOBP-1 [23]. It is therefore possible that aggregation of longer TRIOBP isoforms may play a role in mental illness, but this remains to be investigated.

TRIOBP-5 and/or *6* was also seen to be upregulated in a pancreatic cancer cell line, distinct from another cell line that expressed *TRIOBP-4* in the same study [45]. Curiously, knockdown of *TRIOBP-5/6* in these cells led to reorganization of the actin cytoskeleton and inhibition of filopodia formation [45]. This implies the existence of a more general role for TRIOBP-5/6 in actin dynamics, of potential relevance for cancer. This may occur through its actin binding sites in either repeat region R1 shared with TRIOBP-4, its central coiled coil domain shared with TRIOBP-1, or a combination. One piece of evidence arguing for a TRIOBP-4-like mechanism is that, in a wound healing assay, knockdown of *TRIOBP-5/6* led to reduced cell motility, but this could be rescued through expression of TRIOBP-4 [45]. However, TRIOBP-5/6 was also seen, along with TRIOBP-1, to have its expression inhibited by the metastasis suppressing microRNA *miR-3178*, suggesting that a TRIOBP-1-like role of the longer isoforms also exists, and is of relevance to cancer [46].

5. Conclusions and Unanswered Questions

The *TRIOBP* locus therefore encodes a variety of distinct proteins (Figure 1) with TRIOBP-1 being a structured and ubiquitously expressed protein implicated in mental illness and TRIOBP-4 being a disordered protein with specialized expression pattern essential for hearing. The long isoforms TRIOBP-5 and 6 combine the structures and many of the functions of the shorter isoforms, but with distinct additional roles in the ear, and potentially elsewhere. In spite of this, all the isoforms are linked through their role in stabilizing actin (Table 2, Figure 5).

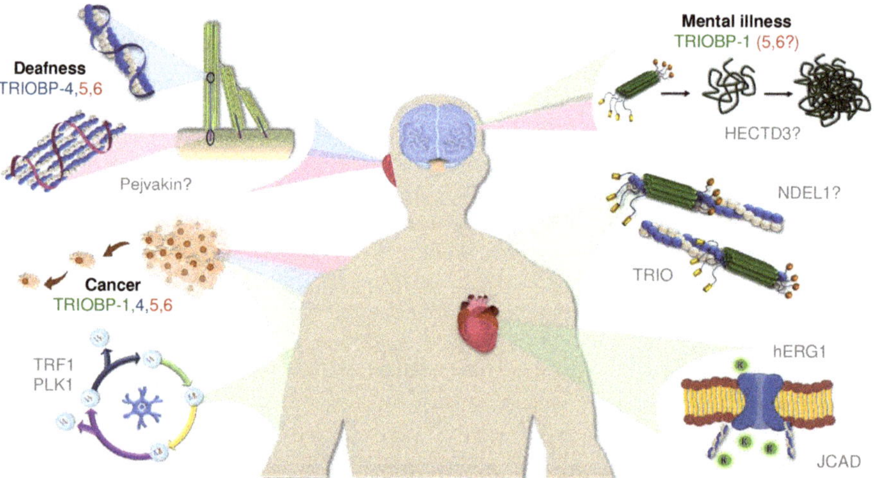

Figure 5. Illustrated representation of the expression and roles of TRIOBP-1 (green), TRIOBP-4 (blue), and TRIOBP-5/6 (red). Clockwise from top: Aggregation of TRIOBP-1 in the brain and mental illness; Role of TRIOBP-1 in F-actin stabilization throughout the body; Linking of actin to ion channel function in the heart by TRIOBP-1; role of TRIOBP-1 in the cell cycle; Importance of all major TRIOBP isoforms in metastasis; Distinct roles of TRIOBP-4 and TRIOBP-5 in the stereocilia of the inner ear and deafness. Protein interaction partners implicated in the various processes are indicated in gray.

Table 2. Normal functions and disease states associated with the TRIOBP isoforms.

Function or Phenotype	TRIOBP-1	TRIOBP-4	TRIOBP-5/6 [1]	Ref.
		Protein Structure		
Principle secondary structure	Helical	Disordered	Disordered	[3,49]
Contains …				
… repeat domains R1 and R1	No	Yes	Yes	[49]
… PH domain	Yes	No	Yes	[3,12]
… coiled-coil domain	Yes	No	Yes	[12,19]
		General function		
Interacts with F-actin	Yes	Yes	Yes	[3,18,50]
Prevents actin depolymerization	Yes	No (?)	?	[12]
Actin bundling activity	No (?)	Yes	Yes (?)	[18,50]
Affects the actin cytoskeleton	Yes	Yes (?)	Yes	[3,45]
Roles in adhesion receptors	Yes	?	?	[13,30]
Implicated in cellular migration	Yes	Yes	Yes	[19,45]
Role in cell cycle progression	Yes	?	?	[21]
		The brain and mental illness		
Expressed in the brain	Yes	No	Yes	[4,5]
Involved in neurite outgrowth	Yes	No	?	[28]
Insoluble (aggregating) in brains of schizophrenia patients	Yes	No (?)	?	[23]
Can aggregate in neurons	Yes	No	Yes	[23]
		Inner ear and deafness		
Expressed in inner ear	Yes	Yes	Yes	[4,5]
Expressed in stereocilia	Yes	Yes [2]	Yes [2]	[18,20,50]
Required in stereocilia for				
rootlet formation	No	Yes	No	[50]
Initial bundling of actin	No	Yes	No	[50]
Sculpting and maintenance	No	No	Yes	[18]
Mouse knockout causes deafness?	(Knockout is lethal)	Yes [3] (profound)	Yes (progressive)	[18,50]
Mutations in human hearing loss	No (?)	Yes [3]	Yes	Table 1
		Cancer		
Upregulated in cancer cells?	Many	Specific	Specific	[43,45]
Potential role in metastasis?	Yes	Yes	Yes	[45,46]
		Role in the heart		
Expressed in the heart	Yes	No	No	[4,5]
Function with hERG	Yes	No	No	[15]

[1] While TRIOBP-5 and 6 likely have at least partially differing roles, no attempt was made to differentiate here due to lack of data. [2] Differences in exact role within the stereocilia. [3] This mouse knockout and these human mutations would also affect TRIOBP-5/6.

The role of TRIOBP-1 in actin dynamics appears to be to bind directly to F-actin [3] and inhibit its depolymerization [12]. This appears to be a general function of TRIOBP-1 in many cell types and organs, with specific roles including the linking of adhesion receptors at the cell surface [13,30], neuronal outgrowth [28], cell migration [19], and signal transduction to mechanical force in the heart [15]. An additional, or possibly alternative, mechanism is that TRIOBP-1 can affect actin in certain circumstances through inhibition of TRIO [13]. TRIOBP-5 and/or 6, which share all the functional domains of TRIOBP-1, are also present in the brain and so can be presumed to participate in many functions there, but not in other TRIOBP-1 expressing tissues such as the heart or liver [4,5]. In contrast, TRIOBP-4 is an actin-bundling protein that likely uses its lack of rigid structure to wrap around actin fibers in the stereocilia, binding using its R1 repeat motif, and bundle them together during rootlet formation and early stereocilia development [45,50]. TRIOBP-5 (and possibly TRIOBP-6) then has a similar, but distinct role, in which it further "sculpts" and maintains the actin core of the stereocilia [18]. TRIOBP-1 is also present in the inner ear and stereocilia [4,5,20], but is seemingly not required for stereocilia formation or hearing [18].

An area where the *TRIOBP* isoforms show greater overlap is in the pathology of cancer. *TRIOBP-1* is expressed in many cancer cells and tissues, while *TRIOBP-4* and *5/6* are more specialized, in partial analogy to their normal expression patterns [42–46], although not necessarily in the same tissue types. Notably, all are implicated in metastasis. Specifically, suppression of TRIOBP-1 and 5/6 expression appears to be a means through which *miR-3178* suppresses metastasis [46], while TRIOBP-4 and 5/6 are implicated the in cell motility of pancreatic cancer cells [45].

While much has therefore been uncovered regarding these proteins, many questions remain. The exact mechanism through which TRIOBP-1 modulates actin is still only partially understood, and its roles in various organs and cell types need further analysis. While both its potential roles in mental illness and cancer are tantalizing, the relationship between it and specific mental illnesses and cancer subtypes needs to be established in larger patient samples. The role of TRIOBP-4 in the stereocilia is understood in more detail, and the role of frameshift and nonsense mutations of *TRIOBP-4* in hearing loss is well established. Nevertheless, the apparent role of more common variants of *TRIOBP-4* in hearing remains to be explored, as does its function in the retina. Perhaps the largest unexplored area of TRIOBP research, however, concerns the other splice variants. Little is known about shorter 3′ variants such as *TRIOBP-2*. For longer variants, putative roles have been found in the inner ear, but their role in the brain and other tissues is unclear, as is the relationship between TRIOBP-5 and TRIOBP-6.

The *TRIOBP* locus therefore provides a fascinating example of how multiple parts of a gene can cooperate in a single function, actin stabilization, through the generation of many different functional splice variants with distinct expression patterns and modes of action. The variety of different human diseases and conditions related to it highlights its importance, however, much work still needs to be done to clarify the exact relationships between these isoforms and with human health.

Author Contributions: Writing—original draft, N.J.B.; Writing—review & editing, B.Z. and B.S.; Visualization, B.Z., B.S., and N.J.B.; Funding acquisition: N.J.B. All authors have read and agreed to the published version of the manuscript.

Funding: This research was funded by the Croatian Science Foundation (HRZZ: Hrvatska zaklada za znanost), grant numbers IP-2018-01-9424, DOK-2018-09-5395, and DOK-2020-01-8580.

Acknowledgments: We thank Shin-ichiro Kitajiri for sharing additional details regarding *TRIOBP* mutations as well as our colleagues who have worked with us on TRIOBP-1, notably Carsten Korth, Antony Yerabham, Maja Odorčić, and Anja Hart.

Conflicts of Interest: The authors declare no conflict of interest.

References

1. Straub, F.B. Actin II. In *Muscular Contraction, Blood Coagulation*; Szent-Györgyi, A., Ed.; S. Karger: Basel, Switzerland; New York, NY, USA, 1943; pp. 23–37.
2. Dos Remedios, C.G.; Chhabra, D. *Actin-Binding Proteins and Disease*; Springer: New York, NY, USA, 2008.
3. Seipel, K.; O'Brien, S.P.; Iannotti, E.; Medley, Q.G.; Streuli, M. Tara, a novel F-actin binding protein, associates with the Trio guanine nucleotide exchange factor and regulates actin cytoskeletal organization. *J. Cell Sci.* **2001**, *114*, 389–399.
4. Riazuddin, S.; Khan, S.N.; Ahmed, Z.M.; Ghosh, M.; Caution, K.; Nazli, S.; Kabra, M.; Zafar, A.U.; Chen, K.; Naz, S.; et al. Mutations in TRIOBP, which encodes a putative cytoskeletal-organizing protein, are associated with nonsyndromic recessive deafness. *Am. J. Hum. Genet.* **2006**, *78*, 137–143. [CrossRef]

5. Shahin, H.; Walsh, T.; Sobe, T.; Abu Sa'ed, J.; Abu Rayan, A.; Lynch, E.D.; Lee, M.K.; Avraham, K.B.; King, M.-C.; Kanaan, M. Mutations in a novel isoform of TRIOBP that encodes a filamentous-actin binding protein are responsible for DFNB28 recessive nonsyndromic hearing loss. *Am. J. Hum. Genet.* **2006**, *78*, 144–152. [CrossRef]
6. Park, S.; Lee, H.; Kim, M.; Park, J.; Kim, S.-H.; Park, J. Emerging roles of TRIO and F-actin-binding protein in human diseases. *Cell Commun. Signal.* **2018**, *16*, 29. [CrossRef]
7. Pei, J.; Grishin, N.V. AL2CO: Calculation of positional conservation in a protein sequence alignment. *Bioinformatics* **2001**, *17*, 700–712. [CrossRef] [PubMed]
8. Sievers, F.; Wilm, A.; Dineen, D.; Gibson, T.J.; Karplus, K.; Li, W.; Lopez, R.; McWilliam, H.; Remmert, M.; Söding, J.; et al. Fast, scalable generation of high-quality protein multiple sequence alignments using Clustal Omega. *Mol. Syst. Biol.* **2011**, *7*, 539. [CrossRef] [PubMed]
9. Jones, D.T. Protein secondary structure prediction based on position-specific scoring matrices. *J. Mol. Biol.* **1999**, *292*, 195–202. [CrossRef] [PubMed]
10. Jones, D.T.; Cozzetto, D. DISOPRED3: Precise disordered region predictions with annotated protein-binding activity. *Bioinformatics* **2015**, *31*, 857–863. [CrossRef] [PubMed]
11. Buchan, D.W.A.; Jones, D.T. The PSIPRED Protein Analysis Workbench: 20 years on. *Nucleic Acids Res.* **2019**, *47*, W402–W407. [CrossRef] [PubMed]
12. Bradshaw, N.J.; Yerabham, A.S.K.; Marreiros, R.; Zhang, T.; Nagel-Steger, L.; Korth, C. An unpredicted aggregation-critical region of the actin-polymerizing protein TRIOBP-1/Tara, determined by elucidation of its domain structure. *J. Biol. Chem.* **2017**, *292*, 9583–9598. [CrossRef] [PubMed]
13. Yano, T.; Yamazaki, Y.; Adachi, M.; Okawa, K.; Fort, P.; Uji, M.; Tsukita, S.; Tsukita, S. Tara up-regulates E-cadherin transcription by binding to the Trio RhoGEF and inhibiting Rac signaling. *J. Cell Biol.* **2011**, *193*, 319–332. [CrossRef]
14. Lan, J.; Zhu, Y.; Xu, L.; Yu, H.; Yu, J.; Liu, X.; Fu, C.; Wang, X.; Ke, Y.; Huang, H.; et al. The 68-kDa Telomeric Repeat binding Factor 1 (TRF1)-Associated Protein (TAP68) interacts with and recruits TRF1 to the spindle pole during mitosis. *J. Biol. Chem.* **2014**, *289*, 14145–14156. [CrossRef] [PubMed]
15. Jones, D.K.; Johnson, A.C.; Roti Roti, E.C.; Liu, F.; Uelmen, R.; Ayers, R.A.; Baczko, I.; Tester, D.J.; Ackerman, M.J.; Trudeau, M.C.; et al. Localization and functional consequences of a direct interaction between TRIOBP-1 and hERG/KCNH2 proteins in the heart. *J. Cell Sci.* **2018**, *131*, jcs206730. [CrossRef] [PubMed]
16. Webb, B.; Sali, A. Comparative protein structure modeling using MODELLER. *Curr. Protoc. Bioinform.* **2016**, *54*, 5.6.1–5.6.37. [CrossRef] [PubMed]
17. Krieger, E.; Vriend, G. YASARA View—Molecular graphics for all devices—From smartphones to workstations. *Struct. Bioinform.* **2014**, *30*, 2981–2982. [CrossRef]
18. Katsuno, T.; Belyantseva, I.A.; Cartagena-Rivera, A.X.; Ohta, K.; Crump, S.M.; Petralia, R.S.; Ono, K.; Tona, R.; Imtiaz, A.; Rehman, A.; et al. TRIOBP-5 sculpts stereocilia rootlets and stiffens supporting cells enabling hearing. *JCI Insight* **2019**, *4*, e128561. [CrossRef]
19. Hong, J.-H.; Kwak, Y.; Woo, Y.; Park, C.; Lee, S.-A.; Lee, H.; Park, S.J.; Suh, Y.; Suh, B.K.; Goo, B.S.; et al. Regulation of the actin cytoskeleton by the Ndel1-Tara complex is critical for cell migration. *Sci. Rep.* **2016**, *6*, 31827. [CrossRef]
20. Kazmierczak, M.; Kazmierczak, P.; Peng, A.W.; Harris, S.L.; Shah, P.; Puel, J.-L.; Lenoir, M.; Franco, S.J.; Schwander, M. Pejvakin, a candidate stereociliary rootlet protein, regulates hair cell function in a cell-autonomous manner. *J. Neurosci.* **2017**, *37*, 3447–3464. [CrossRef]
21. Zhu, Y.; Wang, C.; Lan, J.; Yu, J.; Jin, C.; Huang, H. Phosphorylation of Tara by Plk1 is essential for faithful chromosome segregation in mitosis. *Exp. Cell Res.* **2012**, *318*, 2344–2352. [CrossRef]
22. Li, X.; Lan, J.; Zhu, Y.; Yu, J.; Dou, Z.; Huang, H. Expression, purification, and characterization of Tara, a novel telomere repeat-binding factor 1 (TRF1)-binding protein. *Protein Expr. Purif.* **2007**, *55*, 84–92. [CrossRef]
23. Bradshaw, N.J.; Bader, V.; Prikulis, I.; Lueking, A.; Müllner, S.; Korth, C. Aggregation of the protein TRIOBP-1 and its potential relevance to schizophrenia. *PLoS ONE* **2014**, *9*, e111196. [CrossRef]
24. Lee, S.H.; Lee, Y.J.; Park, S.W.; Kim, H.S.; Han, H.J. Caveolin-1 and Integrin β1 regulate embryonic stem cell proliferation via p38 MAPK and FAK in high glucose. *J. Cell Physiol.* **2011**, *226*, 1850–1859. [CrossRef] [PubMed]

25. Yun, S.P.; Ryu, J.M.; Jang, M.W.; Han, H.J. Interaction of profilin-1 and F-actin via a b-arrestin-1/JNK signaling pathway involved in prostaglandin E2-induced human mesenchymal stem cells migration and proliferation. *J. Cell Physiol.* **2011**, *226*, 559–571. [CrossRef] [PubMed]
26. Lee, Y.J.; Kim, M.O.; Ryu, J.M.; Han, H.J. Regulation of SGLT expression and localization through Epac/PKA-dependent caveolin-1 and F-actin activation in renal proximal tubule cells. *Biochim. Biophys. Acta* **2012**, *1823*, 971–982. [CrossRef]
27. Bradshaw, N.J.; Hayashi, M.A.F. NDE1 and NDEL1 from genes to (mal)functions: Parallel but distinct roles impacting on neurodevelopmental disorders and psychiatric illness. *Cell Mol. Life Sci.* **2017**, *74*, 1191–1210. [CrossRef] [PubMed]
28. Woo, Y.; Kim, S.J.; Suh, B.K.; Kwak, Y.; Jung, H.-J.; Nhung, T.T.M.; Mun, D.J.; Hong, J.-H.; Noh, S.-J.; Kim, S.; et al. Sequential phosphorylation of NDEL1 by the DYRK2-GSK3b complex is critical for neuronal morphogenesis. *eLife* **2019**, *8*, e50850. [CrossRef] [PubMed]
29. Mui, K.L.; Chen, C.S.; Assoian, R.K. The mechanical regulation of integrin–cadherin crosstalk organizes cells, signaling and forces. *J. Cell Sci.* **2016**, *129*, 1093–1100. [CrossRef]
30. Kuo, J.-C.; Han, X.; Hsiao, C.-T.; Yates, J.R., III; Waterman, C.M. Analysis of the myosin-II-responsive focal adhesion proteome reveals a role for β-Pix in negative regulation of focal adhesion maturation. *Nat. Cell Biol.* **2011**, *13*, 383–393. [CrossRef]
31. Xu, S.; Xu, Y.; Liu, P.; Zhang, S.; Liu, H.; Slavin, S.; Kumar, S.; Koroleva, M.; Luo, J.; Wu, X.; et al. The novel coronary artery disease risk gene JCAD/KIAA1462 promotes endothelial dysfunction and atherosclerosis. *Eur. Heart J.* **2019**, *40*, 2398–2408. [CrossRef]
32. Yu, J.; Lan, J.; Zhu, Y.; Li, X.; Lai, X.; Xue, Y.; Jin, C.; Huang, H. The E3 ubiquitin ligase HECTD3 regulates ubiquitination and degradation of Tara. *Biochem. Biophys. Res. Commun.* **2008**, *367*, 805–812. [CrossRef]
33. Lan, J.P.; Luo, Y.; Zhu, Y.Y.; Sun, J.; Lai, X.Y.; Li, J.Y.; Yu, J.; Shi, J.M.; Lin, M.F.; Huang, H. Isolation of Tara protein and its gene cloning. *Zhejiang Da Xue Xue Bao Yi Xue Ban* **2004**, *33*, 486–490.
34. Leliveld, S.R.; Bader, V.; Hendriks, P.; Prikulis, I.; Sajnani, G.; Requena, J.R.; Korth, C. Insolubility of Disrupted-in-Schizophrenia 1 disrupts oligomer-dependent interactions with Nuclear Distribution Element 1 and is associated with sporadic mental disease. *J. Neurosci.* **2008**, *28*, 3839–3845. [CrossRef]
35. Bradshaw, N.J.; Korth, C. Protein misassembly and aggregation as potential convergence points for non-genetic causes of chronic mental illness. *Mol. Psychiatry* **2019**, *24*, 936–951. [CrossRef]
36. Bader, V.; Tomppo, L.; Trossbach, S.V.; Bradshaw, N.J.; Prikulis, I.; Leliveld, S.R.; Lin, C.-Y.; Ishizuka, K.; Sawa, A.; Ramos, A.; et al. Proteomic, genomic and translational approaches identify CRMP1 for a role in schizophrenia and its underlying traits. *Hum. Mol. Genet.* **2012**, *21*, 4406–4418. [CrossRef]
37. Maycox, P.R.; Kelly, F.; Taylor, A.; Bates, S.; Reid, J.; Logendra, R.; Barnes, M.R.; Larminie, C.; Jones, N.; Lennon, M.; et al. Analysis of gene expression in two large schizophrenia cohorts identifies multiple changes associated with nerve terminal function. *Mol. Psychiatry* **2009**, *14*, 1083–1094. [CrossRef] [PubMed]
38. Hennah, W.; Tomppo, L.; Hiekkalinna, T.; Palo, O.M.; Kilpinen, H.; Ekelund, J.; Tuulio-Henriksson, A.; Silander, K.; Partonen, T.; Paunio, T.; et al. Families with the risk allele of DISC1 reveal a link between schizophrenia and another component of the same molecular pathway, NDE1. *Hum. Mol. Genet.* **2007**, *6*, 453–462. [CrossRef] [PubMed]
39. Hennah, W.; Porteous, D. The DISC1 pathway modulates expression of neurodevelopmental, synaptogenic and sensory perception genes. *PLoS ONE* **2009**, *4*, e4906. [CrossRef]
40. Bradshaw, N.J.; Ukkola-Vuoti, L.; Pankakoski, M.; Zheutlin, A.B.; Ortega-Alonso, A.; Torniainen-Holm, M.; Sinha, V.; Therman, S.; Paunio, T.; Suvisaari, J.; et al. The NDE1 genomic locus affects treatment of psychiatric illness through gene expression changes related to MicroRNA-484. *Open Biol.* **2017**, *7*, 170153. [CrossRef]
41. Knight, H.M.; Maclean, A.; Irfan, M.; Naeem, F.; Cass, S.; Pickard, B.S.; Muir, W.J.; Blackwood, D.H.R.; Ayub, M. Homozygosity mapping in a family presenting with schizophrenia, epilepsy and hearing impairment. *Eur. J. Hum. Genet.* **2008**, *16*, 750–758. [CrossRef] [PubMed]
42. Sugaya, M.; Takenoyama, M.; Shigematsu, Y.; Baba, T.; Fukuyama, T.; Nagata, Y.; Mizukami, M.; So, T.; Ichiki, Y.; Yasuda, M.; et al. Identification of HLA-A24 restricted shared antigen recognized by autologous cytotoxic T lymphocytes from a patient with large cell carcinoma of the lung. *Int. J. Cancer* **2007**, *120*, 1055–1062. [CrossRef]

43. Lee, H.; Kim, M.; Park, J.; Tran, Q.; Hong, Y.; Cho, H.; Park, S.; Hong, S.; Brazil, D.P.; Kim, S.-H.; et al. The roles of TRIO and F-actin-binding protein in glioblastoma cells. *Mol. Med. Rep.* **2018**, *17*, 4540–4546. [CrossRef] [PubMed]
44. Ichiki, Y.; Hanagiri, T.; Takenoyama, M.; Baba, T.; Nagata, Y.; Mizukami, M.; So, T.; Sugaya, M.; Yasuda, M.; Uramoro, H.; et al. Differences in sensitivity to tumor-specific CTLs between primary and metastatic esophageal cancer cell lines derived from the same patient. *Surg. Today* **2012**, *42*, 272–279. [CrossRef] [PubMed]
45. Bao, J.; Wang, S.; Gunther, L.K.; Kitajiri, S.-I.; Li, C.; Sakamoto, T. The actin-bundling protein TRIOBP-4 and -5 promotes the motility of pancreatic cancer cells. *Cancer Lett.* **2015**, *356*, 367–373. [CrossRef] [PubMed]
46. Wang, H.; Li, K.; Mei, Y.; Huang, X.; Li, Z.; Yang, Q.; Yang, H. Sp1 suppresses miR-3178 to promote the metastasis invasion cascade via upregulation of TRIOBP. *Mol. Ther. Nucleic Acids* **2018**, *12*, 1–11. [CrossRef] [PubMed]
47. Fardaei, M.; Sarrafzadeh, S.; Ghafouri-Fard, S.; Miryounesi, M. Autosomal Recessive Nonsyndromic Hearing Loss: A Case Report with a Mutation in TRIOBP Gene. *Int. J. Mol. Cell Med.* **2015**, *4*, 245–247.
48. Zou, S.; Mei, X.; Yang, W.; Zhu, R.; Yang, T.; Hu, H. Whole-exome sequencing identifies rare pathogenic and candidate variants in sporadic Chinese Han deaf patients. *Clin. Genet.* **2020**, *97*, 352–356. [CrossRef]
49. Bao, J.; Bielski, E.; Bachhawat, A.; Taha, D.; Gunther, L.K.; Thirumurugan, K.; Kitajiri, S.-I.; Sakamoto, T. R1 motif is the major actin-binding domain of TRIOBP-4. *Biochemistry* **2013**, *52*, 5256–5264. [CrossRef]
50. Kitajiri, S.-i.; Sakamoto, T.; Belyantseva, I.A.; Goodyear, R.J.; Stepanyan, R.; Fujiwara, I.; Bird, J.E.; Riazuddin, S.; Riazuddin, S.; Ahmed, Z.M.; et al. Actin-bundling protein TRIOBP forms resilient rootlets of hair cell stereocilia essential for hearing. *Cell* **2010**, *141*, 786–798. [CrossRef]
51. Tekin, A.M.; de Ceulaer, G.; Govaerts, P.; Bayazit, Y.; Wuyts, W.; Van de Heyning, P.; Topsakal, V. A New Pathogenic Variant in the TRIOBP Associated with Profound Deafness Is Remediable with Cochlear Implantation. *Audiol. Neurotol.* **2020**. [CrossRef]
52. Hoffmann, T.J.; Keats, B.J.; Yoshikawa, N.; Schaefer, C.; Risch, N.; Lustig, L.R. A Large Genome-Wide Association Study of Age-Related Hearing Impairment Using Electronic Health Records. *PLoS Genet.* **2016**, *12*, e1006371. [CrossRef]
53. Wells, H.R.R.; Freidin, M.B.; Abidin, F.N.Z.; Payton, A.; Dawes, P.; Munro, K.J.; Morton, C.C.; Moore, D.R.; Dawson, S.J.; Williams, F.M.K. GWAS Identifies 44 Independent Associated Genomic Loci for Self-Reported Adult Hearing Difficulty in UK Biobank. *Am. J. Hum. Genet.* **2019**, *105*, 788–802. [CrossRef] [PubMed]
54. Yan, D.; Tekin, D.; Bademci, G.; Foster, J.; Cengiz, F.B.; Kannan-Sundhari, A.; Guo, S.; Mittal, R.; Zou, B.; Grati, M.; et al. Spectrum of DNA variants for non-syndromic deafness in a large cohort from multiple continents. *Hum. Genet.* **2016**, *35*, 953–961. [CrossRef] [PubMed]
55. Zhou, B.; Yu, L.; Wang, Y.; Shang, W.; Xie, Y.; Wang, X.; Han, F. A novel mutation in TRIOBP gene leading to congenital deafness in a Chinese family. *BMC Med. Genet.* **2020**, *21*, 121. [CrossRef]
56. Diaz-Horta, O.; Duman, D.; Foster, J.; Sırmacı, A.; Gonzalez, M.; Mahdieh, N.; Fotouhi, N.; Bonyadi, M.; Cengiz, F.B.; Menendez, I.; et al. Whole-exome sequencing efficiently detects rare mutations in autosomal recessive nonsyndromic hearing loss. *PLoS ONE* **2012**, *7*, e50628. [CrossRef]
57. Gu, X.; Guo, L.; Ji, H.; Sun, S.; Chai, R.; Wang, L.; Li, H. Genetic testing for sporadic hearing loss using targeted massively parallel sequencing identifies 10 novel mutations. *Clin. Genet.* **2015**, *87*, 588–593. [CrossRef]
58. Bitarafan, F.; Seyedena, S.Y.; Mahmoudi, M.; Garshasbi, M. Identification of novel variants in Iranian consanguineous pedigrees with nonsyndromic hearing loss by next-generation sequencing. *J. Clin. Lab. Anal.* **2020**. [CrossRef] [PubMed]
59. Pollak, A.; Lechowicz, U.; Pieńkowski, V.A.M.; Stawiński, P.; Kosińska, J.; Skarżyński, H.; Ołdak, M.; Płoski, R. Whole exome sequencing identifies TRIOBP pathogenic variants as a cause of postlingual bilateral moderate-to-severe sensorineural hearing loss. *BMC Med. Genet.* **2017**, *18*, 142. [CrossRef] [PubMed]
60. Shang, H.; Yan, D.; Tayebi, N.; Saeidi, K.; Sahebalzamani, A.; Feng, Y.; Blanton, S.; Liu, X. Targeted next-generation sequencing of a deafness gene panel (MiamiOtoGenes) analysis in families unsuitable for linkage analysis. *BioMed Res. Int.* **2018**, *2018*, 3103986. [CrossRef] [PubMed]
61. Wesdorp, M.; van de Kamp, J.M.; Hensen, E.F.; Schraders, M.; Oostrik, J.; Yntema, H.G.; Feenstra, I.; Admiraal, R.J.C.; Kunst, H.P.M.; Tekin, M.; et al. Broadening the phenotype of DFNB28: Mutations in TRIOBP are associated with moderate, stable hereditary hearing impairment. *Hear. Res.* **2017**, *347*, 56–62. [CrossRef]

62. Thutkawkorapin, J.; Picelli, S.; Kontham, V.; Liu, T.; Nilsson, D.; Lindblom, A. Exome sequencing in one family with gastric- and rectal cancer. *BMC Genet.* **2016**, *17*, 41. [CrossRef] [PubMed]
63. Wang, H.; Pardeshi, L.A.; Rong, X.; Li, E.; Wong, K.H.; Peng, Y.; Xu, R.-H. Novel variants identified in multiple sclerosis patients from southern China. *Front. Neurol.* **2018**, *9*, 582. [CrossRef] [PubMed]
64. Schoonen, M.; Smuts, I.; Louw, R.; Elson, J.L.; van Dyk, E.; Jonck, L.-M.; Rodenburg, R.J.T.; van der Westhuizen, F.H. Panel-based nuclear and mitochondrial next-generation sequencing outcomes of an ethnically diverse pediatric patient cohort with mitochondrial disease. *J. Mol. Diagn.* **2019**, *21*, 503–513. [CrossRef] [PubMed]
65. Pacentine, I.; Chatterjee, P.; Barr-Gillespie, P.G. Stereocilia Rootlets: Actin-Based Structures That Are Essential for Structural Stability of the Hair Bundle. *Int. J. Mol. Sci.* **2020**, *21*, 324. [CrossRef]

Publisher's Note: MDPI stays neutral with regard to jurisdictional claims in published maps and institutional affiliations.

© 2020 by the authors. Licensee MDPI, Basel, Switzerland. This article is an open access article distributed under the terms and conditions of the Creative Commons Attribution (CC BY) license (http://creativecommons.org/licenses/by/4.0/).

Article

Mass Spectrometric Analysis of Antibody—Epitope Peptide Complex Dissociation: Theoretical Concept and Practical Procedure of Binding Strength Characterization

Bright D. Danquah [1], Kwabena F. M. Opuni [2], Claudia Roewer [1], Cornelia Koy [1] and Michael O. Glocker [1,*]

1 Proteome Center Rostock, University Medicine Rostock, 18059 Rostock, Germany; danquahbright@yahoo.com (B.D.D.); claudia.roewer@uni-rostock.de (C.R.); cornelia.koy@med.uni-rostock.de (C.K.)

2 School of Pharmacy, University of Ghana, P. O. Box LG53 Legon, Ghana; kfopuni@ug.edu.gh

* Correspondence: michael.glocker@uni-rostock.de; Tel.: +49-381-494-4930

Academic Editor: Marilisa Leone

Received: 8 September 2020; Accepted: 16 October 2020; Published: 17 October 2020

Abstract: Electrospray mass spectrometry is applied to determine apparent binding energies and quasi equilibrium dissociation constants of immune complex dissociation reactions in the gas phase. Myoglobin, a natural protein-ligand complex, has been used to develop the procedure which starts from determining mean charge states and normalized and averaged ion intensities. The apparent dissociation constant $K^{\#}_{D\ m0g} = 3.60 \times 10^{-12}$ for the gas phase heme dissociation process was calculated from the mass spectrometry data and by subsequent extrapolation to room temperature to mimic collision conditions for neutral and resting myoglobin. Similarly, for RNAse S dissociation at room temperature a $K^{\#}_{D\ m0g} = 4.03 \times 10^{-12}$ was determined. The protocol was tested with two immune complexes consisting of epitope peptides and monoclonal antibodies. For the epitope peptide dissociation reaction of the FLAG peptide from the antiFLAG antibody complex an apparent gas phase dissociation constant $K^{\#}_{D\ m0g} = 4.04 \times 10^{-12}$ was calculated. Likewise, an apparent $K^{\#}_{D\ m0g} = 4.58 \times 10^{-12}$ was calculated for the troponin I epitope peptide—antiTroponin I antibody immune complex dissociation. Electrospray mass spectrometry is a rapid method, which requires small sample amounts for either identification of protein-bound ligands or for determination of the apparent gas phase protein-ligand complex binding strengths.

Keywords: mass spectrometric epitope mapping; gas phase immune complex dissociation; apparent gas phase dissociation constants; apparent gas phase activation energies; ITEM-TWO; native mass spectrometry

1. Introduction

1.1. ESI Mass Spectrometric Analysis of Non-Covalent Complexes

Electrospray mass spectrometric methods have gained broad acceptance for investigation of the constituents of supramolecular complexes and determination of binding surfaces, e.g., for identifying the locations of partial surfaces on antigens which are recognized by an antibody of interest [1]. By contrast, up to now there exists no mass spectrometric method which has gained equal acceptance for investigating gas phase binding strengths of distinct protein-ligand complexes. Previous reports have shown that high pressure mass spectrometry and/or black body irradiation can be applied for analyzing small molecule-ion equilibria and to determine kinetic and thermodynamic properties, such as ion-ligand complex constants in the right order of magnitude [2–4] also for small peptides, the protonated glycine dimer being the smallest possible peptide dimer representative [5]. Gas phase dissociation reactions of Leu-enkephaline dimers [6] and of small proteins, such as ubiquitin [7] had been studied as well. Such investigations included the application of the Eyring–Polanyi equation for bimolecular gas phase reactions [8]. Gas phase dissociation profiles under low pressure collision gas conditions have been determined to estimate relative gas phase binding strengths of DNA duplex structures which correlated to the solution phase stabilities [9] and of antibiotic ligand-peptide complexes [10], respectively. Semi-quantitative analysis of glycan ligand-protein binding has been reported as well to estimate binding strengths by ESI-MS [11].

Recently, and still uncommonly, the combination of fast and robust gas-phase epitope mapping methods [12,13] with mass spectrometry-based determination of quasi-thermodynamic information has been published. The latter was obtained based on desolvated and multiply charged and accelerated protein-protein complex ions in the gas phase [14]. These studies, together with reports on collision induced unfolding reactions of protein ions [15], have enabled the development of a method termed ITEM-TWO (Intact Transition Epitope Mapping—Thermodynamic Weak-force Order) [16] that can simultaneously identify epitopes as well as enables to determine gas phase binding strengths of the respective antibody-epitope peptide interactions. Here, in addition to describing all required experimental in-solution handling steps, we introduce the underlying theoretical concept and explain all necessary mathematical calculations in detail through which the apparent dissociation constants and the apparent activation energies of protein-ligand complex dissociation processes in the gas phase are obtained.

1.2. Theoretical Concept of Dissociation of Protein-Ligand Complex Ions in the Gas Phase

In a typical Q-ToF instrument, dissociation of protein-ligand complex ions in the gas phase is induced by collisions of multiply charged complex ions with noble gas atoms in a collision chamber at a given collision energy. The ligand dissociation reaction encompasses in its most simple approximation a single transition state, as is indicated by a "one stage" chemical reaction where $\mathbf{k}^{\#}$ is the apparent rate constant of product formation (see supplemental information for details).

$$\mathrm{Prot} \times \mathrm{Lig}]^{m^+} \quad \overset{k\#}{\rightarrow} \quad \mathrm{Prot}]^{p^+} + \mathrm{Lig}]^{n^+}$$

To drive the multiply charged protein-ligand complexes' gas phase dissociation reaction, the apparent Gibbs energy of activation, $\Delta\mathbf{G}^{\#}_{\mathbf{mg}}$ (m: mean of charge states, g: gas phase) is required, as is represented by a "one-stage" energy diagram (Figure 1A). From the transition state (TS), the reaction proceeds irreversibly towards the products. $\Delta\mathbf{G}^{\#}_{\mathbf{mg}}$ is the apparent Gibbs energy of activation of the abundance weighted mean charge state of multiply charged and accelerated protein-ligand complex ions in the gas phase. Yet, after electrospraying there is always an external energy contribution ($\Delta\mathbf{G}_{\mathbf{ext}} > 0$) which needs to be considered during dissociation, as the sum of energies affects the experimentally accessible dissociation energy.

To determine the apparent Gibbs energy of activation of protein complex dissociation of "neutral and resting" protein-ligand complexes, the ESI-dependent external energy contributions need to be considered. Thus, the energy diagram of the complex dissociation reaction requires the introduction of $\Delta\mathbf{G}^{\#}_{\mathbf{m0g}}$, which is the apparent Gibbs energy of activation that is needed for the dissociation of a protein-ligand complex in the gas phase without external energy contributions (Figure 1B).

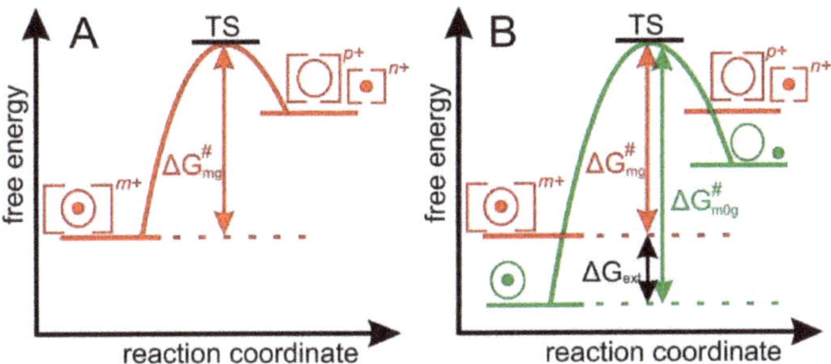

Figure 1. (**A**) Energy diagram showing the apparent Gibbs energy of activation required by charged and accelerated protein-ligand complexes ($\Delta\mathbf{G}^{\#}_{\mathbf{mg}}$) to reach the transition state (**TS**) before dissociating into products ions. (**B**) Energy diagram showing the apparent Gibbs energy of activation required by charged and accelerated protein-ligand complexes ($\Delta\mathbf{G}^{\#}_{\mathbf{mg}} + \Delta\mathbf{G}_{\mathbf{ext}}$) and by neutral and resting protein-ligand complexes ($\Delta\mathbf{G}^{\#}_{\mathbf{m0g}}$) to reach the transition state (**TS**) before dissociating into products ions.

1.3. ESI Mass Spectral Information Extraction

ESI-MS of large biomolecules and macromolecular non-covalent complexes in positive ion mode records series of multiply protonated ions which represent a Gaussian ion intensity distribution of individual charge states for a given molecular/supra-molecular species (complex) [17,18]. For semi-quantitative analyses of ESI mass spectra we postulate that the overall ion characteristics, such as gas phase reactivity of complex dissociation, is well represented by the mean charge state ($\mathbf{m}^{+}$) of the recorded ion series of a molecular/supra-molecular species (Figure 2).

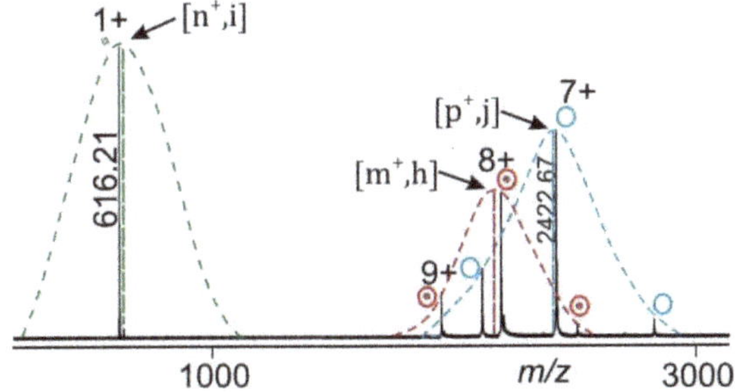

Figure 2. NanoESI mass spectrum of gas phase dissociation of protein complex (e.g., holo-myoglobin; apo-myoglobin—heme complex). Gaussian fits of ion intensities of the related charge state series for each molecular or supra-molecular species (complex) are shown (dashed charge state envelope curves). The arrows point to apices which are determined as maxima of fitted curves. The vertical dashed lines provide heights of charge structure envelopes which represent relative intensities of holo-protein complex ions (h, educts; red), ligand ions (i, product; green), and apo-protein ions (j, product; blue). Locations on the *m/z* axis match with mean charge states of holo-protein complex ions (m^{+}, educts), ligand ions (n^{+}, product) as well as apo-protein ions (p^{+}, product).

Mean charge states of each ion species, e.g., holo-protein ions (educts) and of apo-protein as well as of ligand ions (products) can be separately determined from the mass spectrum. Normalization of ion intensities is achieved by summation of all apex values and setting the sum to 100% (equations and calculations are shown in the supplemental information).

1.4. Data Analysis Procedure

A series of mass spectra in which the collision cell voltage difference is increased stepwise records ion signals with varying intensities of all ion species, i.e., educts and products, as they emerge from the collision induced complex dissociation reaction. Plotting the normalized intensities of the educts of the complex dissociation reaction in the gas phase as a function of collision cell voltage difference (**ΔCV**) provides a sigmoidal shaped curve with Boltzmann characteristics (Figure 3A). The "steep part" of the dissociation reaction dependence (interval 2dx), i.e., the "energy regime" with greatest dependence between educt ion intensities and **ΔCV**, as well as the determination of $\mathbf{\Delta CV_{50}}$ from the Boltzmann fit to the data points is inferred by mathematical procedures which lead to the equation of the tangent line (see supplemental information for calculations).

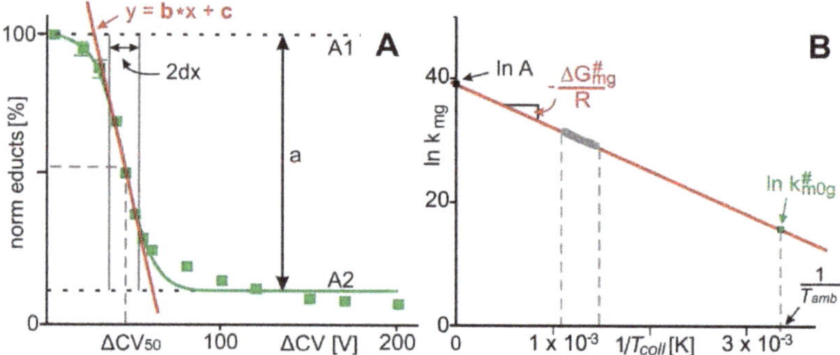

Figure 3. (**A**) Course of normalized ion intensities of complex ions (***norm*** (***educts***)) as a function of collision cell voltage differences (**ΔCV**). Each data point is the mean of several independent measurements. Vertical bars give standard deviations. The curve was fitted using a Boltzmann function. The tangent line equation is taken from the Boltzmann fit. "**a**" describes the difference between the lowest and highest data points on the sigmoidal fit. "**2dx**" is the x-axis interval within which the steepest decline of educts is observed; the center of the **2dx** interval is $\mathbf{\Delta CV_{50}}$. (**B**) Arrhenius plot for the course of protein-ligand complex dissociation in the gas phase. The value for $\mathbf{lnk^{\#}_{m0g}}$ is taken from the point of the line at $\frac{1}{\mathbf{T_{amb}}}$.

Since, in the gas phase of a Q-ToF mass spectrometer, the collision of multiply charged and accelerated complex ions takes place upon reaching elevated energies, the collision temperature ($\mathbf{T_{coll}}$) that is attained by the complex during collision induced dissociation needs to be considered as well. As proposed by a model for collisional activation [19], $\mathbf{T_{coll}}$ can be expressed as the sum of ambient temperature, $\mathbf{T_{amb}}$, plus external temperature contribution, $\mathbf{T_{ext}}$ (see supplemental information for definitions, energy conversion factors [19–21], and equations).

According to the Eyring–Polanyi equation [22], $\mathbf{k^{\#}}$ is directly proportional to an apparent thermodynamic quasi equilibrium dissociation constant, $\mathbf{K^{\#}_{D}}$. The apparent gas phase thermodynamic quasi equilibrium dissociation constant, $\mathbf{K^{\#}_{D\ mg}}$, is also given by the relative ion intensities (cf. Figures 1 and 3). Accordingly, for each experimentally set **ΔCV** value, a corresponding $\mathbf{k^{\#}_{mg}}$ value can be calculated (see supplemental information for equations). From the Arrhenius equation, the apparent energy of activation of protein–ligand complex dissociation $\mathbf{\Delta G^{\#}_{mg}}$ can be determined. Plotting $\mathbf{lnk^{\#}_{mg}}$ as a function of $\frac{1}{\mathbf{T_{coll}}}$ provides the intercept with the y-axis, which is ln **A** (pre-exponential factor), and the slope of the line, which is $-\frac{\mathbf{\Delta G^{\#}_{mg}}}{\mathbf{R}}$ (Figure 3B). Note, at $\mathbf{T_{coll}} = \mathbf{T_{amb}} = \mathbf{298\ K}$ it can be concluded that $\mathbf{\Delta CV} = 0$. Hence, from the Arrhenius plot a value for $\mathbf{k^{\#}_{m0g}}$ is obtained, i.e., the apparent rate constant of dissociation of "neutral and resting" protein-ligand complexes. Similarly, at $\mathbf{\Delta CV} = 0$ the value for $\mathbf{K^{\#}_{D\ m0g}}$, is calculated, i.e., the apparent gas phase thermodynamic equilibrium dissociation constants of protein-ligand complex dissociation, corrected for external energy contributions; i.e., of "neutral and resting" protein-ligand complexes. Therefore, at $\Delta\mathbf{CV} = 0$ the value for $\Delta\mathbf{G}^{\#}_{\mathbf{m0g}}$ is calculated as well, i.e., the apparent Gibbs energy of activation of neutral and resting protein-ligand complexes (see Supplement for equations).

The entire procedure, which is termed "Intact Transition Epitope Mapping—Thermodynamic Weak-force Order (ITEM-TWO)", starts with either generating the protein-ligand complex by mixing the two components in solution or by simply maintaining the natively obtained protein-ligand complex in an electrospray-compatible solution. No further in-solution sample handling steps are needed. A few microliters of complex-containing solution are loaded into a nano-electrospray capillary and all solubilized components including the protein-ligand complex are simultaneously transferred into the gas phase by electrospray. Mass spectrometric data acquisition involves collision induced dissociation of the complex in the gas phase at various applied collision cell voltage differences ($\Delta\mathbf{CV}$). Subsequent in-depth data analysis of intensities of both, resulting product ions and remaining educt ions (survivors) at each of the applied collision energies enables the apparent non-covalent complex stability to be characterized. In this reports supplement, the entire data analysis procedure is described in all detail. In contrast to previous reports the complex dissociation reaction is monitored by investigating the mean charge states and the normalized average intensities of each ion species.

2. Results

2.1. Procedure Development with Myoglobin and Application to RNAse S Dissociation

The mass spectral data obtained from electrospraying myoglobin (the holo-myoglobin complex consists of apo-myoglobin plus heme) were collected by following the ITEM-TWO protocol (see Methods section for in-solution handling and data acquisition steps). The ESI mass spectrum of myoglobin (Figure 4) provides an ion series of multiply charged ions from which a mean charge state of 8.1+ and an average mass of 17,566.95 ± 0.46 Da is calculated for holo-myoglobin. After having electrosprayed the myoglobin solution, and upon having switched on the collision gas and having increased the collision cell voltage difference ($\Delta\mathbf{CV}$) in a step-wise manner (5–20 V/step), dissociation of holo-myoglobin complexes caused appearance of complex-released heme ions (ligands) in the low mass ranges of the mass spectra (Supplemental Table S1). The mass spectra also showed ion signals of apo-myoglobin (mean charge state: 7.2+) in the high mass ranges (average mass: 16,951.46 ± 0.44 Da) with increasing yields (Figure 4). The *m/z* value of the ion that appeared in the low mass range ([heme]+) was 616.21 and corresponded precisely to the calculated values for singly protonated [heme]+ (*m/z* 616.18), resulting in a mass accuracy of 48 ppm.

Mean charge states of holo-myoglobin ions (m+), apo-myoglobin (n+) and heme (p+) as well as apex heights (holo-myoglobin ions (h, educts), heme ions (i, product), and apo-myoglobin ions (j, product) were extracted (Supplemental Table S2) from the triplicate measurements, averaged and normalized. For each $\Delta\mathbf{CV}$ setting one spectrum was generated (Figure 4) and analyzed semi-quantitatively by determining Gaussian fits for all molecular/supra-molecular ion species.

After recording mass spectra under increasing collision cell voltage difference settings ($\Delta\mathbf{CV}$) and after determining Gaussian fits of charge structures for each ion series, the averaged **norm** (**educts**) values were plotted as a function of $\Delta\mathbf{CV}$,which resulted in a sigmoidal shaped course that represented the dependence of educt intensities (starting materials) on $\Delta\mathbf{CV}$ settings (Supplemental Figure S1). All the y-values from the tangent line of the steep decline that fall within the 2dx interval around $\Delta\mathrm{CV}_{50}$ are used for the calculation of $\mathbf{ln\ k^{\#}_{mg}}$ values which are then used in the Arrhenius plot (Supplemental Figure S2). As shown above at $\mathbf{T_{coll}} = \mathbf{T_{amb}} = \mathbf{298\ K}$ there applies $\Delta\mathrm{CV} = 0$ at which the value for $\mathbf{K^{\#}_{D\ m0g}}$ is calculated (Table 1). $\mathbf{K^{\#}_{D\ m0g}}$ is the apparent gas phase thermodynamic quasi equilibrium dissociation constants of heme loss of "neutral and resting" myoglobin in the gas phase. Then, $\Delta\mathbf{G}^{\#}_{\mathbf{m0g}}$ is calculated using the van´t Hoff equation (Table 1).

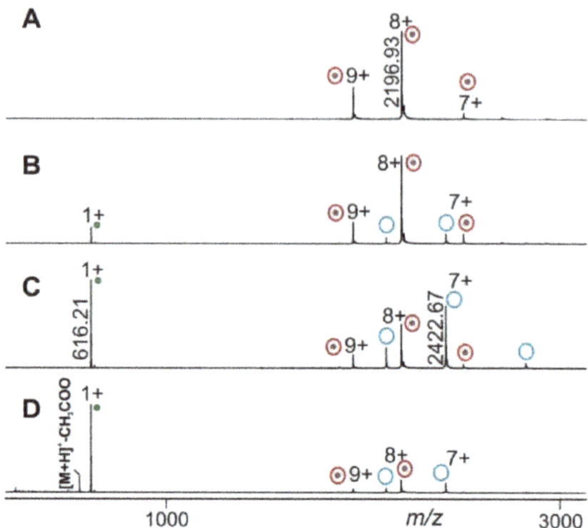

Figure 4. Nano-ESI mass spectra from myoglobin dissociation experiments. Different collision cell voltage differences (ΔCV) are shown. (**A**) 4 V. (**B**) 30 V. (**C**) 60 V. (**D**) 120 V. Charge states and *m/z* values for selected ion signals are given for holo-myoglobin ions (red circles with dots on right ion series), apo-myoglobin ions (blue circles without dots on right ion series) and for the released heme ions (green dots on left ion series). Solvent: 200 mM ammonium acetate, pH 7.

Table 1. Apparent kinetic and apparent quasi-thermodynamic values for gas phase dissociation of non-covalent complexes.

Complex [(a)]	Mean Charge ± std. dev. [(b)]	ΔCV_{50} (V)	dx (V)	$k^{\#}_{m0g}$ (1/s)	$K^{\#}_{D\ m0g}$ (Ø) [(b)]	$\Delta G^{\#}_{m0g}$ (kJ/mol)
myoglobin	8.1 ± 0.01	44.0	8.5	5.1×10^{9}	3.60×10^{-12}	65.3
RNAse S	6.4 ± 0.20	12.5	4.6	7.3×10^{10}	4.03×10^{-12}	65.0
FLAG	24.6 ± 0.30	89.0	19.8	7.9×10^{10}	4.04×10^{-12}	65.0
troponin I	25.9 ± 0.14	26.4	15.9	2.2×10^{12}	4.58×10^{-12}	64.7

(a) "neutral and resting" complex; (b) unitless number.

The gas phase dissociation reaction of RNAse S was investigated in the same manner. Upon electrospraying RNAse S which had been dissolved in 200 mM ammonium acetate solution, pH 7, the collision cell voltage difference was raised in a step-wise fashion and mass spectra were recorded (see the Methods section for in-solution handling steps). The ESI mass spectrum of RNAse S (Figure 5) provides ion series of multiply charged ions from which mean charge states of 6.4+ and average masses of 13,631.68 ± 0.20 Da and 13,544.23 ± 0.60 are calculated for the two most prominent RNAse S species. Commercial RNAse S represents two prominent protein complexes with clearly differentiated ion series, all of which represent related forms of RNAse S (Supplemental Table S3). For determining the overall apparent activation energy of the S-peptide dissociation reaction from RNAse S, all ion series were considered to equally represent the dissociation process as a whole, meaning that all ion signal intensities were subjected to normalization (Supplemental Table S4).

After recording mass spectra under increasing collision cell voltage difference settings (**ΔCV**) and after determining Gaussian fits of charge structures for each ion series, the averaged **norm** (**educts**) values were plotted as a function of **ΔCV**, again generating a sigmoidal shaped course (Supplemental Figure S3). As before, all the y-values from the tangent line that fell within the **2dx** interval around ΔCV_{50} were used for calculating $\mathbf{ln\ k^{\#}_{mg}}$ values. These were subjected to draw the respective Arrhenius plot (Supplemental Figure S4). Then, the value for $\mathbf{K^{\#}_{D\ m0g}}$ was calculated (Table 1) to represent the apparent gas phase quasi thermodynamic equilibrium dissociation constant of S-peptide loss from "neutral and resting" RNAse S in the gas phase. Finally, $\Delta G^{\#}_{m0g}$ is calculated using the van´t Hoff equation (Table 1).

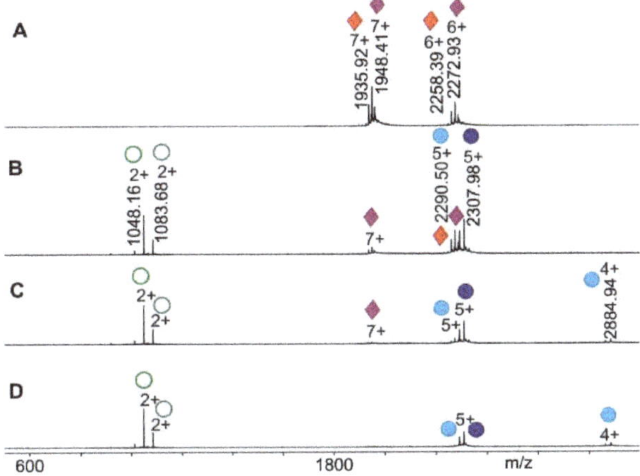

Figure 5. Nano-ESI mass spectra from RNAse S dissociation experiments. Different collision cell voltage differences (**ΔCV**) are shown. (**A**) 3 V. (**B**) 17 V. (**C**) 30 V. (**D**) 50 V. Charge states and *m/z* values for selected ion signals are given for RNAse S ions (red/purple diamonds on right ion series), S-protein ions (light blue/dark blue filled circles on right ion series) and for the released S-peptide ions (light green/dark green circles on left ion series). Solvent: 200 mM ammonium acetate, pH 7.

2.2. Application Examples with Epitope Peptide-Antibody Immune Complexes

The ITEM-TWO procedure was tested with immune complexes which were generated by mixing an epitope peptide-containing solution with a solution that contained its respective monoclonal antibody. The mixture of antiFLAG antibody with seven peptides was investigated. The mass spectrum of this antibody-peptide mixture showed in the high mass ranges three narrowly spaced multiply charged ion triplets at each charge state (Figure 6). The molecular masses of these triplet ions (charge states from 21+ to 27+; mean charge state 24.6+) were determined to be 148,730 ± 92 Da, 149,799 ± 45 Da, and 150,785 ± 61 Da, which represented the antiFLAG antibody, the antiFLAG antibody with one bound FLAG peptide, and the antiFLAG antibody with two bound FLAG peptides, respectively (Supplemental Table S5). The mass differences between each ion signal triplet provided rather inaccurate mass values for the bound peptide and, therefore, unambiguous identification of the epitope peptide out of the mixture of seven peptides was not possible.

Due to the chosen quadrupole settings, the low mass ions of unbound peptides were filtered out. Yet, upon raising the collision cell voltage difference there appeared isotopically resolved ions in the low *m/z* range of the mass spectrum for the FLAG peptide (see inset in Figure 6) which were recorded with high mass accuracy (20 ppm) and enabled unambiguous identification. It is worth noting that the antiFLAG antibody complex only released the FLAG peptide despite the presence of six other peptides in solution.

By increasing the collision cell voltage differences from 4 V to 200 V in a stepwise manner (20–30 V/step), we observed appearance and incremental rise of doubly and triply charged ion signals in the lower *m/z* range together with gradual disappearance of complex ion signals (Figure 6). These relative complex ion intensities, i.e., the heights of apexes of the Gaussian fits, served as the amounts of the various multiply charged ion series of the antibody-peptide complex ions. The height of the apexes of the Gaussian fits of complex-released epitope peptide ion series were used to represent amounts of the released epitope peptides (Supplemental Tables S5 and S6).

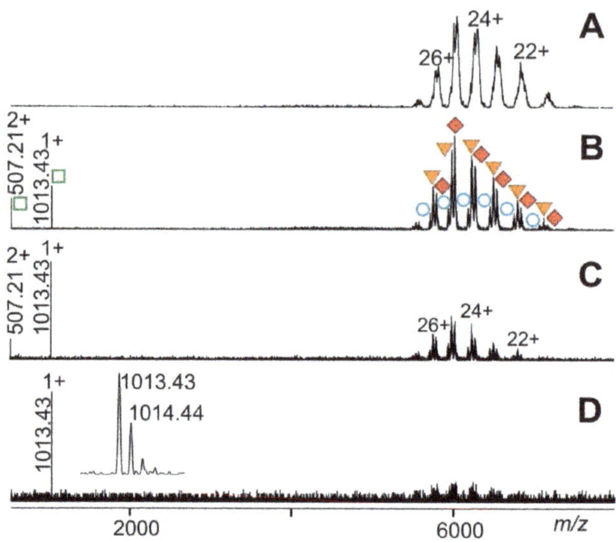

Figure 6. Nano-ESI mass spectra of FLAG-peptide-antiFLAG antibody complex dissociation. Different collision cell voltage differences (ΔCV) are shown. (**A**) 20 V. (**B**) 70 V. (**C**) 120 V. (**D**) 150 V. Charge states and *m/z* values for selected ion signals are given for the immune complexes (antibody plus one FLAG-peptide and antibody plus two FLAG-peptides; filled orange triangles and filled red diamonds, respectively; right ion series), antiFLAG antibody (open blue circles; right ion series), and FLAG-peptide (open green squares; left ion series). The inset shows a zoom of the singly-charged FLAG peptide ion signals. Solvent: 200 mM ammonium acetate, pH 7.

Again, plotting **norm** (**educts**) vs. **ΔCV**, a sigmoidal shaped course was obtained, which represents the dependence of educt intensities on **ΔCV** (Supplemental Figure S5). Next, the x-axis values (**ΔCV**) within the intervals dx above and below $\mathbf{\Delta CV_{50}}$ were used to determine the corresponding y-axis values using the equation of the tangent line. The resulting y-axis values, i.e., **norm** (**educts**), enabled the calculation of $\mathbf{ln\ k^{\#}_{mg}}$ values. Plotting increments of $\mathbf{ln\ k^{\#}_{mg}}$ vs. $\frac{1}{\mathbf{T_{coll}}}$ allowed determination of the part of the apparent dissociation reaction within the "energy regime" located around $\mathbf{\Delta CV_{50}}$ (Arrhenius plot; Supplemental Figure S6). In the same manner as was shown above, the calculated value for $\mathbf{k^{\#}_{m0g}}$ represented the apparent rate constant of dissociation of "neutral and resting" antibody-epitope peptide complexes. Then, $\mathbf{K^{\#}_{D\ m0g}}$ and $\mathbf{\Delta G^{\#}_{m0g}}$ (Table 1) were calculated by applying the respective equations.

At last, we performed ITEM-TWO experiments with an epitope peptide that was derived from human cardiac Troponin I (Tn I) against which was directed a monoclonal antiTroponin I antibody (antiTn I). Generation of the immune complex in solution and subsequent electrospraying of the entire mixture started data acquisition with the respective instrument settings as mentioned (for details see Methods section). The mass spectrum of this antibody-peptide mixture showed in the high mass range three narrowly spaced multiply charged ion triplets at each charge state (Figure 7). The molecular masses of these triplet ions (charge states from 23+ to 28+; mean charge state 25.9+) were determined to be 146,414.75 ± 33 Da, 148,218.75 ± 35 Da, and 150,018.88 ± 33 Da, which were identified to be representing antiTroponin I antibody, antiTroponin I antibody with one bound Troponin I epitope peptide, and antiTroponin I antibody with two bound Troponin I epitope peptides, respectively (Supplemental Tables S7 and S8). Again, upon raising the collision cell voltage difference there appeared isotopically resolved ions in the low *m/z* range of the mass spectrum for the Troponin I epitope peptide (see insert in Figure 7) which were recorded with high mass accuracy (7 ppm).

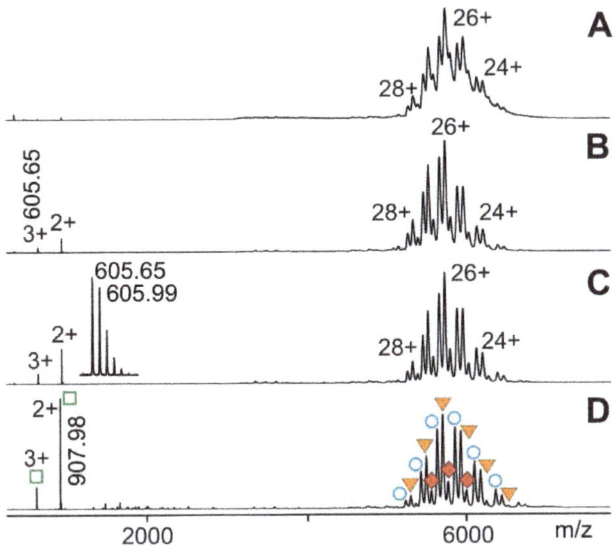

Figure 7. Nano-ESI mass spectra of Tn I-peptide—antiTn I antibody complex dissociation. Different collision cell voltage differences (**ΔCV**) are shown. (**A**) 4 V. (**B**) 16 V. (**C**) 30 V. (**D**) 80 V. Charge states and *m/z* values for selected ion signals are given for the immune complexes (antibody plus one Tn I-peptide and antibody plus two Tn I-peptides; filled orange triangles and filled red diamonds, respectively; right ion series), antiTn I antibody (open blue circles; right ion series), and Tn I-peptide (open green squares; left ion series). Insert shows a zoom of the triply-charged Troponin I epitope peptide ion signal. Solvent: 200 mM ammonium acetate, pH 7.

Again, plotting **norm** (**educts**) vs. **ΔCV** produced a sigmoidal shaped course which represented the dependence of educt intensity on **ΔCV** following Boltzmann characteristics (Supplemental Figure S7). The y-axis values on the tangent line, i.e., **norm** (**educts**), enabled to calculate $\mathbf{ln\ k^{\#}_{mg}}$ values. Plotting increments of $\mathbf{ln\ k^{\#}_{mg}}$ vs. $\frac{1}{\mathbf{T_{coll}}}$ allowed determination of the part of the apparent dissociation reaction within the "energy regime" located around $\mathbf{\Delta CV_{50}}$. This Arrhenius plot again provided **ln***A* (pre-exponential factor) as the intercept with the y-axis and $-\frac{\Delta G^{\#}_{mg}}{R}$ as the slope of the line (Supplemental Figure S8). In the same manner as described above, the value for $\mathbf{k^{\#}_{m0g}}$ was calculated. Similarly, by applying the Eyring-Polanyi equation $\mathbf{K^{\#}_{D\ m0g}}$ was determined. Finally, by using the van´t Hoff equation, $\mathbf{\Delta G^{\#}_{m0g}}$ was calculated (Table 1) representing binding strengths of "neutral and resting" antibody-epitope peptide complexes.

3. Discussion

In solution, the bi-molecular association of the heme group to apo-myoglobin is thought to follow a bimodal process. Association of heme to (partially unfolded) apo-myoglobin is fast (k_A ~10^8 $M^{-1} \cdot s^{-1}$), followed by a slower structural re-arrangement (k ~500 s^{-1}) to generate natively-folded holo-myoglobin (at pH 7) in which the iron atom of the heme group is then primarily coordinated by His93 [23]. Accordingly, in-solution affinity of at least partially unfolded apo-myoglobin to the heme prosthetic group has been characterized as rather strong and the complex consisting of both components was assumed to possess a K_D ~10^{-11} M, whereas natively-folded holo-myoglobin has been determined to form an even stronger complex (K_D ~10^{-13} M to 3×10^{-14} M) [24,25].

In the gas phase, the stability of heme binding to apo-myoglobin has been studied by dissociating holo-myoglobin in the orifice-skimmer region of an electrospray mass spectrometer, i.e., at high pressure, where complex stability was found to correlate with the activation energy of dissociation of the complex in solution [26,27]. In that investigation, heme dissociation kinetics was studied (i) by spraying solutions with pH 5 and (ii) by looking at selected charge state pairs (e.g., 8+ protonated holo-myoglobin and 7+ charged apo-myoglobin). Activation energies ranged from 73 kJ/mol to 106 kJ/mol, depending on myoglobin amino acid sequence mutations. Similar values of activation energies for heme dissociation have been reported for 9+ protonated holo-myoglobin (92 kJ/mol) and for 10+ protonated holo-myoglobin (85 kJ/mol), respectively [19]. Values from our investigations of holo-myoglobin dissociation are somewhat lower but in general stand in good agreement with reported data. The uncertainty of this method has been estimated to be approx. 10% [14]. Hence, myoglobin is considered to be an adequate standard for developing the ESI-MS method by which protein-ligand dissociation reactions may be studied. Interestingly, from ESI-MS ETD studies of multiply charged myoglobin gas phase ions it was concluded that—depending on the complexes′ charge states—the heme group might be coordinated by one of two histidinyl residues, mainly by His93 but also by His64, or by both, suggesting some similarity between in-solution and gas phase complex structures—at least around the heme binding pocket [25]. The existence of relatively defined macromolecular structures during the heme dissociation process (e.g., as transition state) fits our model, so we postulate that dissociation of immune complexes follows in principle a hard spheres model, i.e., entropy contributions at the transition state are small [14]. In fact, in-solution antibody—antigen interactions are enthalpy-driven [28,29]. Non-covalent forces in the gas phase as well as structural properties of other desolvated protein ions [30–32] have demonstrated that higher order protein structures are maintained in the gas phase for a certain period of time [33–35] despite absence of solvation [36]. Hence, similar to myoglobin, the decisive structural properties of antibody-antigen complexes seem preserved in the gas phase, at least to some extent.

The rate limiting factor for irreversible dissociation of immune complexes in the gas phase reaction is the activation barrier that needs to be overcome. With an energy input above a critical threshold, immune complex dissociation proceeds irreversibly but comparatively slowly under CID conditions. At each set energy regime, certain portions of immune complexes reach above threshold conditions which results in mixtures of surviving immune complexes and dissociated products within the timeframe of each single measurement. Thus, despite the de facto irreversible character of the dissociation reaction, apparent equilibrium conditions can be assumed. In contrast to previous work we look at average charge states, to represent a respective protein-ligand complex that has been translated into the gas phase by electrospray, and extrapolate to conditions with no additional external energy contributions, such as multiple charging and acceleration of the complex in the gas phase. Both conditions cannot be realized by experiment, because mass spectrometry experiments are performed with accelerated and (multiply) charged ions. Approximation and extrapolation to "resting and neutral" gas phase complexes is expected to provide a better comparison with in-solution data. As we performed all our work using commercial mass spectrometers, we have no means to change the duration times of the ions in the collision cell. However, by keeping all instrument settings (temperature, pressure, charge states of the complexes, gas identity) constant for the entire duration of the experiments (except of the collision cell voltage difference), we assume that reaction times do not differ too much, when comparing dissociation yields and applying the intensities of all ions for our calculations. Consistent with the literature we observed that at higher collision cell voltage differences the dissociation yields were higher as compared to those which were obtained by applying lower collision cell voltage differences.

As a consequence, from all above considerations it appears well possible to semi-quantitatively compare apparent gas phase binding strengths between complexes and to relate these to in-solution dissociation constants of antibody-antigen complexes, after correcting the energy terms, i.e., by subtracting external energy contributions.

4. Materials and Methods

4.1. Preparation of Myoglobin-Containing Solution

To demonstrate the procedure, electrospray-compatible solutions of myoglobin with neutral pH, in which binding activities are maintained for performing the ITEM-TWO experiments, are prepared. A stock solution was first prepared by dissolving 5.22 mg of myoglobin (lot # 60K7007, Sigma-Aldrich, Steinheim, Germany) in 1 mL of 200 mM ammonium acetate. Next, 50 μL of the stock solution was diluted with 450 μL of 200 mM ammonium acetate. The concentration of the resulting solution was determined to be 0.5 μg/μL using a QubitTM 2.0 Fluorometer (Carlsbad, California, USA). Finally, 100 μL of the 0.5 μg/μL myoglobin solution were further diluted to 200 μL with 200 mM ammonium acetate buffer, pH 7, to obtain a final concentration of 0.25 μg/μL. For each measurement, ca. 3 μL of the 0.25 μg/μL myoglobin solution were loaded into nanoESI capillaries using a microloader pipette tip (Eppendorf, Hamburg, Germany) and were electrosprayed directly.

4.2. Preparation of RNAse S-Containing Solution

A stock solution of ca. 1 mg/mL was first prepared by dissolving the lyophilized powder (0.26 mg) of RNAse S (Lot # 52H7034, Sigma-Aldrich, Steinheim, Germany) in 0.26 mL of 200 mM ammonium acetate, pH 7.0. Then, 100 μL were transferred onto a Microcon centrifuge filter with a 3 kDa cutoff (Millipore Corp., Bedford, MA, USA) together with further 200 μL of 200 mM ammonium acetate solution. This solution was centrifuged for 30 min at 13,000 rpm and 23 °C. The eluate was discarded and 200 μL of 200 mM ammonium acetate solution were added onto the filter. This procedure was repeated for three times. Then, the filter was inverted, placed into a new tube, and centrifuged for 5 min at 4500 rpm at 23 °C. A resulting supernatant of approximately 800 μL was collected and the protein concentration was determined to be 0.78 μg/μL using a Qubit^{TM} 2.0 Fluorometer (Carlsbad, CA, USA) assay. For nano electrospray mass spectrometry 2.56 μL of the purified and concentrated RNAse S solution were diluted to a final concentration of 0.2 μg/μL with 7.44 μL of 10% methanol/200 mM ammonium acetate. For each measurement, ca. 3 μL of the RNAse S solution were loaded into nanoESI capillaries using a microloader pipette tip (Eppendorf, Hamburg, Germany).

4.3. Preparation of FLAG-Peptide-AntiFLAG Antibody Immune Complex-Containing Solution

A volume of 20 μL of 1 μg/μL of mouse monoclonal antiFLAG M2 antibody (product code F 1804, Sigma-Aldrich, Steinheim, Germany) was first re-buffered into 200 mM ammonium acetate buffer, pH 7, using a centrifugal filter (Amicon Ultra cutoff 50 K; Merck Millipore Ltd., Tullagreen, Carrigtwohill Co Cork, Ireland), as described [10,13]. After buffer exchange, 5 μL of antiFLAG antibody solution (0.2 μg/μL; 1.33 μM) was mixed with 1.5 μL of a peptide mixture of seven peptides containing 10 μM each of GPI peptide (ALKPYSPGGPR, 1141.62 Da), FLAG peptide (DYKDDDDK, 1012.40 Da), Angiotensin II (DRVYIHPF, 1045.53 Da), TRIM21A peptide (LQELEKDEREQLRILGE, 2097.11 Da), TRIM21B peptide (LQPLEKDEREQLRILGE, 2065.12 Da), TRIM21C peptide (LQELEKDEPEQLRILGE, 2038.06 Da), and RA33 peptide (MAARPHSIDGRVVEP-NH2, 1632.86 Da) in a molar ratio of 2.2:1 of peptide to antibody. Solvent for the peptides was 200 mM ammonium acetate buffer, pH 7.

4.4. Preparation of Troponin I-Peptide-AntiTn I Antibody Immune Complex-Containing Solution

The human cardiac Troponin I epitope peptide (ENREVGDWRKNIDAL; peptides&elephants, Hennigsdorf, Germany) was obtained as lyophilized powder. The peptide was dissolved in 200 mM ammonium acetate buffer, pH 7, to obtain a peptide concentration of 1.31 μg/μL. The antiTroponin I antibody [MF4] (product code ab38210; Abcam, Cambridge, UK) was obtained dissolved in PBS buffer, pH 7.4. Buffer was exchanged to 200 mM ammonium acetate buffer, pH 7, by loading 21 μL (40 μg) of the antibody stock solution into a centrifugal filter (Microcon with a cutoff of 50 K; Merck Millipore Ltd., Tullagreen, Carrigtwohill, Co. Cork, Ireland). Then 200 mM ammonium acetate buffer, pH 7, were added to reach a volume of 500 μL. The solution was centrifuged at 13,000 rpm for 10 min.

The filtrate was discarded. To the retentate on the filter (ca. 30 µL) 470 µL of 200 mM ammonium acetate buffer, pH 7, were added to reach a total volume of 500 µL. The solution was centrifuged again. This centrifugation/re-filling procedure was repeated eight times. After the last spinning, the filter unit was inverted into a new vial and was centrifuged at 4500 rpm for 5 min to collect the retentate (52 µL). Protein concentration was determined to be 0.33 µg/µL using the Qubit™ 2.0 Fluorometer (Carlsbad, CA, USA) assay. To obtain the immune complex with molar ratio of 2.2:1 of peptide to antibody the antiTroponin I antibody solution (0.225 µM) was diluted 1:2 with 200 mM ammonium acetate and 4 µL were mixed with 1.37 µL of the Troponin I peptide solution 1 which previously had been diluted 1:100 with 200 mM ammonium acetate. The immune complex-containing mixture was incubated at room temperature for at least 1 h. For each measurement, 3 µL of antibody-peptide complex-containing solution were loaded into nanoESI capillaries using a microloader pipette tip (Eppendorf, Hamburg, Germany).

4.5. Production of NanoESI Capillaries

NanoESI capillaries for offline measurements were prepared in-house from borosilicate glass tubes of 1 mm outer and 0.5 mm inner diameters (BF 100-50-10, Sutter Instruments, Novato, CA, USA), using a P-1000 Flaming/BrownTM micropipette puller system (Sutter Instruments, Novato, CA, USA). Capillaries were gold-coated using a sputter coater BalTec SCD 0045 (Bal-Tech, Balzers, Liechtenstein) with the following conditions: current was set to 20 mA for 150 s, table distance was 5 cm, while vacuum was ca. 10^{-3} mbar and Argon gas pressure maintained at ca. 10^{-2} mbar [16].

4.6. Q-ToF 2 Instrument Settings and Data Aquisition

Nano-ESI-MS measurements were performed using a Q-TOF 2 instrument (Waters MS-Technologies, Manchester, UK). The pressure in the source region of the mass spectrometer was manually adjusted to 2.24 mbar using the speedy valve [16]. ITEM-TWO measurements were performed with the following instrumental settings: source temperature, 50 °C; capillary voltage, 1.3 kV; sample cone voltage, 30 V; extractor cone voltage, 3 V; collision voltage, 4 V; pusher time, 124 µs. The Quadrupole and ToF analyzer pressures were typically between ca. 2.0×10^{-5} mbar and 2.50×10^{-7} mbar, respectively. All mass spectra were acquired in positive-ion mode with a mass window of *m/z* 200–4000. The *m/z* axis was calibrated using 50% TFE in 1% orthophosphoric acid. The collision gas was then switched on (1.25 bar) and collision cell voltage differences were increased in a stepwise manner (5–20 V/step) to cause dissociation of the complexes. At each collision cell voltage difference setting, a mass spectrum was recorded for 2 min, each. All scans for a given collision cell voltage difference were combined to generate an average spectrum. Q-ToF MS data were acquired and minimally processed using the MassLynx software version 4.0 (Waters MS-Technologies, Manchester, UK). From each spectrum, the ion intensities (in arbitrary units) were deduced [16].

4.7. Synapt G2S Instrument Settings and Data Aquisition

Nano-ESI-MS measurements were performed using a Synapt G2S instrument (Waters MS-Technologies, Manchester, UK). ITEM-TWO measurements of the Troponin I peptide—antiTroponin I complex were performed with the following instrumental settings: source temperature, 50 °C; capillary voltage, 1.8 kV; sample cone voltage, 110 V; source offset voltage, 110 V; trap gas flow, 8.0 mL/min; cone gas flow, 100 L/h. All mass spectra were acquired in positive-ion mode applying a mass window of *m/z* 200–8000. The *m/z* axis was calibrated using 1 mg/mL sodium iodide dissolved in an isopropanol/water solution (50:50, *v*/*v*). The quadrupole analyzer was used to block transmission of lower molecular weight ions: M1 = 4000 with dwell time of 25% and ramp time of 25%; M2 = 5000 with dwell time of 25% and ramp time of 25%; M3 = 6000. The surviving antibody-peptide complexes were dissociated in the first collision cell (TRAP) by increasing the collision cell voltage difference in a stepwise manner (2–10 V/step). Data were acquired and processed with MassLynx software version 4.1 (Waters MS-Technologies, Manchester, UK). At each collision cell voltage difference setting a mass spectrum was recorded for 1 min, each. All scans for a given collision cell voltage difference were combined to generate an average spectrum. From each spectrum the ion intensities (in arbitrary units) were deduced.

4.8. Mass Spectral Analysis

The Savitzky–Golay method was used for smoothing in five cycles with a window of 10 for the high mass range and in three cycles with a window of 5 for the low mass range, respectively. Fractions of educts and products were derived from heights of ion signals of complex (educt) and its dissociated constituents (products) at all applied collision cell voltage differences (**ΔCV**). At each applied **ΔCV** setting the height of apex of Gaussian fit of the multiply charged ion series of the respective complex as well as of its constituents was determined by recording the intensities and respective *m/z* values of all ion signals for each charge state. Next, these ion intensities were plotted against their respective *m/z* values and fitted to a Gaussian curve (Figure 2) using Origin version 8.1 (OriginLab Corporation, Northampton, MA, USA). Intensity apexes of each molecular species or complex from the heights of the individual ion signals were determined as well. The relative amounts of products, $\mathbf{f(products)}$, and relative amounts of educts, $\mathbf{f(educts)}$, from the nanoESI mass spectra at given **ΔCV** settings were determined. The plots of normalized educt intensities i.e., $\mathbf{norm(educts)}$ vs. **ΔCV** were fitted to a Boltzmann curve ($R^2 \geq 0.99$).

The mass spectrometry data have been deposited to the ProteomeXchange Consortium via the PRIDE [37] partner repository with the dataset identifier PXD021296.

5. Conclusions

ITEM-TWO is able (i) to determine epitopes and (ii) to investigate the epitopes´ binding strengths in the gas phase. Mixing of antigen or epitope peptide and antibody solutions is the only required in-solution handling step when the complex components are dissolved in electrospray-compatible solutions. From normalized ion intensities, the apparent gas phase quasi equilibrium dissociation constants ($\mathbf{K^{\#}_{D\ m0g}}$) can be deduced from which apparent dissociation activation energies for neutral and resting immune complexes in the gas phase ($\mathbf{\Delta G^{\#}_{mOg}}$) can be calculated. As suitable electrospray mass spectrometry equipment has become amply available, our ITEM-TWO method should be easily adaptable by mass spectrometry laboratories all around the world.

Supplementary Materials: The following are available online, Equations and theoretical explanations. Figure S1: Course of normalized ion intensities of holo-myoglobin ions (*norm* (*educts*)) as a function of collision cell voltage differences (ΔCV), Figure S2: Arrhenius plot for the course of myoglobin dissociation in the gas phase, Figure S3: Course of normalized ion intensities of RNAse S ions (*norm* (*educts*)) as a function of collision cell voltage differences (ΔCV), Figure S4: Arrhenius plot for the course of RNAse S dissociation in the gas phase, Figure S5: Course of normalized ion intensities of FLAG-peptide—antiFLAG antibody ions (*norm* (*educts*)) as a function of collision cell voltage differences (ΔCV), Figure S6: Arrhenius plot for the course of FLAG-peptide—antiFLAG antibody complex dissociation in the gas phase, Figure S7: Course of normalized ion intensities of Troponin I peptide—antiTroponin I antibody complex (*norm* (*educts*)) as a function of collision cell voltage differences (ΔCV), Figure S8: Arrhenius plot for the course of Troponin I peptide—antiTroponin I antibody complex dissociation in the gas phase, Table S1: Ion intensities, charge states, and m/z values for myoglobin at various collision cell voltage difference settings, Table S2: Apex heights and mean charge states of educt and product ion signals upon gas phase dissociation of myoglobin, Table S3: Ion intensities, charge states, and m/z values for RNAse S at various collision cell voltage difference settings, Table S4: Apex heights and mean charge states of educt and product ion signals upon gas phase dissociation of RNAse S, Table S5: Ion intensities, charge states, and m/z values for FLAG-peptide—antiFLAG antibody complex at various collision cell voltage difference settings, Table S6: Apex heights and mean charge states of educt and product ion signals upon gas phase dissociation of FLAG-peptide—antiFLAG antibody complex, Table S7: Ion intensities, charge states, and m/z values for TroponinI peptide—antiTroponinI at various collision cell voltage difference settings, Table S8: Apex heights and mean charge states of educt and product ion signals upon gas phase dissociation of TroponinI immune complex.

Author Contributions: Conceptualization: B.D.D. and M.O.G.; methodology: B.D.D., K.F.M.O., and M.O.G.; validation: K.F.M.O., C.R., and C.K.; formal analysis: B.D.D., C.R., and C.K.; investigation: B.D.D. and M.O.G.; resources: M.O.G.; data curation: B.D.D.; writing—original draft preparation: B.D.D. and M.O.G.; writing—review and editing: B.D.D., K.F.M.O., C.R., C.K., and M.O.G.; visualization: B.D.D. and M.O.G.; supervision: M.O.G.; funding acquisition: B.D.D., K.F.M.O., and M.O.G. All authors have read and agreed to the published version of the manuscript.

Funding: The authors thank the German Academic Exchange Service (DAAD) for providing a postdoctoral fellowship (Re-invitation Programme for Former Scholarship Holders) for KO (91548123) and a doctoral scholarship for BD (No. 91566064). The WATERS Synapt G2S mass spectrometer has been bought through an EU grant [EFRE-UHROM 9] made available to MOG.

Acknowledgments: We express our thanks to Michael Kreutzer for providing his expertise on bioinformatics and to Peter Lorenz and Hans-Juergen Thiesen for providing the FLAG peptide.

Conflicts of Interest: The authors declare no conflict of interest.

References

1. Opuni, K.F.M.; Al-Majdoub, M.; Yefremova, Y.; El-Kased, R.F.; Koy, C.; Glocker, M.O. Mass spectrometric epitope mapping. *Mass Spectrom. Rev.* **2016**, *37*, 229–241. [CrossRef]
2. Sunner, J.; Magnera, T.F.; Kebarle, P. Ion–molecule equilibria measurements by high pressure mass spectrometry. Some recent advances in concepts and technique. *Can. J. Chem.* **1981**, *59*, 1787–1796. [CrossRef]
3. Kebarle, P. Equilibrium studies of the solvated proton by high pressure mass spectrometry. Thermodynamic determinations and implications for the electrospray ionization process. *J. Mass Spectrom.* **1997**, *32*, 922–929. [CrossRef]
4. Armentrout, P.B.; Rodgers, M.T. Thermochemistry of Non-Covalent Ion–Molecule Interactions. *Mass Spectrom.* **2013**, *2*, S0005. [CrossRef] [PubMed]
5. Armentrout, P.B.; Heaton, A.L.; Ye, S.J. Thermodynamics and Mechanisms for Decomposition of Protonated Glycine and Its Protonated Dimer. *J. Phys. Chem. A* **2011**, *115*, 11144–11155. [CrossRef] [PubMed]
6. Meot-Ner, M.; Dongre, A.R.; Somogyi, A.; Wysocki, V.H. Thermal decomposition kinetics of protonated peptides and peptide dimers, and comparison with surface-induced dissociation. *Rapid Commun. Mass Spectrom.* **1995**, *9*, 829–836. [CrossRef]
7. Price, W.D.; Schnier, P.D.; Jockusch, R.A.; Strittmatter, A.E.F.; Williams, E.R. Unimolecular Reaction Kinetics in the High-Pressure Limit without Collisions. *J. Am. Chem. Soc.* **1996**, *118*, 10640–10644. [CrossRef]
8. Li, C.; Ross, P.; Szulejko, A.J.E.; McMahon, T.B. High-Pressure Mass Spectrometric Investigations of the Potential Energy Surfaces of Gas-Phase SN2 Reactions. *J. Am. Chem. Soc.* **1996**, *118*, 9360–9367. [CrossRef]
9. Pan, S.; Sun, X.J.; Lee, J.K. Stability of complementary and mismatched DNA duplexes: Comparison and contrast in gas versus solution phases. *Int. J. Mass Spectrom.* **2006**, *253*, 238–248. [CrossRef]
10. Jørgensen, T.J.D.; Delforge, D.; Remacle, J.; Bojesen, G.; Roepstorff, P. Collision-induced dissociation of noncovalent complexes between vancomycin antibiotics and peptide ligand stereoisomers: Evidence for molecular recognition in the gas phase. *Int. J. Mass Spectrom.* **1999**, *188*, 63–85. [CrossRef]
11. Kitov, P.I.; Kitova, E.N.; Han, L.; Li, Z.X.; Jung, J.; Rodrigues, E.; Hunter, C.D.; Cairo, C.W.; Macauley, M.S.; Klassen, J.S. A quantitative, high-throughput method identifies protein–glycan interactions via mass spectrometry. *Commun. Biol.* **2019**, *2*, 268. [CrossRef]
12. Yefremova, Y.; Opuni, K.F.M.; Danquah, B.D.; Thiesen, H.-J.; Glocker, M.O. Intact Transition Epitope Mapping (ITEM). *J. Am. Soc. Mass Spectrom.* **2017**, *28*, 1612–1622. [CrossRef]
13. Danquah, B.D.; Rower, C.; Opuni, K.F.M.; El-Kased, R.; Frommholz, D.; Illges, H.; Koy, C.; Glocker, M.O. Intact Transition Epitope Mapping—Targeted High-Energy Rupture of Extracted Epitopes (ITEM-THREE). *Mol. Cell. Proteom.* **2019**, *18*, 1543–1555. [CrossRef] [PubMed]
14. Yefremova, Y.; Melder, F.T.I.; Danquah, B.D.; Opuni, K.F.M.; Koy, C.; Ehrens, A.; Frommholz, D.; Illges, H.; Koelbel, K.; Sobott, F.; et al. Apparent activation energies of protein–protein complex dissociation in the gas–phase determined by electrospray mass spectrometry. *Anal. Bioanal. Chem.* **2017**, *409*, 6549–6558. [CrossRef] [PubMed]
15. Donor, M.T.; Mroz, A.M.; Prell, J.S. Experimental and theoretical investigation of overall energy deposition in surface-induced unfolding of protein ions. *Chem. Sci.* **2019**, *10*, 4097–4106. [CrossRef]

16. Danquah, B.D.; Yefremova, Y.; Opuni, K.F.; Röwer, C.; Koy, C.; Glocker, M.O. Intact Transition Epitope Mapping—Thermodynamic Weak-force Order (ITEM—TWO). *J. Proteom.* **2019**, *212*, 103572. [CrossRef] [PubMed]
17. Przybylski, M.; Glocker, M.O. Electrospray Mass Spectrometry of Biomacromolecular Complexes with Noncovalent Interactions—New Analytical Perspectives for Supramolecular Chemistry and Molecular Recognition Processes. *Angew. Chem. Int. Ed.* **1996**, *35*, 806–826. [CrossRef]
18. Loo, J.A. Studying noncovalent protein complexes by electrospray ionization mass spectrometry. *Mass Spectrom. Rev.* **1997**, *16*, 1–23. [CrossRef]
19. Donor, M.T.; Shepherd, S.O.; Prell, J.S. Rapid Determination of Activation Energies for Gas-Phase Protein Unfolding and Dissociation in a Q-IM-ToF Mass Spectrometer. *J. Am. Soc. Mass Spectrom.* **2020**, *31*, 602–610. [CrossRef] [PubMed]
20. Wells, J.M.; McLuckey, S.A. Collision-Induced Dissociation (CID) of Peptides and Proteins. *Biol. Mass Spectrom.* **2005**, *402*, 148–185. [CrossRef]
21. Douglas, D.J. Applications of collision dynamics in quadrupole mass spectrometry. *J. Am. Soc. Mass Spectrom.* **1998**, *9*, 101–113. [CrossRef]
22. Chang, R. *Physical Chemistry for the Biosciences;* University Science Books: Mill Valley, CA, USA, 2005.
23. Hargrove, M.S.; Barrick, D.; Olson, J.S. The Association Rate Constant for Heme Binding to Globin Is Independent of Protein Structure. *Biochemistry* **1996**, *35*, 11293–11299. [CrossRef] [PubMed]
24. Culbertson, D.S.; Olson, J.S. Role of Heme in the Unfolding and Assembly of Myoglobin. *Biochemistry* **2010**, *49*, 6052–6063. [CrossRef] [PubMed]
25. Enyenihi, A.A.; Yang, H.; Ytterberg, A.J.; Lyutvinskiy, Y.; Zubarev, R.A. Heme Binding in Gas-Phase Holo-Myoglobin Cations: Distal Becomes Proximal? *J. Am. Soc. Mass Spectrom.* **2011**, *22*, 1763–1770. [CrossRef]
26. Hunter, C.L.; Mauk, A.G.; Douglas, D.J. Dissociation of Heme from Myoglobin and Cytochromeb5: Comparison of Behavior in Solution and the Gas Phase. *Biochemistry* **1997**, *36*, 1018–1025. [CrossRef] [PubMed]
27. Schmidt, A.; Karas, M. The influence of electrostatic interactions on the detection of heme-globin complexes in ESI-MS. *J. Am. Soc. Mass Spectrom.* **2001**, *12*, 1092–1098. [CrossRef]
28. Ito, W.; Kurosawa, Y. Development of an artificial antibody system with multiple valency using an Fv fragment fused to a fragment of protein A. *J. Biol. Chem.* **1993**, *268*, 20668–20675.
29. Ito, W.; Iba, Y.; Kurosawa, Y. Effects of substitutions of closely related amino acids at the contact surface in an antigen-antibody complex on thermodynamic parameters. *J. Biol. Chem.* **1993**, *268*, 16639–16647.
30. Hoaglund, C.S.; Valentine, S.J.; Sporleder, C.R.; Reilly, J.P.; Clemmer, D.E. Three-Dimensional Ion Mobility/TOFMS Analysis of Electrosprayed Biomolecules. *Anal. Chem.* **1998**, *70*, 2236–2242. [CrossRef]
31. Konijnenberg, A.; Butterer, A.; Sobott, F. Native ion mobility-mass spectrometry and related methods in structural biology. *Biochim. Biophys. Acta (BBA) Proteins Proteom.* **2013**, *1834*, 1239–1256. [CrossRef]
32. Seo, J.; Hoffmann, W.; Warnke, S.; Bowers, M.T.; Pagel, K.; Von Helden, G. Retention of Native Protein Structures in the Absence of Solvent: A Coupled Ion Mobility and Spectroscopic Study. *Angew. Chem. Int. Ed.* **2016**, *55*, 14173–14176. [CrossRef]
33. Hopper, J.T.S.; Oldham, N.J. Collision induced unfolding of protein ions in the gas phase studied by ion mobility-mass spectrometry: The effect of ligand binding on conformational stability. *J. Am. Soc. Mass Spectrom.* **2009**, *20*, 1851–1858. [CrossRef]
34. Zhong, Y.; Hyung, S.-J.; Ruotolo, B.T. Ion mobility–mass spectrometry for structural proteomics. *Expert Rev. Proteom.* **2012**, *9*, 47–58. [CrossRef] [PubMed]
35. Breuker, K.; McLafferty, F.W. Stepwise evolution of protein native structure with electrospray into the gas phase, 10–12 to 102 s. *Proc. Natl. Acad. Sci. USA* **2008**, *105*, 18145–18152. [CrossRef] [PubMed]
36. Heck, A.J.R. Native mass spectrometry: A bridge between interactomics and structural biology. *Nat. Methods* **2008**, *5*, 927–933. [CrossRef] [PubMed]

37. Perez-Riverol, Y.; Csordas, A.; Bai, J.; Bernal-Llinares, M.; Hewapathirana, S.; Kundu, D.J.; Inuganti, A.; Griss, J.; Mayer, G.; Eisenacher, M.; et al. The PRIDE database and related tools and resources in 2019: Improving support for quantification data. *Nucleic Acids Res.* **2018**, *47*, D442–D450. [CrossRef] [PubMed]

Sample Availability: Samples of the peptide compounds are available from the authors on request.

Publisher's Note: MDPI stays neutral with regard to jurisdictional claims in published maps and institutional affiliations.

© 2020 by the authors. Licensee MDPI, Basel, Switzerland. This article is an open access article distributed under the terms and conditions of the Creative Commons Attribution (CC BY) license (http://creativecommons.org/licenses/by/4.0/).

MDPI
St. Alban-Anlage 66
4052 Basel
Switzerland
Tel. +41 61 683 77 34
Fax +41 61 302 89 18
www.mdpi.com

Molecules Editorial Office
E-mail: molecules@mdpi.com
www.mdpi.com/journal/molecules

www.ingramcontent.com/pod-product-compliance
Lightning Source LLC
LaVergne TN
LVHW070112170726
843515LV00007B/1675